GREGOR MENDEL

GREGOR MENDEL

The First Geneticist

◆

Vítězslav Orel

Emeritus Head
The Mendelianum
Brno
Czech Republic

Translated by Stephen Finn

Oxford   New York   Tokyo
OXFORD UNIVERSITY PRESS
1996

Oxford University Press, Walton Street, Oxford OX2 6DP

Oxford   New York
Athens   Auckland   Bangkok   Bombay
Calcutta   Cape Town   Dar es Salaam   Delhi
Florence   Hong Kong   Istanbul   Karachi
Kuala Lumpur   Madras   Madrid   Melbourne
Mexico City   Nairobi   Paris   Singapore
Taipei   Tokyo   Toronto
and associated companies in
Berlin Ibadan

Oxford is a trade mark of Oxford University Press

Published in the United States
by Oxford University Press, New York

A catalogue record for this book is available from the British Library

Library of Congress Cataloging in Publication Data
Orel, Vítězslav.
Gregor Mendel : the first geneticist / Vítězslav Orel. — 1st ed.
Includes bibliographical references and index.
1. Mendel, Gregor, 1822–1884.   2. Geneticists—Austria—Biography.
3. Genetics—History.   4. Mendel's law—History.   I. Title.
QH31.M45067   1996   575.1'092—dc20   94–29077

ISBN 0 19 854774 9

Typeset by EXPO Holdings, Malaysia

Printed in Great Britain by
Bookcraft (Bath) Ltd,
Midsomer Norton, Avon

To my wife
Olga

Preface

The beginnings of research in the life sciences date back to the ideas put
forward by philosophers and natural scientists in the seventeenth century
in an endeavour to explain how animals and plants reproduced themselves.
At the same time they tried to devise a system of classification of natural
forms, both animate and inanimate. Francis Bacon (1561–1626) was convinced that man would never master nature or influence its development
before he understood the covert processes which lay behind all natural phenomena, and which he supposed to be governed by particles too tiny for
our senses to perceive. Factors of corpuscular size and their motion were
also used by René Descartes (1596–1650) to try to explain the enigma of
generation. Carl Linnaeus (1707–78) worked out a system of classification
of all living organisms, including man, and he also gave botanists the idea
of experimenting with the artificial pollination of plants.

The results of a study of living nature led late-eighteenth-century scholars to take a new view of both animate and inanimate objects. J.B. Lamarck
(1744–1829) proposed a system for examining both these worlds, and it
was he who invented the name biology for the branch dealing with living
things. The concept of biology in connection with a study of living objects
was also proposed by adherents of the German *Naturphilosophie*. After the
turn of the nineteenth century, the study of living things was no longer
exclusively the domain of physiology, this science being gradually supplemented by specialized disciplines such as histology, embryology, palaeontology, and later also biochemistry. Research into the ancient mystery of
generation led scholars to a recognition of the structures of plant and
animal tissue, and to the discovery of the cell. In this connection the physiologist R. Wagner (1805–64) from the University of Göttingen suggested in
1853 that experiments should be performed to investigate the enigmas of
both generation and heredity.

At the start of the nineteenth century Lamarck tried to explain the development of organisms from the simplest to the most complex, including
man. It was not, however, until Charles Darwin (1809–82) came on the
scene that naturalists learned the mechanism of the origin of species.
Darwin was himself well aware that natural selection operates on the basis
of hereditary variability, and admitted that the laws governing inheritance
were still unknown. Independently of progress in the natural sciences,
farmers had at the turn of the century developed methods of breeding more

productive animal races and plant varieties. They introduced recording of pedigrees and progeny testing, and applied techniques of crossing in animals and hybridization in plants, but came no nearer to an understanding of the problem of heredity. Even H. Nathusius, an acknowledged expert on animal breeding, writing in 1872, complained that 'the laws of heredity have not yet fallen from the tree of knowledge, that which, as legend would have it, led Newton on the right path to the conception of the Law of Gravity'.

It was not until the end of the last century that experimental biology appeared as a branch of science in its own right. Through the latest findings in the field of cytology and the development of the theory of evolution, biologists became aware of the very pressing problem of heredity, but they came up with no methodological approach to its solution. The explanation was not revealed to the world until the year 1900, with the unearthing of a paper published by Mendel in 1866 in Brno (Brünn), capital of a province of the Austro-Hungarian Empire called Moravia (from 1918 to 1992 a part of Czechoslovakia, now in the Czech Republic).

After 1900 Mendel's *Experiments in plant hybridization* was published repeatedly in various languages and stimulated geneticists to develop the new science of heredity, which since 1906 has been known as 'genetics'. The recognition Mendel's work achieved abroad led the Augustinians in Brno to search for documents relating to the activities of their former abbot, and a Mendel museum was gradually set up at the monastery. With the development of the science of heredity, geneticists kept returning to Mendel's work for instruction and inspiration. Various views were expressed regarding the significance of Mendel's theory in connection with the progress of knowledge on evolution. The theory has also been misused in support of Nazi racism, and was subsequently rejected by Marxists in the Soviet Union. It was this ideology which in 1950 led to the closure of the Mendel Museum at the Brno monastery, which was itself forcibly shut down.

After 1950 research moved forward at all levels of organization of living systems, leading to a rehabilitation of genetics in countries with communist regimes, and of Mendel's research work in the country where he was born and worked. On the occasion of the hundredth anniversary of Mendel's discovery, 1965, the Mendel museum was re-opened by the Moravian museum under the name 'Mendelianum'. The monastery community was not able to return until after the events of November 1989.

When the *Pisum* paper was rediscovered, the name of its author, Gregor Mendel, was almost unknown to world science. Today he ranks among the great discoverers in the natural sciences, such as Nicholas Copernicus

(1473–1543), who laid the foundations of modern astronomy, Isaac Newton (1642–1727), who made a decisive contribution to the development of physics, and Charles Darwin (1809–82), who explained the origin of new species through the action of natural selection. Since almost the beginning of this century, geneticists have acknowledged Mendel as the founder of the science of heredity. In the second half of the century, new information has been brought to light, and has served to awaken new interest among historians of science. Recently critical voices have been raised among the historians, philosophers, and sociologists of science, suggesting that Mendel's research work might have been overestimated.

A systematic study of the documents and publications relating to Mendel and the part played by his research on the origin and development of genetics shows that attitudes to Mendel have varied, right from the initial publication of his works. This can be seen particularly in journals and books written in different countries since the start of the 1950s. An examination of what are frequently contradictory views may lead one to ask a number of methodological questions, most of which have never before been posed from such angles or in such depth. They can be summarized briefly as follows: Where did Mendel get the motivation to study natural sciences, especially plant hybridization? What led to his creation of such an original method of research? How did he manage to carry out such time-consuming and laborious experiments in a monastery? What was the connection between his research, the enigma of generation, and the development of the cell theory? How did he understand the development of living forms in nature, and what was his attitude to Darwin's Theory of Evolution? How did his contemporaries in provincial Brno receive his theory, and why was there no significant reaction from readers of the journal abroad? What was decisive for the acceptance of his theory after 1900, and when was his theory incorporated into the body of biological sciences?

The answers to these questions should be sought in the evolution of the cultural and social milieu in which Mendel grew up and later worked, and in connection with world progress in science, with which he became acquainted at university, in the company of Brno natural scientists, and while studying scientific literature. A more detailed analysis of this development shows how the problem of heredity gradually emerged in the natural sciences from the start of the nineteenth century, and appeared in parallel with a growing consciousness of the inheritance of traits in agricultural breeding practice. At the same time one may point to the complex problem of generation, which included heredity.

A more detailed explanation of the effect of Mendel's research on the origin and development of genetics is now sought. This book may help the reader to understand better the extreme views which are published from time to time, on both Mendel the scientist and Mendel the man. Readers are also offered a critical assessment of interpretations of Mendel's motivation toward scholarship and research in his cultural and social climate, and also of how he was able to adopt such an original approach to experimentation and the formation of an original theory.

Brno, Czech Republic   V.O.
1994

Contents

1
An enigmatic discoverer

We have only to read Mendel's paper to see how much further he saw into the nature of heredity than did any of his contemporaries.
—C. ZIRKLE (1951, p. 51).

In 1900 scholars in various countries pointed out the great significance of Mendel's publication *Experiments in plant hybridization*, published in Brno in 1866. (A year before that publication, Mendel had delivered two lectures on the subject at monthly meetings of the Natural Science Society.) In 1900 biologists suddenly became interested in the man who had performed these experiments, and who was almost totally unknown. Even in Brno, the town where the scientist had worked, little was known about him sixteen years after his death. A study of his paper led researchers to seek more information on the life and work of this newly famous experimenter. The first scanty details which appeared in the early years of this century were filled out in later years. On the occasion of each new anniversary, discussions were held not only about Mendel's scientific work, but also about his origin, his studies, and his life at the monastery and in public affairs. For the most part there was nothing new, and certain misconceptions were passed on over and over again. Later, however, new material turned up, and the amount of information available on Mendel and his work began to grow.

In comparison with other outstanding natural scientists, such as Darwin, our data on Mendel remain unsatisfactory to this day. He did not keep a diary, and virtually no research notebooks from his experiments have survived. His research was conceived and performed in the seclusion of a monastery on the edge of a provincial town, and in the world outside he mixed only with colleagues at the schools where he taught, or with members of the learned societies he belonged to. Sometimes, as a committee member, he chaired their meetings. As abbot, he associated with experts from the Agricultural Society, and helped deal with some of the problems of agriculture in the province. As a member of the Moravian Diet, he attended the meetings of special commissions, where he was able to offer the benefit of knowledge and experience gained during his studies, during his work as a teacher, and later as administrator of the monastery landholdings. He was a popular teacher, and kept in touch with his pupils after they left school.

One direct source of information we do have is a small number of letters from Mendel to his parents and relatives, revealing his relationship with his family and with his native village. One cannot but subscribe to the generally held view that he was of the introspective type—a conclusion borne out by what he wrote on the title-page of a meteorology textbook by Kunzek (1850): 'He who does not know how to be alone is not at peace with himself.' The life he chose was that of a monk in a quite exceptional monastery, where he was able, alone and without distraction, to seek scientific truth. When he was twenty-eight years old, he wrote a curriculum vitae in connection with his application for permission to take his teacher's examination (Mendel 1850); in it he expresses the wish to devote himself to the study of natural sciences. Father Gregor wished to present himself at the examination as a natural scientist, and thus gain the qualification he needed to teach, even though he had never studied at university before. This is a unique pointer to Mendel's thought processes at the time he decided to fulfil his ambition to become a teacher, and thus to a certain extent ensure his financial independence.

The two papers on plant hybrids offer only the slenderest of clues to the motivation for Mendel's scientific work (Mendel 1866*a*, 1870*a*). More detail on the subject can be gleaned from his letters to C. Nägeli (1817–91), not made available until twenty-one years after his death (Correns 1905). Information regarding Mendel's specialized interests can also be found in the publications of learned societies in Brno. He emerges as a natural scientist with a clear interest in research and a grasp of the applications of science in practical life. The obituaries draw attention mainly to the merits of the late abbot and to his work in public institutions. Little attention is paid in them to his scientific interests and research. It should, however, be remembered that every memorial occasion emphasizes those merits which are most apparent in the contemporary context.

In the year when *Experiments in plant hybridization* was rediscovered, the scientists who started to study it with such interest knew only the basic facts regarding its author's origin, education, and research work. C. Correns (1900*a*) and E. Tschermak (1901) published skeleton biographies of the newly famous scientist. Soon afterwards, Mendel's nephews Alois and

Ferdinand Schindler began to publish recollections of their uncle and information on his origins. Alois Schindler (1902) published in expanded form the lecture he gave about Mendel on the occasion of the unveiling of the first memorial plaque, at the village fire-station in Hynčice (Heinzendorf), his birthplace. The villagers put it up in memory of the benefactor who paid for the firefighting equipment there. The published text of the lecture formed the basis of many other publications in different countries, and among those who drew on it was the English natural scientist W. Bateson, an influential proponent of Mendel's work in English-speaking countries. When he published the results of his experiments on trait transfer in animals, he appended a brief life of Mendel (Bateson 1902). Later, in 1904, Bateson visited Brno in search of more detail on Mendel's work and his life story. He was disappointed to find no records of the experiments, and gave up his original idea of writing a biography. Some details relating to Mendel were published by G.S. Saunders (1906) in the *Report* on the first Conference on Genetics, held in London in 1906.

Bateson's visit to Brno aroused the interest of A. Matoušek, procurator of the Augustinian monastery, in collecting documents relating to the monastery's former abbot, now a celebrated scientist (Orel and Marvanová 1966). He built up a small Mendel museum in the monastery, which was visited by the participants in the 1922 centennial celebration of the researcher's birth. From 1905 the Brno natural scientist H. Iltis (1882–1952) studied Mendel's personality and scientific work. He found especially valuable documents relating to the period of study at Vienna University. After publishing a number of articles on the subject, he wrote a biography which was subsequently translated into English and Japanese (Iltis 1924). This was to be the chief source of information about the founder of genetics for authors of biographical studies. Some new facts were revealed in serial form in the monthly *Messenger*, published in Kentucky, USA, in 1938–41 (see Dodson 1979). Another researcher in Brno, O. Richter (1943), brought to light new documents relating to Mendel, publishing them in a booklet in which, *inter alia*, he criticized Iltis' presentation of Mendel as a freethinking natural scientist and an admirer of Darwin. Richter underlined Mendel's qualities as a priest. The article was published during the Nazi occupation of Czechoslovakia, and it depicted Mendel as a scientist who thoroughly rejected Darwin's Theory of Evolution (Richter 1941).

In the 1930s geneticists were preoccupied with clarifying the relations between the theories of evolution and heredity, and paid little attention to Mendel the man. This situation altered significantly in 1948, when Lysenko in the Soviet Union and a few other biologists elsewhere in Europe began once more to write about the inheritance of acquired characteristics, and in this connection criticized Mendel's scientific work (see Medvedev 1969; Joravsky 1970; Gaissinovitch 1988; and Soyfer 1989). In 1950, the newly-installed communist authorities in Czechoslovakia closed most of the monasteries in the country, among them the Augustinian monastery in Brno. The Mendel museum in the monastery was also closed. At the critical time, naturalists from the Moravian Museum managed to save most of the documents and instruments dating from Mendel's time. It was only after the fall of Lysenko in the Soviet Union, together with the demise of his disciples in Czechoslovakia, that it became possible to prepare the ground for re-establishing a Mendel museum.

But by the time the golden jubilee of genetics was commemorated at Ohio University by the Genetic Society of America in September 1950, prominent geneticists from many countries were evaluating the research from the viewpoint of biochemistry and immunology, and it was the eve of the birth of molecular genetics. L.C. Dunn (1951b) went on to say that many of the principles of genetics had turned out to be of a general character, and that its rules applied not only to plants, as Mendel had found, but also to animals of all species, including man, and to the whole world of microorganisms which had been revealed since Mendel's day. On his departure for the USA before the Second World War, Iltis took with him copies of documents relating to Mendel which are now treasured exhibits in the Museum of Natural History at the University of Illinois at Urbana-Champaign (Hoffmeister and Henriksen 1979). He now presented the picture of Mendel he had acquired during his many years of research in Brno. Iltis' (1951) opening sentence drew attention to the problems involved: 'When I was asked to speak about Gregor Mendel's heritage, I immediately saw the difficulties of this task. Mendel's was a sober mind, his thoughts were concerned mainly with concrete facts, and he had little inclination for sentimentalism of any kind.' Iltis was right, in that the limited number of surviving documents permit no more than an educated guess as to what went on in the mind of the student, teacher, scientist, and abbot. But beneath the apparently sober mind and the interest in concrete facts, thought processes were in motion which led Mendel to a recognition of scientific truth in the research programme he had resolved to devote his life to.

There were new biographies of Mendel in the 1950s, most of them based on the one by Iltis. A life of Mendel was published in Dutch by Meijknecht (1950), and two editions of Krumbiegel's German biography (1957, 1967) appeared. Later biographies on Mendel have been published in German by Sajner (1974), in English by George (1975) and by Orel (1984), and in Japanese by Orel and Cetl (1973) and by Nakazawa (1978). Biographies of

Mendel in more popular form have appeared in German (Heinen 1943), in English (Sootin 1959), and in Russian (Volodin 1968). In 1955 the Mendelianum in the Moravian Museum in Brno published with captions in English and in Czech 120 documents and photographs relating to Mendel (Marvanová *et al.* 1965).

On the centenary in 1965 of the first delivery of Mendel's historic paper, geneticists from many lands met in Brno to acknowledge his contribution in the light of the latest developments in genetics (Sosna 1966). B.L. Astaurov (1966, pp. 57–62), president of the All-Union Society of Geneticists and Selectionists, newly founded in the USSR under the name of N.I. Vavilov, stated in his address: 'The real penetration into the very essence of the phenomena discovered by Mendel required such a radical change of biological thinking as can be compared with the transition from classical concepts of Newtonian mechanics to the ideas of modern quantum physics.' A misunderstanding of these radical changes in the thinking of some biologists was among the reasons why Lysenko and his disciples came to reject the science of genetics and the validity of Mendel's theory. The depth of ignorance of Mendel's scientific work was illustrated by I.M. Lerner (1966, pp. 189–97), referring to a remarkable paper by R.A. Fisher (1936), the first to point out that, in spite of the immense publicity Mendel's paper had received, it had 'rarely been examined with sufficient care'.

The 1965 Mendel centennial also led prominent geneticists H. Stubbe (1963), L.C. Dunn (1965*b*), A.H. Sturtevant (1965), and F.A.E. Crew (1966*a*) to undertake a critical review of the history of genetics, and in this context to reconsider Mendel's role in the origin and early development of the science. The Moscow publishing house Nauka published a further Russian translation of both Mendel's papers on plant hybridization, along with the first Russian translation of Mendel's letters to Nägeli and a reprinting of the introduction to the 1935 edition of Mendel's papers by N.I. Vavilov (Gaissinovitch 1965*a,b*). Later, Gaissinovitch (1967, 1988) published a more detailed analysis of Mendel's research in the context of the previous history of the natural sciences. Twenty years later he returned to the subject, going into considerable detail about the wayward route to an understanding of Mendel's theory taken by the development of genetics in the USSR. R.C. Olby (1966), in his book *The origins of Mendelism*, stimulated the interest of historians of science in re-evaluating Mendel's research into plant hybridization and plant physiology. The book was dedicated to the founder of the Mendelianum, J. Kříženecký, who published a source book on Mendel at the Leopoldina Academy in Halle in 1965, and another at the Czechoslovak Academy of Sciences in Prague republishing the *Pisum* paper and twenty-seven papers from the rediscovery era (Kříženecký 1965*a,b*).

Since 1965 new documents and information on Mendel and the early days of genetics have been published by geneticists and historians of science in the series *Folia Mendeliana*, throwing new light not only on Mendel's scientific and professional work, but also on the cultural and social environment from which he came. The information contained in Iltis' 1924 biography and in the books published since 1965 can now be filled in, particularly with new data on the scientist's work and his interest in study and research. Additional information has been provided by Matalová (1981) in a review of the published primary sources relating to Mendel's biography. The reconstruction and analysis of his experiments with *Pisum* by Fisher (1936) and Orel (1971a) and with other plant species by Cetl (1973) indicate that Mendel had a long-term research programme, starting with *Pisum* and ending with crossing races of bees, undertaken in an attempt to expand his theory.

2
Heredity before Mendel

For he that knows the way of nature will more easily observe her deviations: and on the other hand he that knows her deviations will more accurately describe her ways.

—FRANCIS BACON (1620, pp. 237–8).

The effects of heredity had been apparent since time immemorial, and man had taken advantage of them ever since he first kept animals or cultivated crops, deliberately choosing for breeding those individuals he liked best or which gave the best yields. He must soon have realized that the characteristics of parents tended to be passed on to their offspring, and that his own species was no exception. But it remained a mystery why one offspring was like its mother and another like its father, while the only resemblance some of them bore was to one or other of the grandparents. Variable traits such as height often seem to appear in the children as the average of those of their parents. This observation gave rise to the notion of 'blending' inheritance subscribed to by ancient philosophers and natural historians. Hippocrates of Cos (460–377 BC), considered to be the founder of medical science, propounded a theory according to which minute particles from every part of the body entered the seminal substance of the parents, and by their fusion gave rise to a new individual exhibiting the traits of both of them.

Hippocrates' theory of pangenesis still dominated the ideas of natural scientists in the nineteenth century, culminating in Darwin's attempt to explain heredity. A more concrete conception of heredity was developed by Aristotle (384–322 BC). He supposed that every part of the new organism was contained within the semen, which was formed by sanguinous nutriment. It acquired the active power to shape a new embryo. The menstrual blood of a woman passively contained each and every part of her body, which was shaped into a new organism by the action of the principle of motion of the sperm. In this conception, the sperm produced qualitative changes in the matter of the female organism, and thus Aristotle was the first to attribute to the mother an essential role in the process of generation. These ideas formed the starting-point for ancient speculation on the nature of heredity.

The pragmatic Romans accepted the speculative views of the Greeks, as could be seen in the writings of the influential physician Galen (AD 131–201). They paid more attention to the problems of growing plants and breeding farm animals. In this connection they described local breeds of animals and varieties of cultivated plants, and even some aspects of breeding practice, including progeny testing.

In the early centuries of the Christian era St Augustine (AD 345–430) paid serious attention to questions of generation in his writings. He tried to bring the teaching of Aristotle into line with Christian theology, accepting the Old Testament cosmology as a revelation that the world was created (*creatio ex nihilo*). According to St Augustine, God had endowed matter with certain powers of self-development, leaving free the operation of natural causes in the production of plants and animals. But heredity only emerged as a problem of natural science much later.

HEREDITY IN LATTER-DAY SCIENCE

By combining the findings of astronomy and physics, the French philosopher and natural scientist Descartes presented a new view of the world, based on the principles of mechanics. Natural scientists started to believe that everything in nature could be investigated and explained in terms of the existence and regrouping of minute particles of matter. Even living organisms were considered as structures subject to the laws of mechanics. After Isaac Newton offered a mathematical explanation of the force of gravity, scientists began an objective investigation of the macrocosm of the Universe, and later of the structure of the Earth. The living world, too, soon caught their attention, and one of the questions they considered was the 'enigma' of generation. The English physician William Harvey (1578–1657) linked his speculation on the origin of birds' eggs with the mysteries of generation and heredity, assuming that new individuals arose through a fusion of the formless substance of both parents. This notion of *epigenesis* was in conflict with the long-held idea of the pre-existence of life-forms.

In the seventeenth century naturalists began to use the microscope to study the development of the embryo, leading to apparent confirmation of the idea of pre-existence. Leeuwenhoek (1632–94) supposed, on examining the sperm of animals and man under the microscope, that he could see *Samentierchen* (minute animals). He thought that these were preformed embryos whose nutritional needs were catered for by the egg. Around the same time Malpighi (1628–94), on studying the development of avian embryos, likewise supposed he had found a preformed embryo in the egg, which he said began development on coming into contact with sperm. These notions gave rise to the contradictory theories of *preformation* known as the ovistic and the spermistic respectively.

A new concept of generation was introduced by the French physicist and zoologist Réamur (1683–1757). He supposed there were organic molecules in the seminal substances of the parents, into which a new order was introduced after their fusion, through the action of a special force which thus gave rise to a new individual. The inheritance of traits was connected with this process. These ideas were further developed by the mathematician and astronomer Maupertuis (1698–1759). Influenced by Newton's idea of attraction and repulsion, he investigated the transmission of the traits of parents to their offspring. He also studied the inheritance of polydactyly in man, which Réaumur, too, had found interesting. He assumed that there existed in the seed of parents some sort of fluid elements which came from all parts of the body, and on fusing together formed a new individual with traits from both parents. Maupertuis tried to explain the normal development of an embryo, and the origin of monsters (teratology), a problem dealt with in the second half of the eighteenth century by other naturalists. His views contradicted the idea of preformation—a position also reached by Buffon, though for somewhat different reasons.

Wolf (1734–94) undertook a systematic observation of the process of growth and development of plants and later of the chicken embryo in the egg, and supposed he had discovered a 'law of organic bodies'. In his thesis *Theoria generationis*, Wolf (1759) concluded that the developing parts of the chicken embryo were formed from the structureless soluble yolk grains on the basis of a special force he called the *vis essentialis*. He saw the basis of the development of an organism in the process of nutrition; for him the essence of conception was perfect nutrition provided by the extreme nutritive value of semen. Gaissinovitch (1990) shows how Wolf even argued in the unpublished notes on his thesis that semen was 'necessary to form the initial basis of the embryo' and served to create 'the first fermenting movement'. Wolf went on to carry out more work which refuted the preformation theory, and tried to prove the existence of epigenesis, even trying to offer a confused explanation of variation and heredity.

Botanists began experimenting with plant crossing in connection with the demonstration of sex in plants, and they considered their results from the viewpoint of the idea of generation and later of fertilization. Most natural scientists saw a supernatural order in nature, with the organisms created at the start of the Universe simply replacing themselves, and forming an un-broken chain from the lowest forms up to man. It was considered a closed system, devised according to the Creator's plan. Then an examination of fossil remains revealed forms of plants and animals which no longer existed, and natural scientist began looking for an explanation. When, in 1735, Linnaeus published his system of classification of living nature, he believed the number of species to be the same as it was at the Creation. But subsequent experi-mental crossing of plants convinced him that hybridization gave rise to com-binations of parental traits. He thought the genus rather than the species to be the basic unit of creation, and now admitted the possibility of new species appearing in nature and disappearing from it. He formed an open system, interpreting it in harmony with the Creator's design.

Most natural scientists accepted Linnaeus' classification system, and began to consider the question of whether new species came into being and, if so, how. Hybridization became a problem connected, *inter alia*, with the classification of forms of plants. In 1749, J.G. Gmelin, professor of chemistry at the University of Tübingen, gave J.G. Kölreuter (1733–1806) the idea of experimenting with the artificial fertilization and crossing of plants, supposing that he would refute experimentally the ancient and still current dogma of the constancy of species. Kölreuter (1761–6) published a three-part report on experimental plant crosses involving thirteen genera and fifty-four species. His results have been analysed by E. Mayr (1986). Kölreuter introduced hybridization as a method of investigation of the old enigma of generation. As an essentialist, he believed that the uniform, fluid semen material of two parents, which blended in an intermediate condition in the progeny, was 'designed by the Creator for joining'. As an outstand-ing observer in nature he was convinced that when a species was fully grown and began to develop flowers, the male and female fluid semen again had to segregate (Olby 1986). According to this conception he was able to confirm experimentally the controversial sexuality of plants, and described the uniformity of hybrids and the results of reciprocal crosses and the segregation of parental traits in the hybrid progeny. He put exceptions to his theoretical expectation down to the irregular mixing of semen from both parents. Kölreuter saw his explanation as agreeing with the Aristotelian theory of generation through the semen of both parents, in contradiction to the still widely held theory of preformation. In publishing the results of his experiments Kölreuter became a supporter of the theory of epigenesis, according to which the newly-formed germ is homogeneous, and differentiates only as it develops.

Another unique feature of Kölreuter's theory was his explanation of the variable degree of sterility of plant hybrids as being due to a disturbed balance of the semen stuffs of the two parents, and to the influence of the environment. In repeated fertilization of hybrids with paternal pollen, Kölreuter wished to explain how the hybrids reverted to the original paternal form. His conclusion was that his experiments had not refuted the dogma of the constancy of species. These were the first carefully performed experiments with artificial fertilization of various forms of plants, in which the author described in the hybrid progeny segregation of three types of offspring, those like the hybrid, and those like each of the parental forms. He sought an explanation on the level of contemporary alchemy, in analogy with salt formation.

After Kölreuter's death voices were raised in opposition to the idea of sex in plants, so in 1822 the Berlin academy of sciences offered a prize for new research elucidating the question. The winner was A.F. Wiegmann (1828), an apothecary from Brunswick, who crossed experimentally various species of peas. He described hybrid forms which were similar to the maternal or paternal plants, or combined the traits of the two. But he also observed traits which bore no resemblance to those of either parent, and failed to reach the nub of the problem. In 1830 the academy of sciences at Haarlem in The Netherlands offered a similar prize. The conditions laid down for its award stipulated that the work should explain the manner in which hybridization could be exploited in the breeding of plants.

The prize was won in 1837, by the German botanist F.C. Gärtner (1772–1850). In 1849 he published an extensive monograph on plant hybridization, reviewing the published findings of other researchers as well as reporting his own experimental results. In performing over 10 000 artificial fertilizations in 700 plant species, yielding 250 different hybrids, he demonstrated, in agreement with Kölreuter's findings, that hybrids exhibited decreased fertility. Gärtner also rejected the notion that natural hybridization gave rise to new plant species, thus confirming the constancy of species, and in accordance with German *Naturphilosophie* he asserted that form and essence were the same. His starting-point was the inborn character of the traits of the plant and the fact that their preservation and reproduction were assured by fertilization. When pollination with foreign pollen took place, in Gärtner's view, the traits of the plant were altered, since the pollen had a life-giving and, at the same time, a form-building force. From this point of view, according to Gärtner (1849, p. 250) 'the explanation of the formation and creation of bastards from elements and characters of the parent stock is as important for the physiology of the plants as for systematic botany'.

In his monograph Gärtner distinguished between hybrids from various points of view, explained in more detail in Chapter 5. For the most part he considered, like Kölreuter before him, hybrid plants in whose progeny alternative traits appeared to be the result of a process of reversion of hybrid forms to the stock species. However, he also pointed to the occurrence of new trait combinations, which could be used for creating new varieties of cultivated plants. In some species Gärtner also drew attention to the appearance of constant hybrid forms, capable of propagating themselves unchanged as new species. Later, M. Wichura (1854, 1865) published the results of crossing experiments with various species of willow, which Mendel was later mistakenly to assign to the category of constant hybrids.

In 1861 the French academy of sciences offered a prize for an explanation of the fertility of hybrids and the constancy of their traits. Two essays were later submitted. In the first A.A. Godron (1863) argued the fertility of hybrids within the species and the infertility of those between species. In the other, C. Naudin (1863), who won the first prize, offered an explanation of the uniformity of hybrid plants, and the fact that the same results are obtained from reciprocal crossings of parental forms. He supposed that hybrids display either a preponderance of the traits of one or other of the parents, or an average of the parental traits, and also observed segregation of parental traits in the offspring of hybrids, offering the explanation that it was due to the action of a 'specific essence'.

At the time that Wolf was active the theories of preformation and epigenesis were still competing. The subsequent dispute was heavily influenced by Blumenbach (1752–1840), a representative of German *Naturphilosophie* and professor at the University of Göttingen. In his essay *On the formation force and generation affair (Über den Bildungstrieb und das Zeugungsgeschäft)*, published in three editions in 1771–91, Blumenbach rejected any form of preformation, and as an experienced anatomist and anthropologist emphasized that the embryo formed from the very start by the action of a special 'formation force'. His natural history textbook (Blumenbach 1830), published in twelve editions between 1780 and 1830, had a profound influence on his contemporaries. Among these was the Czech physiologist Jan Evangelista Purkyně (1787–1869). (In German and English papers he spelled his name Purkinje.) In 1825 he dedicated one of his first papers to Blumenbach: his remarkable study on the origin of the avian egg, demonstrating a 'germinal vesicle' in the yolk, later shown to be the nucleus of the cell. Purkyně (1834), in his extensive speculations on generation, mentioned the transfer of traits from parents to offspring, though without using the term heredity (Orel *et al.* 1987). He supposed that a process of 'involution' reduced the parental traits to a mere quality in the germs of the parents and that, following their fusion, a process of evolution produced the embryo of a new individual, bearing the traits of the parents. Purkyně was also studying animal tissue under the microscope, and in 1837 he pointed to the analogy between plant cells and 'globules' in the tissue of animals. Two years later this analogy was generalized by Schwann (1810–82) as the cell theory. The idea of the cell as a common unit of structure and function in animals and plants became the starting-point for new efforts to explain the enigma of generation, and later that of fertilization.

In 1842–53, Professor R. Wagner, Blumenbach's successor at Göttingen university, edited in cooperation with leading European physiologists a *Dictionary of physiology*, comprising a series of monographs expounding the latest knowledge. The last of which, entitled *Generation (Zeugung)*, was by Wagner's pupil R. Leuckart (1853), at the time professor of zoology at Giessen. He connected the enigma of generation with heredity, but could find no explanation for it, nor a method of investigating the problem experimentally. Dissatisfied, Wagner (1853) wrote an eighteen-page postscript dealing with the problem from an historical and methodological point of view, and recommending experimental research into heredity. He expected the explanation of the problem of heredity to resolve the enigma of generation. Shortly afterwards Wagner acquainted himself with the latest literature, demonstrating the penetration of sperm into the mammalian egg, and immediately added a second, four-page postscript. He inclined towards the view that the embryo came into being with the participation of both parents, attributing the idea to Purkyně, who had learnt it from research by Barry (1840). (Attention was drawn to his contact with Purkyně by Wood (1989–90).) Purkyně made a detailed exposition of his ideas in lectures entitled *Physiological morphology*, most probably written in Prague shortly after 1850 (Orel and Janko 1988). He had already been in correspondence with Wagner, and around this time received a visit from him. Purkyně set out from the assumption that in the process of fertilization a fusion occurred between the germ substances of the two parents. He supposed the participation of a 'living formative force and a creative endowment', which were 'involuted', and from which under certain conditions the evolution of an embryo began. Kříženecký (1987) drew attention to Purkyně's visit to the Augustinian monastery in Brno in 1850. He undertook an excursion in the surroundings of Brno with Mendel's Fr. Klácel, discussing with him scientific ideas of the time relating to the phenomena of life, which interested them both. At that time Klácel too had experimental knowledge of his own about the artificial fertilization of plants, and he was interested in the scientific concept of evolution.

Later, in an article on the animal cell in 1860, Purkyně distinguished in each organic part 'two manners of being, the *external* material, by which it persists physically and chemically in being and presents itself to the senses, and the *internal*, germinal, life-forming, involuted, through which in the course of time and according to the laws of life, it develops to maturity of propagation, and finally extinction'. Attention has recently been drawn to this by Janko and Orel (1989–90). Special germinal matter in the context of heredity was a theme later developed by Weismann.

Shortly after 1850, the physiologists Franz Unger (1800–70) and Carl Nägeli (1817–91) began their physiological investigations into the problem of hybridization of plants at the Universities of Vienna and Munich respectively. They were seeking the basis of the variability of traits and the origin of new species. They set out from the enigma of fertilization, looking for an explanation in accordance with the cell theory as it was understood at the time. Their conception of research shows an attempt to make use of the methods of physics and chemistry, and hybridization is considered in quite a different light to that in which Kölreuter and Gärtner regarded it.

Since ancient times heredity had also been included in the observation of the varieties of the living world. The explanation was sought in the reflection of some large purpose in nature predestined by the will of God. In England scientists and theologians also saw God in the light of nature, and J. Ray (1627–1705), the father of British natural history, described the great variety of perfections and adaptations in animals and plants.

Ideas on heredity had also appeared in the first speculations regarding evolution. G.L.L. Buffon (1707–88), a contemporary of Linnaeus, placed the variability of traits of animals and plants in the same context as his speculations regarding the evolution of living things. He rejected the idea of constancy of species and pointed to their temporary existence, and the influence of external environment on heredity. It was on his teaching that Lamarck based his theory of transformation of organisms and of evolution from the lowest forms through to man himself. Lamarck (1809) attributed to living matter a tendency to grow in complexity, thus explaining the continual improvement of organisms. His name is associated with the idea of the inheritance of traits acquired through the action on organisms of the environment (as elucidated by R.W. Burkhardt 1977). Larmarck's ideas had a major impact on natural science not only in the nineteenth century, but also throughout the first half of the twentieth. German naturalists, under the influence of *Naturphilosophie*, conceived the idea of the evolution of the whole of living nature, including the evolution of individual organisms, whose investigation was to be treated as *Entwicklungsgeschichte*, translated into English as 'developmental history'. Orel *et al.* (1987) published the manuscript of Purkyně's lecture, explaining his understanding of the concept. This is dealt with in Chapter 5, p. 167. On Purkyně's suggestion, his pupil Valentin (1835) published a book on the topic. It was in connection with this attempt to clear up the enigma of generation and *Entwicklungsgeschichte* that the issue of heredity emerged as an independent problem.

The mysterious teleology in the evolution of living nature was explained without resort to divine intervention by Charles Darwin (1859) in his masterpiece *The origin of species*. His theory states that the progeny of parents includes individuals with random modifications of traits, which may by a process of natural selection become the basis not only of new traits, but even of new species. In his preoccupation with the concept of natural selection Darwin failed to explain the basis of heredity, admitting that 'the laws governing inheritance are for the most part unknown'. Following the publication of Darwin's theory natural scientists began to consider the problem of heredity, and did so in connection with the latest findings regarding the structure of the cell and cell nucleus. By the end of the 1860s the issue of heredity was making its appearance alongside that of evolution itself, and the preceding speculation about the hybridization of plants and the crossing of animals was fading into the background.

HEREDITY IN BREEDING PRACTICE

We shall never know when it was that a farmer first began selecting and crossing animals and plants in order to obtain higher-yielding forms. There is just the occasional mention in the literature of isolated efforts to select deliberately individual specimens for agriculture, or of a combination of selection and crossing to produce new plant varieties. An example is the letter written by the noted chemist J. Priestley (1797) to Sir John Sinclair, where he draws the latter's attention to a remarkable improvement achieved in plants by the American breeder Joseph Cooper (1759–1840), whose practice was 'the same as that adopted by Mr. Bakewell in England with respect to animals'.

The pioneers of plant breeding through hybridization knew nothing of the arguments over the sexuality of plants which were raging among botanists, and did not consider questions of the constancy of species or theories of generation. They based their practices on the observed variability of plant traits and their combining by crossing. A natural curiosity led them to carry out field experiments. At that time few pure naturalists took into account the empirical knowledge acquired in agriculture. Francis

Bacon, the intellectual founder of scientific philosophy, had already considered the possibility of changing plant and animal 'kinds' through the accumulation of variation during domestication and in agricultural plant and animal production. In his essay *A new Atlantis* he gave free rein to his imagination on the subject of the creation of totally new forms of animals and plants in agricultural practice; but his views seem to have had little influence, if any, on animal and plant breeders.

Isolated reports of breeding new varieties by means of crossing are to be found even in the seventeenth century. One facet of travel to newly-discovered regions in the sixteenth and seventeenth centuries was that natural scientists were sent out to investigate exotic flora and fauna. New plant species and varieties were found, and some of them were introduced to European countries. In newly established botanical gardens investigations were carried out into methods of plant propagation, allowing the dissemination of more attractive or more productive forms. The German naturalist G.A. Agricola (1716) published a remarkable book dealing with plant propagation. It is well illustrated with copperplate engravings to explain the various techniques of plant propagation. According to Agricola 'The plant soul is material and divisible, and as such can be divided into innumerable particles, so that it is preserved in the smallest particles in its function, which can be proved *a posteriori* through its effect.' Agricola's deliberations led him to the conclusion that 'vegetabilia or trees and shrubs' could be 'mutually transmuted and changed'. He therefore drew attention to the propagation of fruit trees and vine varieties from seed.

The French naturalist M. Duhamel (1700–82) was most probably the first to apply natural science systematically to growing plants. In his book *La physique des arbres* Duhamel (1758) tried to explain the variability of cultivated plants. In catalogues of seeds and plants he found descriptions of the traits of previous varieties not referred to at a later date. His conclusion was that most of the fruits which gardeners called new were nothing more than composites of previous varieties. The explanation he offered was that this occurred 'through a mixture of the pollen powder, through mutual fertilization'. On that occasion Duhamel referred to a similar process in crossing different races of animals, using for plant hybrids the term 'metis', formerly reserved for crossed animals.

A contemporary of Duhamel's in England was P. Miller (1692–1771), regarded as a leading expert on the question of raising cultivated plants. From 1722 to 1770 he was head of an apothecaries' company near Chelsea Gardens in London. His book *The gardener's and florist's dictionary or complete system of horticulture* was published in eight English-language editions during that period. Before its publication not more than a thousand species of cultivated plants were known. By the time he died more than 5000 plant species had been described. Miller also wrote about the process of plant fertilization and hybridization, stating that female plants could not bear fruit before being fertilized by the flower of the male plant. Miller (1751) called pollen grains flour or dust. He considered the discussion which was under way at that time on whether 'the impregnation process from *farina fecundans*, or male dust' entered the uterus 'in substance or effluvia' to be irrelevant, since Robert Boyle had shown that 'all effluvia are subtle particles of matter'. The mixture of traits in the progeny of hybrids Miller ascribed to the effluvium of particles from the dust into the fertilized egg. This was Miller's understanding of the origin of new varieties arising from hybridization. The German version of Miller's dictionary (1751) is preserved in the university library in Brno.

Kölreuter was the first pure naturalist to carry out systematic experiments with plant hybridization. Attention has recently been drawn to his interest in the application of hybridization in plant breeding by Mayr (1986). In the third part of his essay on sex in plants Kölreuter (1763, p. 176) wrote: 'I could wish that I, or someone else, might one day be lucky enough to produce hybrids of trees, the use of whose timber might have great economic effect. Among other good properties such trees might have one would be that if the original trees needed, for example, a hundred years of full growth, the hybrid would achieve the same in half that time. At least, I do not see why they should behave differently from other hybrid plants.'

Towards the end of the eighteenth century in the newly founded parks and botanical gardens in different European countries, introduced plants were grown and their economic traits were compared with those grown traditionally in fields and forests. In that period a unique park was organized in the agricultural economy of the large estates of the Duke of Liechtenstein's family in Lednice, south Moravia. After studying the organization of parks in France and in The Netherlands, T. Walashek (1753–1834), also known later by his title von Wahlberg, founded an extensive park where he grew a great collection of field and forest plant species and varieties. This collection was known in his lifetime as *herbarium vivum*. In the view of Wahlberg, the comparison of different plant varieties in certain soil and climatic conditions was the best way to find more productive forms to be cultivated. Wahlberg published the results of his investigation in an extensive book in Vienna in 1810. It was the first book to critically evaluate, in detail, the economic traits of a great collection of cultivated plants, in the modern spirit of the gene reservoir.

By the end of the eighteenth century progressive plant cultivators in England had begun a systematic investigation into the hybridization of cultivated plants to obtain increased plant production. But it was Thomas Andrew Knight (1759–1838) who began to publish information on this new trend in plant growing, on the basis of his own experiments. Setting out from the idea of the limited duration of plant varieties, Knight (1797) recommended constant selection of new varieties, and also considered the possibility of using artificial pollination. His most influential paper was published under the title 'An account of some experiments of the fecundation of vegetables' (Knight 1799). In the introduction attention is drawn to the analogy between animal and plant breeding, and it is also emphasized that the same means would produce the same consequences. His experiments soon confirmed his expectations. He set out from the animal breeders' idea of superfetation. According to this idea, the situation in plants was that in the fertilization process 'the explosion of the two vesicles of farina (taken from different plants) at the same moment may afford seeds (as I have supposed) of common parentage'. Ten years later Knight (1809) returned to the analogy that subsists between plants and animals, and considered the animal breeder's idea of superfetation for the purpose of explaining the role of parents in the transmission of traits.

In his deliberations on creating new varieties of fruit trees with the application of artificial pollination, Knight was aware that 'several years must elapse before the success or failure of this process could possibly be ascertained'. He therefore came upon the idea of carrying out hybridizing experiments with annual plants, in order to obtain experience of his new breeding system more quickly. Among these plants 'none appeared so well calculated to answer my purpose as the common pea, not only because I could obtain many varieties of this plant, of different forms, sizes, and colours, but because the structure of its blossom, by preventing the ingress of adventitious farina, has rendered its varieties remarkably permanent'. In his experiments, which Knight had begun as early as 1787, he examined the transmission of individual traits of seeds and observed a uniform appearance of the first hybrids and a greater variation of the trait in the next generation. The crucial problem here was whether the largest and smallest quantities of farina produced any difference in effect. Knight's publication stimulated other scientists to carry out experiments in plant-crossing. The publications most often cited are those of J. Goss (1824) and A. Seton (1824), relating to experiments with *Pisum*. Along with the dominance of seed traits they also described the segregation of parental traits in the hybrid progeny.

Being acquainted with the experiments of Goss and Seton, Knight (1823) returned to his motivation for experiments in plant hybridization, stressing that in his experiments with peas he wished to obtain such information as would enable him to 'calculate the probable effects of similar operations upon other species of plants'. In this context he believed that it would not be easy to suggest an experiment of cross-breeding upon this plant, of which one has not seen the result, 'through many generation'.

Knight's most influential paper on hybridization, published in 1799, did not escape the attention of plant breeders on the Continent. In the following year a German translation appeared in Leipzig (Knight 1800), bearing a remarkable footnote in which the anonymous translator states that 'The art of forming new varieties from two plants through blending of seeds is already known to Germans from the discoveries of Kölreuter, and later from various writings and observations of J. Hedwig in the years 1793–7'. The latter experimented with the crossing of different varieties of cucumbers, though his work was interrupted by his premature death (Hedwig 1797). In the footnote to the German paper it is stressed that Knight's paper has been written for the benefit of agriculturalists, and therefore the results of further experiments were anticipated. The journal in which the paper was published is deposited in the university library in Brno, with the stamp of the Agricultural Society. It was thus available to those who were interested in breeding plants in Brno (Orel 1978*b*).

Knight also influenced the respected French pomologist A. Sageret (1763–1851) who after 1800 began experimenting on the crossing of melons and studying the transfer of individual traits to the hybrids and their progeny. Sageret (1826) showed the dominance of traits in hybrids and the segregation of the parental traits in the hybrid progeny, and also the combination of parental traits. Like Knight, he gave as his motivation the goal of verifying in a few months what would in fruit trees be the work of many years.

Knight's interest in breeding new plant varieties came from his previous experience with sheep and cattle breeding for meat production according to the methods of Robert Bakewell (1725–95), whose achievements encouraged both animal and plant breeders at the end of the eighteenth century (Mylechreest 1988).

Bakewell's fame was founded chiefly on the creation of a new breed of meat-producing Dishley sheep (Wood 1973). Continental farmers, however, were more interested in keeping sheep for wool. Even in the seventeenth century it was known that the best-quality wool was that obtained from Spanish sheep, and in the course of the eighteenth century they came to be exported to various European countries. Because of the new conditions under which these animals were bred, their owners had to provide fodder all the year round and sheds for the winter, and there was also the new problem of reproduction within the relatively small flock of imported animals. Experts began to publish manuals and handbooks on the breeding of imported sheep; mention was made of the crossing of imported rams with local ewes and of consanguineous mating, at the time publicly regarded as violating the prohibitions of religious dogma.

The noted French naturalist L.G.M. Daubenton (1716–1800), on the instigation of government authorities, began in 1767 to experiment with the crossing of sheep breeds in order to improve the quality of wool produced. Ten years later he read to the Academy of Science a paper reporting the results he had achieved. In it he evaluated methods of crossing and the wool quality of the parental animals and of the progeny. This lecture was included in his book on sheep husbandry (Daubenton 1782).

After 1800 most of the books published on agriculture on the Continent dealt with the breeding of wool-producing sheep. The high prices commanded by wool from Spanish sheep aroused interest in improving the quality of that produced by local breeds by crossing them with imported Spanish rams. The finest wool, which was the type that fetched the highest prices, was then produced by imported Spanish sheep in Saxony. After 1800, however, greater and longer-lasting success was achieved in the Habsbury provinces of Moravia and Austria. The best known breeder in the early nineteenth century was Ferdinand Geisslern (1751–1824), who came to be known as the 'Moravian Bakewell'. The economic advantages of his breeding practices, as in England, stimulated interest among farmers in applying hybridization to plant-breeding practices in Mendel's homeland, Moravia (Orel and Wood 1981).

At the start of the nineteenth century Brno, capital of the province of Moravia, became the centre of the textile industry of the Habsburg monarchy (Freudenberger 1977) (Fig. 2.1). With an eye to this, those who were chiefly concerned with textile development tried to organize the furtherance of natural science on the basis of learned societies. The learned naturalist Christian Carl André (1763–1831) was soon to be one of the foremost figures. In 1806, the Moravian Society for the Improvement of Agriculture, Natural Science, and Knowledge of the Country (henceforth referred to as the Agricultural Society) was inaugurated, combining the stagnating Moravian Agricultural Society, established in 1770, with very active private societies of naturalists. André (1815) drew up a programme of scientific development, emphasizing the importance of developing basic and applied research in the natural sciences, and in this connection pointed out the significance of great scientific discoveries. The examples he gave were of the discoverers Copernicus and Newton, and he expressed a conviction that the members of the new society might, by dint of carefully conceived research work, lay the foundations for some similarly great achieve-

21

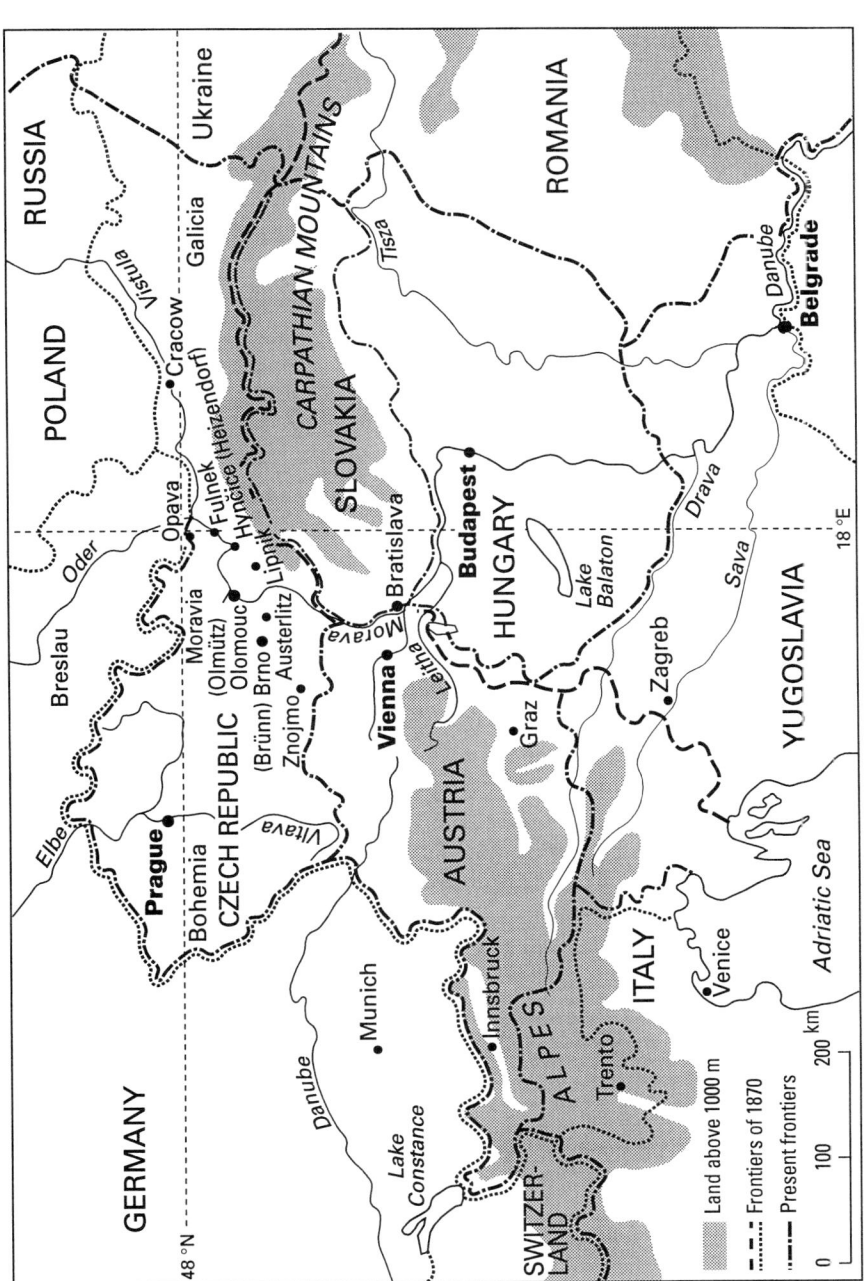

Fig. 2.1 Map of the Habsburg monarchy, 1870, showing places connected with Mendel.

ment, though it might not see the light of day until a much later date, when the whole of civilization would be indebted to Moravia. At the time his choice of words seemed extravagant. Soon afterwards André was to devote more and more time and attention to the question of sheep breeding, considering it a scientific activity, and in this connection he came into contact with the question of heredity. Later H.F. Salm and C.C. André (1814) founded in Brno, in affiliation with the Agricultural Society, an 'Association of Friends of, Experts on, and Supporters of Sheep Breeding, for the achievement of the further well-founded advancement of this branch of the economy, and the manufacturing and commercial aspects of the important woollen industry which are based upon it'. Throughout this book this association is referred to as the Sheep Breeders' Society. This was the first animal-breeding association to be set up in continental Europe, and every year it held meetings of sheep breeders from throughout central Europe and also from as far afield as Mecklenburg in Prussia. The agenda included methods of artificial selection and the transmission of traits of wool from parents to offspring—in other words, heredity (Orel 1974a).

Before coming to Brno, André, a former student of the University of Jena, had worked in Saxony as an educationist and author of natural scientific and economic publications. He also became a founder member of the first mineralogical society, established in Jena in 1798, and was an enthusiastic student of this nascent branch of science. In developing methods of sheep breeding, André emphasized that this was the start of a new science, for which a new terminology would have to be created (Franke and Orel 1983). André had been prepared for his understanding of the enigma of reproduction before coming to Brno through writing his textbook on zoology. In his extensive introduction to that work André (1795), on the basis of his experience in animal breeding, explained the participation of both parents in the origin of the new individual, rejecting the old notion of preformation. In journals published in German in thousands of copies, André reported the discussions of the society regarding methods of scientific breeding and heredity, though he used the latter term only exceptionally. He regularly published reviews of publications on agriculture and the natural sciences associated with it, paying special attention to animal and plant breeding.

Economic achievements in England and Moravia aroused interest among plant breeders in using methods of artificial pollination in order to achieve new and higher-yielding varieties of fruit trees. In 1816 André drew attention to the success of Knight in breeding new varieties of vine and fruit trees, and recommended that his method should be used for creating new varieties of fruit trees. In the same year he founded in Brno a Pomological and

Oenological Association, which later became known simply as the Pomological Association. Shortly afterwards the committee established a stock nursery for the creation of new varieties, combining useful traits from existing ones (Orel 1978*b*). André repeatedly published reports on the activities of the Horticultural Society of London, established in 1804, of which Knight was president. Their author was André's friend G.C.L. Hempel, secretary of the Pomological Association at Altenburg, near Leipzig, founded in 1803, who was also an overseas member of the London Horticultural Society. In 1818 it was even suggested that permanent cooperation should be initiated between London, Brno, and Altenburg.

A most remarkable article appeared in 1820. André had asked Hempel, whom he considered an expert in plant breeding, to explain the possibility of applying the methods of artificial fertilization to breeding new varieties of cereal. In an extensive paper Hempel (1820) described the experience of Knight in selecting new varieties of fruit trees. He was sure that 'higher scientific pomology' was moving towards the point where the breeder would be able to create a new variety according to a previously conceived ideal type, combining traits of fruit size, shape, colour, and flavour. But he emphasized that it would first be necessary to explain the laws of hybridization which applied to sexually reproducing plants such as grain crops. Only then, he thought, would hybridization in breeding practice open up the way to the creation of entirely new varieties of all species of cultivated plants, with much higher yields than farmers could at present imagine. He did not expect this problem to be solved in the near future, stating that a new type of natural scientist would have to emerge, able to perform demanding experiments in plant hybridization. He even tried to characterize him: he would be 'a researcher with a profound knowledge of botany and sharply defined powers of observation, who might, with untiring and stubborn patience, grasp the subtleties of these experiments, take a firm command of them, and provide a clear explanation'. This was a pretty fair description of Mendel, born two years later in Moravia (Orel 1974*b*).

At the time Hempel's predictions were published, the president of the Brno Pomological Association was André's friend Jan Sedláček von Harkenfeld (1760–1827), who was then beginning to use methods of hybridization in breeding new varieties of fruit trees and vines. Shortly before he died he outlined a plan for breeding vines in order to achieve new varieties which would not only be 'constant in vegetation and suitable for our climate', but whose grapes would also 'unite in themselves many superior characteristics, so that the wine pressed from them would be of higher quality'. Perhaps, he adds, they would also 'be such noble varieties as until now we do not have even from abroad' (von Harkenfeld 1826).

After Sedláček's death his place at the association's helm was taken by C.F. Napp (1792–1867), newly elected abbot of the Augustinian monastery in Brno (Orel 1975*a*). It was supposed that he would continue to 'promote pomology and wine-growing through experiments, observation and instruction, enrich science, and propagate useful findings'. He was the right choice for the job. Before being elected abbot, Napp had organized a nursery garden on the previously neglected monastery farm at the village of Šardice, about 80 km south-east of Brno. By 1829 it had 3400 seedlings, of which 700 were characterized as improved. According to the report of the Pomological Association, Napp also wrote a handbook to instruct village people on how to improve fruit trees in that neglected part of the province. The stock nursery for breeding and disseminating new varieties of fruit and grapes established by Sedláček inside the association soon became unsatisfactory, so in the 1830s Napp created another one in the monastery, later described by Hungarian expert F. Schams (1836) as 'an institute created for practical experiments' and 'a jewel among nurseries'. Over a hundred varieties were investigated there (Weiling 1991) from the viewpoint of yield. Keller, the head of the institute, wished to examine individual grape varieties in separately produced wines.

After 1820 the method of artificial pollination began to be used in many European countries for breeding new varieties of fruit trees and ornamental plants, and later also for other plant species. Much attention was paid to the breeding of plants by the members of the Horticultural Society in Bavaria, based in Frauendorf. The society published a journal which was subscribed to by some of the members of the Brno Pomological Association. Its president, Napp, used a report on the contents of this journal during meetings. A German translation of the English papers by J. Goss and A. Seton (Anon. [G.] 1837) appeared in the journal, describing the experiments on crossing of peas which they had performed at the suggestion of Knight. The German translation did not mention the English authors. The text also falsely gave the year of the start of the experiments as 1831 instead of 1821. The reader was certain to get the impression that the experiments had been carried out in Bavaria.

At the time much attention was also being devoted to the application of hybridization in plant breeding in The Netherlands and in Prussia, and prizes were being offered for elucidating the effect of hybridization.

The chief organizer of the development of methods of scientific breeding of sheep and plants in Moravia was C.C. André, who laid the foundations for an exchange of experience with experts from abroad (d'Elvert 1870, II pp. 119–20). In 1820 he was forced to leave the territory of the Habsburg monarchy because of his liberal views, and moved to Stuttgart, where he no longer had the opportunity to devote his attention to questions of breeding animals and plants. But he had set off a chain reaction in Moravia, and his former colleagues continued to support the development of methods of breeding practice.

HEREDITY IN UNIVERSITY TEACHING

The introduction of rational methods in agriculture called for a knowledge of the natural sciences and its application to the production process. So in the late eighteenth and early nineteenth centuries the subject of agricultural science appeared on university curricula. One of the first universities to introduce this branch of study was Edinburgh. Later an enthusiastic propagator of the rational methods of British agriculture was A. Thaer (1752–1826), professor of agricultural science at the University of Berlin from 1810. His meritorious activities were reported by André in his journal, one of whose readers was Thaer himself. André had suggested setting up a professorship of agricultural science in Moravia as far back as 1808. He himself was the most suitable candidate, but he was not accepted, most probably because of his liberal views. The first professor of agriculture in Moravia was established at Olomouc in 1811, the second in Brno in 1816. But progress in methods of sheep and plant breeding was not affected by university teaching until the next generation of professors. In 1823 a former colleague of André's, J.K. Nestler (1783–1841), began teaching the subject in Olomouc, and the Brno chair of agriculture was taken by F. Diebl (1770–1859), an associate of Napp's, in 1825. Nestler paid greater attention to sheep breeding, Diebl to plant breeding (d'Elvert 1870, II, pp. 203–15, 280–9, 315–31; Orel 1975*a*, 1977; Weiling 1976, 1982). Both professors took an active part in the activities of breeding societies, and published articles dealing, *inter alia*, with hybridization and heredity. At other schools in the Habsburg monarchy, teachers of agriculture dealt with questions of animal and plant breeding only marginally.

Nestler (1829) published his lectures on breeding under the title 'The influence of generation on the characteristics of the progeny'. An editor's note pointed out that this was 'part of the model lectures on agricultural science and natural history in which that worthy author develops the most important aspects of rational animal breeding, motivated by their various relations even in 1827'. The title does not mention the word inheritance, but the reader knew that this was a study of the principles according to which traits were transferred from parents to offspring. Right at the beginning of his text Nestler stated that 'fruitful generation with heredity of all the substantial traits in the progeny is possible only between two kinds (in the natural historical sense) belonging to the same species'. The lectures describe empirical knowledge derived from selection practice, but treated from a wider natural scientific understanding of the phenomena of variation, crossing, and hybridization. Attention is also paid to the hybridizing experiments of Kölreuter. Examples are given from the animal and plant kingdoms, and man is also mentioned. Finally the influence of consanguineous crossing on hereditary diseases is considered, and in this connection the author points to the significance of progeny testing in animal breeding.

Nestler described the animal and plant traits determining production in agriculture, and mentioned the endowment (*Anlage*) of traits, but did not get as far as distinguishing the determinants governing heredity. The lectures deal with the concept of blending inheritance, but on the other hand mention is made of the observed inheritance of alternative traits, such as the presence of horns in cattle. Nestler refers to the sudden appearance of new traits as sports, or freaks, of nature, which he considered inexplicable at that time. He also tries to differentiate between the influence of the environment and that of heredity in the manifestation of the production traits. The heredity of acquired characteristics is described as problematical. In describing the application of consanguineous crossing in breeding practice, Nestler defends this method against its opponents, who were still in the majority.

The publication of Nestler's lectures on animal and plant breeding evoked a renewed discussion among sheep breeders regarding the theoretical bases of the process of scientific breeding, which had been a subject of discussion in the new journal *Mittheilungen*, published by the Agricultural Society since 1821. Studying natural scientific literature, Ehrenfels (1831) tried to explain the origin of life from 'the lifeless chaos of matter' by means of a 'generation force'. He also referred to it as a 'genetic force', which in interaction with climate and nutrition was considered 'the main lever of nature in the formation of matter'. From this the sheep breeders deduced that it was possible to modify the traits of animals in order to form organisms with a higher production-capacity. These speculations based on empirical findings from breeding practice encouraged them to elucidate the theory of breeding and to seek a new breeding method in order to improve production.

The discussions among breeders in Brno regarding generation and artificial selection reached a climax in 1836 and 1837. The president of the Sheep Breeders' Association, E. Bartenstein, asked Professor Nestler to choose the topic which would be most benefit to and most desirable for sheep breeders. According to Nestler 'the most essential thing of all for improved sheep breeding, as well as being an urgent question of our time' was 'the ability to inherit'. He thus provoked a discussion among the participants in the annual meeting on a crucial problem in breeding practice, heredity. The participants expected a solution to the problem to come from an analysis of pedigree records and the data from progeny testing of traits determining production capacity (Orel 1978c).

The two professors of agricultural science and natural history in Moravia worked together with fruit and plant breeders, and every year they published numerous specialized articles explaining the latest scientific findings and their application to agricultural practice. The agricultural schools in the monarchy used the textbook by L. Trautmann (1814), a professor at the University of Vienna. It came out in four editions. Another widely diffused book was that written by agriculturalist J. Burger (1819), which was translated into Swedish and Polish. There is no mention in either of these books of the breeding of new plant varieties through hybridization. The number of students studying agricultural science in Moravia increased rapidly, a trend which was reflected in the publication of a five-volume book by F. Diebl (1835–44) under the appropriate title *On the science of agriculture for agriculturalists, especially those devoting themselves to a study of this science*. The second volume, on plant production, describes the anatomy of flowers and the technique of artificial pollination as a method of obtaining new, more productive varieties. Both Nestler and Diebl taught agricultural science and natural history. This led to a textbook on which they cooperated, called *General natural history* (Nestler and Diebl 1836).

In the journal of the Agricultural Society there appeared an article by Baron von Witten (1828) called 'On wheat varieties', which had previously been published in the journal of the Prussian Horticultural Society. It was considered interesting, and in the next year Diebl (1829) published a critique of it explaining his own view on the classification of cultivated plants. This was based on a knowledge of agriculture and of natural scientific literature, especially concerning the practice of selecting cereals. According to him the naturalist considers only constant traits which are inherited. Under special conditions these traits do not appear, but they may reappear. Their origin cannot be explained from the viewpoint of contemporary natural history. Plant physiology, however, allows us to accept the view that such traits arise through hybrid fertilization under certain conditions, through a force still unknown to us. Most cultivated plants have their subspecies, and investigation shows that their distinctive traits are not constant, and undergo changes attributable to the factors mentioned above. Natural historians only establish the existence of varieties, but the plant breeder investigates them scientifically with the goal of creating new, more productive forms.

Another remarkable article appeared in Brno some time later. The author (Anon 1834) pointed out the possibility of prudent pairing of individual plants with different traits. He emphasized that by using artificial selection it was possible to obtain a combination of selected traits and reproduce them permanently in new varieties. At the end of the article there is a reference to a different technique of pollination in beans and peas. The author may again have been Diebl. This self-taught man had, through tremendous hard work, attained a broad knowledge, and was acknowledged not only as a skilful farm manager, but also as a breeder and later as a teacher. Diebl (1839) considered the development of agricultural production in an article called 'The necessary struggle of a farmer for the improvement and preservation of agricultural products'. In his view 'man raised himself from his original animal position in nature and placed himself over the animal kingdom'. His needs began to increase, and their satisfaction began to refine his existence. He considerably improved the products of nature for his own purposes. But improved plants and animals still had a tendency to revert to their original state. Man found that the improvement of forms had its limitations. But now natural science was offering new possibilities of pushing back the barriers to further improvement, and a new process of plant and animal breeding was leading to the gradual overcoming of nature. In this progressive spread of new knowledge Diebl gave pride of place to the local grass-roots gatherings of learned societies, where agriculturalists met natural scientists, which led to the diffusion of more and more new knowledge, and also encouraged new discoveries in natural science.

In the 1830s Professor Diebl worked in close conjunction with Abbot Napp on the committee of the Agricultural Society, and in particular in the Pomological Association, of which Napp was president. Napp and Diebl (1838) announced a prize essay sponsored by the Pomological Association on the subject of a new, improved variety of currant bush. They expected a presentation of the improved fruit and an essay describing the manner in which the variety was obtained. In 1839 the essays of the gardeners M. Frey (1839) and J. Twrdy (1839) were published in the journal of the Agricultural Society. There is a footnote to Frey's paper describing the method of artificial fertilization. Its author can only have been Professor Diebl, who had already published his ideas on the breeding of cereal varieties using artificial selection.

In 1840 the Agricultural Society, on whose committee Abbot Napp occupied a position of great influence, decided to offer an extraordinary prize of 1000 guilders, and to award the gold medal of the society, to the author of a chronicle of the development of natural science over the preceding hundred years in connection with the improvement of agriculture. The anonymously published announcement stated that many of the findings of natural science had still to be exploited in practice, and that the work was intended to evoke interest in a more detailed study of the subject and in particular of ways in which its findings could be used by farmers. Suggested models for the work were the history of the natural sciences published by the Frenchman G. Cuvier (1769–1832) or the history of natural science by the Englishman H. Davy (1778–1829). The author was expected to deal with agricultural production in its entirety, but without going into detail.

Not until 1850 was a manuscript submitted anonymously by N.K. Fraas (1810–75), professor of agricultural botany at the University of Munich. A commission of ten experts appointed by the committee of the Agricultural Society studied the work at daily sessions over a period of two months. Their report to the committee stated that it represented a summary of knowledge acquired in the course of progress in natural and agricultural science, with an extensive bibliography of literature published in Germany, France, England, Italy, Holland, Sweden, and elsewhere. It was thus of enduring value and a tribute to both the author and the society which had given the stimulus for its creation (Hackler 1851). Fraas (1852) published his work in Prague, in two volumes. The introduction to the commission's report highlights the author's pointing to inconsistencies between the way in which certain problems are understood by 'pure' natural scientists and by agricultural experts. Fraas emphasizes that naturalists have failed to grasp that agriculture is a synthesis of the application of natural science with that of economic principles. Though knowing nothing of research into the cell and the penetration of the pollen tube in the fertilization of a plant, or of the systematics of plants, farmers have for centuries made use of the sexuality of plants in the artificial pollination of various varieties. New varieties, even species, of agricultural crops are created, and no-one can now say where they originated from. Agriculturalists have thus refuted the mistaken view of natural scientists that species are constant, and have shown that hybrids are fertile.

The report on Fraas' work was presented to the committee of the Agricultural Society, where the views on artificial pollination and the breeding of new plant varieties cannot have escaped the attention of Abbot Napp. The report was published in Brno at the time Napp sent Mendel to study at Vienna University.

HEREDITY AS A RESEARCH PROJECT

Research into heredity received its first substantial impetus from sheep breeding. C.C. André (1812) (Fig. 2.2), dealing with the scientific concept of breeding, had described artificial selection as a method of increasing the

Fig. 2.2 Christian Carl André (1763–1831), who gave the main impulse for the promotion of science in Moravia after 1800. He also began to organize animal and plant breeding on a new scientific level.

production capacity of wool sheep. Four years later his son Rudolf André (1816) published an important manual on scientific methods of sheep breeding, emphasizing that this was a new method. Two years after this, the elder André (C.C. André 1818) reached the conclusion that if animal propagation in the closest kinship was continued unrestricted it would lead to a weakening of the breed. Here he saw the operation of 'a natural physiological law', which he wished to investigate in more detail, adding: 'To do justice to this problem I should have to write a book on it.'

In the same year André the elder asked Count E. Festetics (1764–1847) to formulate the main principles of the use of interbreeding in breeding practice. In the following year Festetics (1819) published the first rough formulation of 'genetic laws', already explicitly using a terminology (in German *genetische Gesetze*) that was to be more generally introduced into genetics in 1906. He states: (1) the characteristic traits of healthy parents are inherited by the progeny. (2) sometimes the traits of one of the grandparents appear in the progeny. (3) on other occasions the progeny exhibit traits which are quite different, and, if they do not correspond to the aims of the breeder and are heritable, are undesirable. (4) the main requirement for the use of inbreeding is scrupulous selection of the stock animals. In conclusion Festetics recommends the circumspect use of consanguineous mating to maintain the required production traits of animals (Orel 1989, 1994).

Count Festetics was an acknowledged expert on sheep breeding in Hungary, and as a participant in the annual meetings of breeders held in Brno he returned to Rudolf André's proposal for the testing of wool quality: 'It will be judged as marking the beginning of a new epoch in the science of breeding, that in 1819 grades of wool fineness were established and defined with mathematical precision (Festetics 1820).'

An important milestone along the road to understanding the principles of scientific sheep breeding was the annual meeting in 1836. The president of the Sheep Breeders Association at that time, Bartenstein (1837), reached the conclusion that even if sheep breeders knew much more of the principles of selection since the founding of the society than they had before that, they realized that there was 'much more to be investigated'; indeed, they were only just arriving at the stage of rethinking their ideas. According to him it was necessary to take 'a deep look at the great and mysterious works of nature'. Given such a bold start, one might 'through the weakness of mortals' easily and understandably be led astray. Thus the president of the society, Bartenstein, turned to university professor Nestler to suggest a new topic for discussion at the annual general meeting of the Association in 1836. Nestler pointed out the importance of clearing up the question of which traits were transmitted more readily to the offspring, which were transmitted with difficulty, and which were passed on predictably, and under what circumstances.

Sheep breeders, along with Nestler, supposed that they could contribute to new findings, and to the monitoring of heredity through *a posteriori* analysis of the pedigree and progeny testing records of animals. According to Teindl *et al.* (1836), the discussion was pursued by Abbot Napp, who asserted that 'heredity of the characteristics from the producer to the produced (*seitens der Erzeuger auf die Erzeugten*) consisted above all of mutual elective kinship of the paired animal'. As a result a ram should be chosen for the ewe which corresponds to it in its inner and outer organism; this process must be the result of a substantive physiological study.

The meeting mentioned above also provoked a subsequent discussion on heredity in the pages of the Agricultural Society's journal. In the next year the prominent Austrian sheep breeder Ehrenfels (1837) called the discussion on heredity which arose very important, and said that the result was that it was carried on at the annual general meeting the next year. Its notable conclusions were summarized by Napp with rigorous brevity. He emphasized that it was not only a question of the breeding process, but that the question to be answered was 'what is inherited, and how?'. Professor Nestler (1837) reviewed the debate, and in an article entitled 'Heredity in sheep breeding' expressed his satisfaction over the progress which had been achieved, though he was unable to add anything new to the discussion. But a *volte-face* had occurred in his conception of the matter. Whereas in 1829 he had published his lectures on scientific breeding without using the term heredity in the title, by 1837 he considered heredity to be a new scientific problem with serious economic implications that was crying out for experimental investigation. At this latter date he also explained how he had arrived at a definition of the problem of heredity. At an exhibition of pedigree sheep a prize-winning ram from the breeding farm of Count R.E. Wrbna provoked the question of what price to put on such an animal, and concluded: 'It cannot be sold at any price if its advantages are inherited by its offspring; if they are not, then its price is no more than that of its wool, meat, and skin.' Nestler went on: 'Thus Count Wrbna has posed the question of heredity to me, and through me to the Sheep Breeders' Association.' At that time the question seemed to many participants, according to Nestler, a strange one. Some even declared it not worth discussing, which Nestler considered a great mistake. He was convinced that he had 'sown the seed of the question in the proper soil' and that it could 'gradually be developed into the luxuriant fruit of science', if the embryo was cared for.

In 1840 the fourth congress of German-speaking farmers and foresters was held in Brno, with Abbot Napp chairing a discussion on fruit-tree breeding, where he defended the hypothesis that artificial fertilization was a method of creating new varieties, against the view of some other participants that hybridization was merely a random process. Nestler edited the proceedings. According to Napp: 'In the mean-time nothing certain can be said in advance as to why production through artificial fertilization remains a lengthy, troublesome and random affair (Nestler 1841, p. 337).' In fact he was thus completing his formulation of the problem of heredity begun in 1837 with a third point—what was the role of chance in heredity?

In the 1830s the discussion of methods of sheep breeding in Brno was at its height. Before that the members of the Sheep Breeders' Association had exchanged their experience at the annual meetings and on the pages of journals. This gave rise to mutual cooperation, which J. Teindl (1822), described as follows: 'Through mutual endeavour a desire for the progressive improvement of domestic animals is being awakened in landlords as a matter of priority; instead of uncertain procedures, more correct principles derived from experience and exact observation are being established, and candid warnings against the false trails and blind alleys to be avoided are being disseminated, so that the general standard of oviculture is everywhere being simultaneously improved.' At the congress of farmers in Brno in 1840 mentioned above Nestler, at the instigation of Napp, noted (Nestler 1841, p. 216) that: 'Moravia can claim special credit for having become, through Geisslern's school at Hoštice and from 1814 through the stimulus provided by the meritorious activity of Sheep Breeding Society in Brno, a source of modern rational sheep breeding.' After André left Moravia, his former colleague Nestler continued to develop the principles of scientific breeding. In 1841 he died suddenly; but by then the import of cheap wool from British colonies had begun to have an impact, and sheep breeding in Moravia had lost much of its economic significance. In the 1840s no one showed any further interest in performing experiments for the improvement of sheep-breeding methods. Increased attention was, however, being paid in Brno to the use of plant hybridization, mainly owing to the efforts of Professor Diebl and Abbot Napp.

The enigma of heredity in connection with sheep breeding reappeared in 1853 in the theoretical speculations of the pure naturalist Professor Wagner (1853, p. 1007) at Göttingen; he was seeking an explanation of what were then the key questions of generation and heredity. He mentioned the corporeal peculiarities of organisms which were passed on to the next generation and beyond, and perspicaceously called for a critical examination of the observed facts and a careful elucidation of the accumulated statistics (*Zusammenstellungen*). He believed that 'more exact ascertainment of the numerical data could furnish a reliable clue'. Wagner defined six areas of interest in research into heredity, some of which had already been explained by Professor Nestler in his published lectures on the basis of empirical knowledge arising from sheep breeding. Wagner was sure that research into heredity could be carried out, but he saw it as being both an expensive and time-consuming undertaking.

In 1851 he himself tried to experiment with crossing various amphibians with differing qualitative traits. In the end he chose frogs and fish. He was aware that experiments must be carried out on a large number of animals.

But he did not manage to obtain specimens of the same age after artificial fertilization, and he abandoned his efforts, realizing that it was not feasible to carry out such experiments in the laboratory of his institute. He expressed the opinion that they could be performed on large stud- or sheep-breeding farms. He also admitted the possibility of doing experiments with crossing exotic animals at London Zoo. In connection with attempts to achieve artificial insemination in dogs he suggested that experiments might be carried out at veterinary schools. Professor Wagner had no idea that experiments had been carried out on a large scale in Moravia in connection with sheep breeding, and that on the basis of the results obtained Professor Nestler had, along with Abbot Napp, already defined a research task relating to the basis of heredity.

A noteworthy conclusion of Wagner's (1853, p. 1018) was that 'teaching on plant hybridization rests on a similar, more enlightened basis'. He refers to the latest publications on plant fertilization. Thus Wagner was the first pure naturalist who explicitly drew attention to the possibility of independent research into heredity through plant-hybridizing experiments.

The variance of views on research into hybridization and the use of hybridization methods in breeding plants was clearly pointed out by learned forester R. Geschwind (1829–1910), who lived and worked in Bohemia, Slovakia, and Hungary. He first applied his knowledge of botany to the successful breeding of roses (Geschwind 1885). His achievements in this field led him to propose the founding of nurseries for breeding new varieties of forest trees. In a remarkable paper Geschwind (1864) describes his experience with breeding woody species. The introductory sentence emphasizes that the art of producing plant bastards by means of the artificial cross-fertilization of plants had become established in horticulture a hundred years previously, over a broad area, and would transform horticulture. Hence, it is no wonder that men of science have welcomed the progress of hybridization with open arms, because of the opportunity it provides of examining the mystery of its laws (Orel 1986; Weiling 1975a).

Geschwind's experiments were in fact applied research. He showed a familiarity with the literature on research by botanists, the methods of breeders, and the work of physiologists investigating hybridization and the fertilization enigma. He also knew that some naturalists were already engaged in theoretical research into the problem. Forty-four years previously, Hempel had foreseen a new type of researcher, who would explain the laws of hybridization in the distant future. By the time Geschwind's paper was published, just such a researcher was already writing up in Brno his theoretical conclusions drawn from experiments into plant hybridization.

The problem of hybridization as seen by botanists and plant breeders was treated in an extensive paper by Ch.F. Hornschuh (1848). Its citation in Gärtner's monograph did not escape Mendel's attention. Hornschuh wrote that the problem of the transformation of species was the weakest link in plant science, and that no one had yet explained the transformation of species or refuted the idea. It was still claimed that many forms of vetch were derived from forms of peas and lentils. Hornschuh drew attention to the fact that the rigorous segregation of forms of plants came to the fore without the occurrence of any intermediate forms. In his view these questions were only just starting to become the subject of research by a few isolated scholars, whilst others considered them almost ridiculous. Mendel was among the scholars for whom the problem was ripe for scientific research.

What, then, was the milieu from which this 'new type of researcher' came, and what qualifications, what motivation, did he have for undertaking such research?

3
The early years

His sorrowful youth taught him early the serious aspects of life, and it also taught him to work.

—MENDEL (1850) in his curriculum vitae.

Mendel was born on 22 July 1822, the only son of a peasant farmer in the small village of Hynčice, consisting of seventy-two households. The entry in the baptismal register of the village church gives his date of birth as 20 July, but Mendel himself always stated that he was born on 22 July, and the biographical sources accept this date (van der Pas 1972). It was a part of the rural tradition that an only son followed in his father's footsteps. At the time the village was situated in Moravian Silesia, a province of the Habsburg empire. The area was known historically as Kravařsko, from the fact that in the thirteenth and fourteenth centuries it was ruled over by the house of Kravař. The estates consisted of a strip of land 24 km long and around 3 km across, giving way at the northern end to a range of hills. The forested ridge known as Pohoř rises above the village. Under the Kravař family most of the inhabitants were ethnic Czechs. In the thirteenth century settlers began to arrive there from various German provinces, and the two ethnic communities gradually became intermingled.

In the sixteenth century most of the inhabitants were non-Catholics. Between 1618 and 1620 the administrator of the Moravian Brethren's school in the nearby town of Fulnek was John Amos Comenius (1592–1670), the famous educationist. Following the Battle of the White Mountain in 1620, the Kingdom of Bohemia, including the Margravate of Moravia, was subjected to the Habsburg monarchy. The repressive re-establishment of Catholicism forced Comenius into exile, first in Poland, then in Sweden, and finally, for the longest period, in The Netherlands. As an acknowledged expert on education, who had published a great deal of material on the subject, he was invited to several other lands to reorganize their educational systems. While still in Moravia, Comenius recommended, after studying the works of Francis Bacon (1581–1626), that natural history should be included in a programme of general education. Though his proposal was not accepted, Comenius revived the idea on a number of occasions. In 1641 he was invited to England to modernize the educational system, but the English Revolution soon broke out, and he was obliged to return to The Netherlands. When the Royal Society was later founded in London, Comenius (1668) dedicated his book *The way of light* to 'the torchbearers of this enlightened age, members of the Royal Society'. The book expressed a desire that advances in natural science should be to the benefit of man, in accordance with its author's conception of the unity of theology, philosophy, and scientific knowledge. Time has shown how well his ideas survived in the last place in which he was active in Moravia, Fulnek and its vicinity, the area in which Mendel was born.

At the turn of the nineteenth century Hynčice was part of the estates of Countess Walpurga Truchsess-Zeil (1762–1828), who was a supporter of the freemasons, and under the influence of the ideal of the Enlightenment endeavoured to improve the standard of education of those who lived and worked on her estates. She considered education an essential part of economic, cultural, and social development. In 1792 the Countess founded a private institute of education at her seat in Kunín, modelled on the highly reputed Philanthropinum at Schnepfenthal in Saxony. This establishment largely owed its origin to the spread of Comenius' ideas of general education. C.C. André taught there before his emigration to Moravia in 1798. The institute at Kunín was administered by J. Schreiber (1769– 1850), whose pioneer teaching did not escape André's attention. In the first volume of his journal *Patriotisches Tagesblatt*, published in 1800, an anonymous article described the well-managed school at Kunín and its 'diligent and skilled' teacher, Schreiber. The article refers to boys and girls being educated at Kunín for a practical life, and characterizes the school as 'a kind of Philanthropinum', the name given to Salzmann's school at Schnepfenthal in Saxony, where André had been a tutor and teacher. In the following year Schreiber (1801) described the education programme as 'a training for diligence and permanent occupation for acquiring useful knowledge and noble feeling'. The pupils were educated according to the book by M.K. Traugott-Thieme (1795) *First nutrition for the healthy reason of man*. Among its axioms were 'money and property can be taken from me, but never the art of scientific knowledge'. In a footnote to Schreiber's article André highly commended Schreiber's skills as a pedagogue, saying he was an example to others in the province, and praised the principles upon which education in the school was founded. He states, *inter alia*: 'We must bless the founders, keepers and supporters of such institutes.'

On the Countess's orders talented boys and girls from her landholdings were admitted to the institute, and she paid their bed and board. By supporting education she helped introduce a new element of culture to rural life and work. At first the provincial authorities were full of praise for the educational efforts of the institute, but with the rise of absolutism in the monarchy in reaction to the French Revolution it was labelled excessively liberal and subjected to investigation. Accusations of 'spreading alien notions' were levelled at it, a phrase which encompassed the teaching of natural science. Schreiber was held responsible for this 'scandal', and in 1802 was forced to leave. That was not, however, the end of the institute's tribulations: further investigations followed, and in 1814 it was closed. By then Schreiber was parish priest in Dolní Vražné (Gross Petersdorf), the parish to which Mendel's native Hynčice belonged.

FAMILY ENVIRONMENT AND EARLY EDUCATION

Anton Mendel, father of the researcher, was born in Hynčice in 1789. He took part in the final stages of the Napoleonic Wars as a soldier in the Austrian army, and thus had the opportunity to see foreign lands and different ways of life, experience he later put to good use on his smallholding. In 1818 he married a girl from his village, Rosine Schwirtlich, five years his junior. She was a gardener's daughter, a fact Mendel's biographers have seen as instrumental in stimulating her son's interest in growing plants. Most of the villagers in those days considered themselves ethnic Germans. In the early years of the twentieth century Mendel's nephew, A. Schindler, traced his uncle's family backwards for seven generations, and found that their names were registered in Czech or in German according to the ethnic origin of the parish priest. According to him (Křiženecký 1965b, pp. 134–60) the available registers show the name Mendel to have appeared in the family in 1550. The Mendels were most probably fugitives from Württemberg at the time of the peasants' revolt there. The family of Mendel's father was from the neighbouring village of Veselí, whose inhabitants at the time of Mendel's birth were mostly ethnic Czechs. Schindler discovered that about three-quarters of Mendel's forefathers were of German origin, the other quarter of Czech origin. Thus even his ancestry contained the 'magic' ratio of three to one which was to be characteristic of his experiments. His father seems to have been very hard-working. He rebuilt his originally wooden house in brick, and the main features of it are preserved to this day. The two rooms of the living quarters now house a memorial exhibition to the founder of genetics. The rest of the farmstead still retains the salient features present in Mendel's parents' day.

There was a garden around the house, where Mendel's father grew fruit trees and had beehives. In this late feudal period of the Habsburg empire, Anton Mendel was obliged to work three days a week for his landlord, half of that period using a team of horses. The villagers were smallholders, and not one of them had more than the two horses owned by Mendel's father. They eked out their earnings by carting lime from a nearby kiln. One of their important sources of food was fruit, so great attention was paid in the village to growing the best varieties. There was no shortage of work on such a farm, and an only son would have to do his share from an early age.

In 1820 a daughter, Veronica, had been born to the Mendels. Johann followed two years later, and a second girl, Theresa, seven years after that. Two other siblings died soon after birth, and are usually not even mentioned in biographies. The inhabitants of Hynčice were anxious for their children to be educated. A. Schwirtlich, Mendel's great uncle, had taught the village children in 1780–8, on his return from military service (Kříženecký 1965b, pp. 134–8). At the time the villagers built a school so that the children would not have to go all the way to Vražné. From 1796 the village teacher was Thomas Makitta, described by Mendel's nephews as a capable teacher. Classes were large, and children had to be given different lessons according to their age, which was no small problem. Following the Countess's instructions, Makitta and Fr. Schreiber included in the syllabus basic natural history. To a limited extent, Schreiber was able to continue an activity whose worth he had proved at the Kunín Philanthropinum, where he founded a fruit-tree nursery and grew in it new varieties he obtained from France. The Countess had them distributed among the villagers, but they did not trust their masters, and, not realizing the value of these new varieties, failed to take advantage of the generous offer. Marvanová (1971) has pointed out Fr. Schreiber's astute perception of the rustic mentality, choosing as he did a subtler manner of distributing the new varieties. The fruit-tree nursery at Kunín was to be 'strictly' guarded, and by order of the Countess anyone stealing fruit trees would be punished. The guards were instructed to make plenty of noise if anyone appeared in the nursery, but not actually to catch any thieves. Within a few days the seedlings had been stolen. The villagers grew them for many years with great care, and their descendants continued to benefit from the new varieties.

In Kunín Fr. Schreiber and his pupils gathered seeds in their thousands to produce seedlings, the best of which were propagated for planting in villages. When he came to Vražné Schreiber transformed the presbytery garden into a fruit-tree nursery, giving grafts of his best varieties to his parishioners. Among the recipients was Anton Mendel, and at the village school Schreiber taught Anton's son Johann the basic techniques of fruit-tree improvement. When, in 1816, the Pomological Association was formed in Brno, Schreiber became a founder-member (Orel and Vávra 1979). Soon afterwards he was elected a corresponding member of the Brno Agricultural Society, a considerable honour. His interest in progress in fruit-tree breeding methods led Schreiber to attend the annual meetings of the Pomological Association. In 1822 he was present at the meeting which discussed the successes achieved by T.A. Knight in breeding new varieties of fruit trees.

Schoolteacher Makitta and Fr. Schreiber made a practice of informing parents if their children were gifted, and they soon took notice of young Mendel. On their own initiative they sent two pupils to the Piarist school in Lipník (Leipnik), some 25 km from the village. The intention was to verify their talents and to see whether they were capable of studying at a *Gymnasium*. This was how Johann Mendel came to enter the third grade of the Piarist school in 1833. He proved an able pupil, and in the following year was moved to the *Gymnasium* in Opava (Troppau), which is 36 km from his native village. For his parents the decision was not an easy one to make; it involved giving up any idea of Johann taking over the running of the farm, and it was by no means certain to what extent they would be able to maintain him during his studies. They were still paying off a loan they had received to build their house, and were able to pay only for bed and half board. Whenever the occasion arose they would send him bread and other products from their farm. Johann only came home during the school holidays, a period in which he was able to meet Fr. Schreiber, whose broad interest in natural history and agriculture may well have aroused Mendel's interest in natural science.

Throughout his studies in Opava, Mendel achieved excellent results. He was thus able to give lessons to his less gifted schoolfellows, which helped him pay his way. The headmaster of the school was F. Schaumann, a member of the Augustinian community in Brno, whom Mendel was to meet again in the monastery. The young Mendel spent six years in the alien environment of Opava, a period of privation, as he describes in a curriculum vitae written fifteen years later: 'Owing to several successive disasters, his parents were completely unable to meet the expenses necessary to continue his studies, and it therefore happened that the respectfully undersigned, then only sixteen years old, was in the sad position of having to provide for himself entirely. For this reason he attended the course for

School Candidates and Private Teachers at the district teachers' seminary in Opava. Since, following his examination, he was highly recommended in the qualification report, he succeeded by private tutoring during the time of his humanities studies in earning a scanty livelihood (Mendel 1850).'

A natural history museum had been set up at the Opava *Gymnasium* in 1814 on the suggestion of André, and this encouraged teachers' interest in a further study of the natural sciences and in improving their knowledge for purposes of teaching. The collection and library had been built up gradually, and one of the subjects covered was meteorology, in which Mendel later showed an interest which endured for the rest of his life. Studies at the *Gymnasium* lasted six years, and those who wished to enter a university had to undergo a further two years' 'philosophical study', which in the Habsburg Empire at the time was taught either at a Lyceum attached to a university, or at independent Philosophical Institutes. The nearest institution of this kind for Mendel was at Olomouc (Olmütz).

Farmers are to this day dependent on weather conditions, but in the days of Mendel's childhood, before improved agrotechnology was introduced, this was doubly so. As a student at Opava, Mendel saw the fluctuations in harvest yields caused by weather conditions, and the quotation from his curriculum vitae underlines the fact. During his studies his father was also unfortunate enough to sustain a serious injury during his obligatory work for the landlord in the forest, and his ability to work was impaired. At Whitsuntide the seventeen-year-old lad became seriously ill, and stayed at home for the rest of the year. In spite of this he moved up into the next class without resitting any examinations, and he again proved one of the best pupils.

A fragment of poetry written by Mendel while at the *Gymnasium* has survived (Iltis 1966, pp. 36–7; Marvanová 1966). The lines pay homage to the power of the printed word, which opens the way to learning. He expressed the conviction that scientific knowledge would rid mankind of the superstitions it was prone to. He also modestly indicates that he would like to contribute to the expansion of that knowledge. The poem clearly reflects the ideals of the Enlightenment, which Fr. Schreiber had spread in his native village. Mendel's enthusiasm for study enabled him to overcome his material problems, and was decisive for his future.

A FATEFUL DECISION

'When he graduated from the *Gymnasium* in the year 1840, his first care was to secure for himself the necessary means for the continuation of his studies. Because of this, he made repeated attempts in Olomouc to offer his services as a private tutor, but all his efforts remained unsuccessful because of a lack of friends and references. Sorrow over these disappointed hopes and the anxious, sad outlook which the future offered him affected him so powerfully at that time that he fell sick and was compelled to spend a year with his parents to recover' (Mendel 1850). This bleak account indicates that on leaving the grammar school Mendel sought an opportunity for further study. At the same time P. Křížkovský, a member of the Augustinian monastery, later to become a well-known composer, had also just left grammar school and was trying to make a living as a teacher.

In 1841 Mendel's name appeared on the roll of students at the Philosophy Institute in Olomouc, where he would have to study for two years to be entitled to enrol at the university to study law, theology, or medicine. But his father was no longer able to work because of his old injury, and he asked his son for the last time to consider taking over the farm. At the age of eighteen, Mendel decided to continue his studies. He found himself without friends, in a city where most of the inhabitants spoke Czech, a language he had only just begun to learn. The problem of making a living, which had been serious enough in German-speaking Opava, now loomed very large indeed. He went through a grave mental crisis, and became ill during the exams which closed the first term, after passing in mathematics and Latin philosophy with top grades, as Weiling (1991) recently pointed out.

When he had recovered his strength in a family environment Mendel returned to Olomouc, and over the next two years completed his course. Meanwhile, his elder sister had married A. Sturm, and they had taken the farm over from her father. The text of the contract of sale, dated 7 August 1841, has survived (Iltis 1966, p. 39). The property was valued at 400 convention-coins, and the parents were to be paid an old-age pension. Part of the contract refers to the vendor's son: 'The purchaser shall pay to the son of the seller, Johann by name, if the latter as he now designs should enter the priesthood, or should he in any other way begin to earn an independent livelihood, the sum of 100 fl., say one hundred gulden convention-coins, and also annually, so long as Johann is still engaged in his studies, shall pay the father the sum of 10 fl. convention-coins as an aid to the cost of study, and shall also defray all the expenses connected with the first mass. But should the son Johann be prevented by any mishap from taking priest's orders, or should he in any other way be hindered from earning an independent livelihood, he shall be entitled to free quarters in the reservation rooms, and in every field after his father's death to the use of one measure of arable.'

In this contract Mendel's parents provided for the bare minimum Johann would require in order to continue his studies and eventually become a priest. They sensibly allowed for the unlikely event of his being forced to seek refuge in his old home. The sum of 100 florins mentioned in the contract was only enough to cover part of the expenses associated with his studies in Olomouc. Mendel indicated that he had a certain limited opportunity to teach privately, and in the end he just managed to finish his studies. Alois Schindler, son of the younger sister, Theresa, was later to reveal how his mother helped out by offering Johann part of her dowry. He remained grateful for the rest of his life, supporting all three of her sons in their studies.

Olomouc University was founded in 1569. The standard of teaching was high, since lessons were given by the university lecturers. Documents show Mendel to have been examined not only in religious studies, but also in theoretical and practical philosophy, ethics, mathematics, physics, and pedagogics. He did not, however, take part in the course of natural history and agriculture. This can be explained by the fact that the lecturer in this subject, Professor Nestler, was ill at the time, and died in 1841. His place was taken by Professor J. Helcelet (1812–76), whom Mendel met for the first time in Brno some time later. The subjects which may be supposed to have played a role in his natural science studies and research were physics, mathematics, and logic. Weiling (1991) pointed out that of the twenty or so hours of classes he attended per week, seven and eight respectively were devoted to mathematics in the first year and to physics in the second. Both were areas of special importance to Mendel's later scientific career. The mathematics teacher was Professor J. Fux, who had published a textbook (Fux 1839) approved for grammar schools throughout the Empire. It included the rudiments of combinatorics. Physics was taught by Professor F. Franz, who had previously taught at the Brno Philosophy Institute, during which period he was accommodated in the Augustinian monastery. Mendel's physics textbook had been written by Professor A. Baumgartner (1823) while he was the incumbent of the chair at Olomouc. In 1824 he moved to the University of Vienna. His textbook was republished repeatedly, finally in cooperation with Professor Ettingshausen (Baumgartner and Ettingshausen 1843), and was still used by university students at the time Mendel studied in Vienna.

Mendel's talent for and interest in the study of physics, together with the fact of his meeting Professor Franz, were decisive in forming his future career. His curriculum vitae exposes the feelings of a young student seeking the means by which to continue his studies. Seven years after the event, he wrote: 'It was impossible for him to endure such exertion any further. Therefore, having finished his philosophical studies, he felt himself com-

pelled to enter a station in life that would free him from the bitter struggle
for existence. His circumstances decided his vocational choice. He requested
and received in the year 1843 admission to the Augustinian monastery of
St Thomas in Brno' (Mendel 1850). Schindler said that out of thirteen candidates, four were shortlisted for the novitiate. The choice was facilitated by
Professor Franz, a member of the Premonstratensian Order at Nová Říše in
Moravia. The Augustinian abbot in Brno, C.F. Napp, had asked him to keep
an eye open for likely candidates for admission to the Brno monastery. In a
letter dated 14 July 1843, Franz recommended only one student, Johann
Mendel, who, he said, had 'during the two-year course in philosophy
almost invariably had the most exceptional reports', and was 'a young man
of very solid character, in my own branch almost the best' (Iltis 1966,
pp. 42–3). Franz added that though the candidate's Czech was poor, he was
prepared to work on it. This recommendation sufficed to secure Mendel's
admission to the monastery, on 7 September 1843, without his even having to attend the interview that most would-be novices had. His parents'
written permission, dated 19 September 1843, has also survived. An examination showed Johann to be medically fit to join the order, and on 9
October he began his novitiate, taking the name Gregor. Novices were
entrusted to the care of A. Keller (1783–1853), a member of the
Agricultural Society and an active member of the Pomological Association.
His interest in plant improvement can be illustrated by an article describing
alternative traits of six different varieties of melons (Keller 1828). He
showed a special interest in the economic aspects of breeding new varieties
of fruit trees and vines. Thus, right from the start, the newcomer was in the
company of an acknowledged expert on agriculture, pomiculture, and
viticulture.

Mendel was twenty-one when he entered the monastery, the history of
which is described by A. Neumann (1930). He must soon have realized
what a rare milieu it offered: so long as he fulfilled his clerical duties, he
was free to devote himself to private study. He writes in his curriculum
vitae that he had now achieved the social security which is 'so beneficial to
any kind of study'. According to the orders of his superiors he studied the
classical subjects prescribed for his probationary year, and in his spare time
he 'occupied himself with the small botanical and mineralogical collection
which was placed at his disposal in the monastery' (Olby 1985, p. 198). He
was able to devote his full attention to a study of natural science, for which
he said he had 'a special liking, which deepened, the more he had the
opportunity to become familiar with it'.

The study of natural science had originated in the monasteries in the
Middle Ages, but there were immense differences in the attitudes different

orders took to it. The degree to which scientific study was tolerated, or even encouraged, depended largely on the personalities in each monastery, and above all on who was in charge. A major role was also played by the overall political and economic development in different countries. The exceptionally good conditions created in the Brno Augustinian monastery during the abbacy of F.C. Napp are a case in point.

The monastery in Brno, next to the church of St Thomas, was founded in 1350 by the Margrave of Moravia, Johann Heinrich (1322–75), brother of Emperor Charles IV (1346–78) who in 1348 had founded Prague University, the first in Central Europe. The Margrave and his descendants were to be buried in the monastery, and the monks were obliged to pray for their souls. The material needs of the monastery were provided for by a number of villages allotted to it as a source of feudal income. It also received from Charles IV the famous 'Black Madonna' painting, which legend asserted to have been the work by Luke the Apostle. The Přemysl family had acquired it in the twelfth century from the Emperor Frederick Barbarossa. It was installed on the altar of the cathedral in Brno in a ceremony attended by Charles himself. The picture was reputed to have supernatural powers, a fact which drew large numbers of visitors to the cathedral. There was a special liturgy for its adoration, and possession of this rare asset brought the community both considerable income and certain ecclesiastical privileges, one of them being the fact that the monastery was directly subordinated to the director general of the order in Rome, or his delegate. Since 1721 the head of the monastery had been appointed for life. Later he received the extraordinary title of abbot. In time the monastery's power and influence increased still further, and the community gained in prestige in the public eye. Prominent families frequently left it property, which steadily bolstered its wealth. Because it was a major landowner, the abbot had a seat in the provincial Diet, giving the monastery still further social privileges. During the seventeenth century the monastery and its outbuildings, situated in the centre of the city, were rebuilt to provide accommodation for forty-two monks, and a large library was established. In the eighteenth century the members of the order began to pay greater attention to the study of theology and philosophy.

But the privileged position of the monastery was threatened by the altered policy of Emperor Josef II (1780–90) towards the Church. In 1782 more than half the monasteries in the Empire were abolished, and those which remained were to serve not only the Church, but also the State. This meant that the monks were obliged to work in parishes and hospitals, and to teach religion in schools. The Augustinian monastery was not dissolved, but in 1783 the monks were forced to leave their splendid, centrally-

Fig. 3.1 The former Queen's Monastery in Old Brno which became the seat of the Augustinian Order in 1873. This photograph is from the period when Mendel lived in the monastery.

situated building to take over a building on the outskirts of the city recently vacated by Cistercian nuns. The convent and its church had been built in the first half of the fourteenth century (Fig. 3.1). The buildings were in poor condition, and the cost of moving, together with the expense of adapting their new home, left the once rich Augustinian community in debt.

In 1807, before the monastery had a chance to recover from its financial problems, an order came from Emperor Franz I (1792–1835) for its members to take on the teaching of mathematics and biblical studies in the newly established Philosophy Institute and the Brno theology college. The remaining teaching staff were to be provided by the Benedictines in Rajhrad and the Premonstratensians in Nová Říše. The monks were not prepared for these new tasks, and it was not long before the supervisory authorities instructed them to prepare evidence that they intended to take up their new obligations. This constituted a threat to the very existence of this handful of surviving monasteries, and their superiors began looking for young men with the interest and the ability to study and to become teachers. It was not until 1819 that the first Augustinian teacher of biblical studies, P.F. Neděle (1778–1827) took up his post (until then the monastery had been obliged to pay members of other orders to carry out the work). Neděle had studied at the grammar school in Litomyšl in Bohemia, and then at the University of Olomouc, where he took the degree of doctor of philosophy and theology. He soon became involved in the Czech national revival movement, and translated pedagogical and didactic literature from German to Czech. He was well qualified as a teacher, and popular with his students, but the reaction of the bishop's consistory was less favourable. Neděle was accused of being too liberal in his teachings, and of harming the interests of the Church, and in 1821 was removed from his teaching post. He became prior of the monastery, assuming charge of the library. He took the teaching ban very badly, and it may have been a contributory factor in his illness and early death, at the age of forty-eight.

The place of the demoted Neděle was taken by a passionate philosophy and theology student from Olomouc University, F.C. Napp, who at the time was preparing to take his doctoral examination in those subjects. It was his ambition to become a university professor. When he arrived in Brno he began to teach biblical studies and the rudiments of oriental languages. He took over the job of librarian, and catalogued the extensive monastery library, in the course of which he gained a picture of the available literature in a number of different branches of science. His election in 1824 as abbot for life brought a major change in his lifestyle, but Napp was so attached to his teaching post that he continued to occupy it. Early in 1827, however, the bishop took it away from him, saying that he must devote his attention to an improvement in monastery discipline. Napp considered this move unfair both to himself and the monastery; this was the start of a decline in relations between the Augustinian monastery and the consistory which was to continue during Mendel's abbacy. Napp's many activities were desribed by Neumann (1930), to which attention was recently paid by Czihak and Sladek (1991). In 1932 the Olouc University awarded Napp the degree Doctor *honoris causa* in the field of theology.

Napp used his broad knowledge and considerable organizational skill to good effect. He accepted with enthusiasm the imperial directive for the monastery to prepare its monks for the teaching profession: the interpretation he put upon it was that they should assist in spreading the word on the latest scientific findings. He started to look for talented young monks able to study various branches of science. In order to do this, the community first needed to create the necessary financial infrastructure. They had to pay off the debts they had incurred in the process of moving premises, and above all to complete the adaptations their new home badly needed. They applied for exemption from contributions to the ecclesiastical fund set up by Josef II. The application stated as a reason the monastery's difficult financial situation and the added expense of preparing new members of the community for teaching work. One way in which an improvement could be brought about in the monastery's finances was by introducing rational

methods of management to its estates. With this in mind, Napp himself
studied agricultural literature and associated natural science publications.
As a member of the Lords' Diet and of the committee of the Agricultural
Society, he soon became prominent in his efforts to rationalize agriculture.
To a certain extent it might be said that, under new social conditions,
Napp's public work in this area took up where André's considerable
successes had left off (Orel 1975a).

In order to modernize production methods on the monastery farms, Napp
enlisted the assistance of an acknowledged agricultural expert, F. Diebl, for
whom he helped to secure the post of professor of agricultural science in
Brno (Orel 1975a; Weiling 1968; 1976, 1982). According to his instructions, the abbot introduced crop-rotation, fodder-crop cultivation, and sheep
breeding (Balcárek 1977). Documents show that in 1825 wool production
was the most profitable side of the monks' agricultural business. Out of a
total of 7057 guilders' profit at the Šardice farm, no less than 5153 guilders
were earned from sheep. Napp was also instrumental in the introduction of
leguminous fodder crops, carrying out field trials and publishing the results
in the Agricultural Society journal (Napp 1829, 1832a, b). At the Šardice
farm and in the Brno monastery gardens he established fruit-tree nurseries
for spreading new, improved varieties, and he even wrote a manual on
fruit-tree growing, which unfortunately has not survived.

When, in 1826, insect pests invaded the monastery's estates, Abbot Napp
examined the situation and published his observations on the spread of the
insects and the extent of the damage. At that time he was cooperating with
F. Diebl, who pointed out the taxonomy and manner of reproduction of the
insects. Both asked readers of the Agricultural Society journal to observe the
appearance of insects in the following year (Napp and Diebl 1826). Two years
later the two of them summarized the results of their observations in a further
paper, explaining that the high temperatures in September and October had
been a major factor contributing to insect numbers (Napp and Diebl 1828).

When, in 1827, Napp was elected president of the Pomological Association,
he supported the improvement of methods of fruit- and vine-growing in the
province, publishing the results of his experiments with fruit trees
(Napp 1831). As president of the Pomological Association he began to
work with Diebl, the secretary of the Association, organizing annual meetings in Brno and publishing extensive reports. They also offered prizes for
solutions to important pomicultural and viticultural problems of the day
(Napp and Diebl 1838). For economic reasons Napp also took part in
the meetings of the Sheep Breeders' Association. Here too, he gained
expertise, becoming a member of the commission which examined the
participants in courses for shepherds. He took part in discussions aimed

at seeking theoretical explanations for complex problems encountered in breeding practice. In 1836, for instance, Napp pointed out that herecity was a problem for whose solution it was necessary to look chiefly to physiological experimentation. Today we can only guess as to whether Napp was interested in breeding practice simply from a landholder's point of view, or from that of an agricultural expert intent on seeking theoretical explanations for the things breeders knew empirically.

André, in organizing the promotion of natural sciences in Moravia, had proposed that research should be done on the rich flora of Moravia, and to this end he recommended that a botanical society should be established. It was a branch of science to which attention turned relatively late in the day. The first notable botanical expert in Moravia was the Augustinian A. Thaler (1796–1843), who taught mathematics at the Philosophy Institute (Hrabětová-Uhrová 1974). In 1830 Napp gave him permission to establish an experimental garden measuring 37 m × 5 m under the refectory windows, and he used it to grow rare Moravian plants. Thaler also founded a herbarium, which Mendel was later to use in the early stages of his private study of natural science.

On entering the monastery Mendel came into contact with other personalities who devoted their attention to science. Among them was the mathematics teacher F. Schaumann, then in his early fifties, whom he knew from Opava, where he had been headmaster of the grammar school when Mendel was a pupil there. X. Wieser was in his forties and taught at the theology faculty in Olomouc, and Dr A. Alt, in his late thirties, was a mathematics teacher at the grammar school, and subsequently headmaster. Another mathematics teacher was thirty-two-year-old F. Gabriel. Romance languages were taught by B. Fogler, who was in his forties, and with whom Mendel later taught at the *Realschule*. P. Křížkovský, also in his forties, entered the monastery a couple of years after Mendel. Napp took him on as a talented musician, and he soon made a name for himself as a composer. He helped raise the standard of music teaching in Brno. Napp took a special liking to him, and towards the end of the abbot's life Křížkovský was not only his confessor, but also his closest companion.

Among the Brno Augustinians, Matouš Klácel (1808–82) deserves special attention (Orel 1972*c*). In the year Mendel entered the monastery he took over care of the experimental garden from Thaler. He was already well known as a philosopher who had published a number of studies, but also studied natural science. Seeing science as an individual whole, Klácel was convinced that the development of one branch of science affected that of the rest. His views were based on the teaching of G.W.F. Hegel, which had an evolutionary slant and an idealistic content. Klácel was influenced by the German pantheistic *Naturphilosophie*, an idea which he summed up in a caption to his photograph: 'Deliberate university is the goal of all love and science.' Klácel was acknowledged in Brno as a natural scientist versed in botany, mineralogy, and to some extent astronomy. This fact is reflected in his membership of the Brno Agricultural Society.

In 1841–3, Klácel published in the journal of the Prague National Museum articles showing his enthusiasm for scientific progress. In an article called 'The development of science' Klácel (1941) pointed to the importance of progress in the social sciences, too. An article called 'On death' (Klácel 1843) proposed substituting for the Czech word *příroda* the word *přeroda*. This change of prefix was intended to emphasize the constant process of renewal in nature; the similar word *přeroda* means a regeneration, or transformation. In another article, entitled 'A proposal' (Klácel 1842), he developed further *Naturphilosophie's* notion of evolution, stating that 'every age has much which is transient, which the dialectic sifts and polishes, until its kernel is revealed'. Klácel associated these ideas with a picture of nature in the process of developing, and also with the cultural and social development of society, as a member of the small Czech nation, at the time dominated by the German-speaking minority. He therefore emphasized the importance of the development of Czech nationalist ideas and the teaching of Czech in schools. This brought the wrath of the German authorities and his superiors in the Church upon his head.

In 1842 Klácel was accused of spreading pantheistic notions, and a year later of pantheism and the spread of Hegelian ideas, in contradiction to the Catholic faith. Abbot Napp defended his monk to the very limits of prudence. Klácel (1847) wrote a three-part essay *The philosophy of rational good*, submitting it to the censors in Vienna. The first two parts were passed, forming the basis of the book *Ethics*, published in Brno. But in the third part the censors found 'harmful sentences', and this was the reason for Klácel's dismissal from his post as a philosophy teacher. Napp was forced to accept the fact that Klácel would no longer be allowed to teach. It was at the end of a year of such tribulations for Klácel that Mendel arrived in the monastery. The two became firm friends, and Mendel admired his fellow monk's broad knowledge of various natural scientific disciplines, at a time when he himself was just beginning to study natural science. In his curriculum vitae Mendel mentions the advice of experienced men. We can assume that the foremost among these was Brother Klácel.

In 1848 Klácel was among the organizers of the pan-Slavonic congress in Prague, and he expected to be able to resume his teaching duties. In a letter written at the time he asks Mendel to take over the experimental garden. After the congress was broken up by force, Klácel became librarian of the monastery, and was able to carry on his studies. He continued to experiment with plants in the botanical garden. He later mentioned what these experiments involved in a lecture on Darwinism given in the USA, where he had lived since 1869 (Matalová 1979). He recalled having observed the combination of the traits of parental plants in hybrid progeny, probably peas and potatoes, attributing it to evolution as conceived by Hegelian philosophy. The lecture on Darwinism followed Klácel's study of Darwin's works, not only *The origin of species*, but also his writings on the domestication of animals and plants. He linked his previous *Naturphilosophie* with Darwinian ideas, placing it in the context of the evolution of man.

Another member of the monastic community who was a student of both philosophy and natural science was Tomáš Bratránek (1815–84). Like Klácel, he based his ideas on *Naturphilosophie*. He joined the Augustinian community in Brno in 1834, on leaving the Olomouc Philosophy Institute. Abbot Napp gave him the opportunity to study at Vienna University, where he took a doctorate in philosophy. On completing his studies in Vienna, Bratránek began his career as assistant to Professor I. Hanuš at the University of Lemberg (now Lvov in the Ukraine, but then part of the Habsburg Empire). From 1843 to 1851 he taught philosophy in Brno in place of Klácel, and for a short time he also taught natural history. After Klácel's demotion he realized that there was no room for free philosophy in the Empire, and in 1851 moved to the University of Cracow as professor of German literature (Loužil 1972). Bratránek (1853) published his remarkable book *The aesthetics of plants*, in which he expressed his *Naturphilosoph* conviction that an understanding of the metaphysical essence of what went on in nature was possible only through a kind of empathy, an intuitive identification of the researcher with the object of research. This view was typical of *Naturphilosophie*, and Goethe was among its adherents. Bratránek (1874) published in Cracow three volumes of Goethe's correspondence with leading natural scientists, chiefly Count C. von Sternberg and the von Humboldt brothers. In the first volume he compared the poet's ideas on metamorphoses in nature with Darwin's theory of evolution. In contradiction to E. Haeckel's statement that there are in Goethe's work some Darwinian ideas, Bratránek quotes the opinion of the zoologist O. Schmidt, dated 1871, as follows: 'For Darwin, the type is a random and, from the viewpoint of purpose, fixed form, i.e. a product. For Goethe, on the contrary, it is the necessary basis for its development and random variation, i.e. a prerequisite.'

While in Cracow, Bratránek maintained contact with his fellow monks in the Brno monastery. In 1863 he was elected a member of the newly formed Natural Science Society, being proposed by Mendel. Abbot Napp considered Bratránek the most outstanding scientific figure in the monastery, and after

Napp's death he was the leading contender to succeed him. But he preferred to stay on at the university, thus opening the way for Mendel. In
1881, he returned to Brno to live in retirement.

In 1834 Bratránek described the conditions for study provided in the
monastery (Sajner 1974, pp. 27–8). They would surely have been substantially unchanged when Mendel arrived there. Each of the monks had
spacious and tastefully furnished quarters, usually consisting of two rooms.
According to his station, everyone had his daily duties as a priest and as a
worker, doing either parish work or teaching. The remainder of his time
was devoted to study. The monks rose at six for morning mass and prayers,
followed by breakfast, after which the members of the community went
their separate ways. Those who had no parish work or teaching could go
and study. A large library was at their disposal, and if they needed any
literature not kept there, they could request its purchase. In this milieu
Mendel studied theology and philosophy, which were compulsory, and
natural science, which was his private interest and for which there were
botanical and mineralogical collections available in the monastery. His
curriculum vitae notes that private study was very difficult, but that he had
taken such a liking to the study of nature that he would 'not spare any
effort to fill the gaps ... through self-instruction' (Mendel 1850).

After a year's probation, his novitiate, Mendel began his study of theology,
which lasted until 30 June 1848. In 1846 he also attended the lectures in
agricultural science which were then compulsory for theology students, and
the new lectures on pomiculture and viticulture by Professor Diebl at the
Brno Philosophy Institute. In these subjects he passed three examinations
with distinction (Fig. 3.2). In the course of his studies he learned the technique of artificial pollination of plants for breeding new, higher-yielding
varieties. The lectures were aimed at teachers and priests working in the
country, and the idea was that in passing their knowledge on to pupils and
parishioners they would help raise the economic and social standing of country people. Mendel may well have recalled what he had learned as a boy from
the teacher and parish priest in his native village, and learned to understand
it on a higher level. He also mentions these studies in his curriculum vitae, in
an effort to show that he had the necessary background knowledge to justify
his being admitted to the teachers' examination.

By tradition, after completing his studies in Brno, Mendel 'received permission from his prelate to prepare himself for the philosophical rigorosum'.
The curriculum vitae makes only passing mention of this. But the year
1848 had arrived, and with it a revolutionary movement that began in
Vienna and was soon to sweep through the whole of the Empire. Before
long its effects were felt in Brno, including the Augustinian monastery.

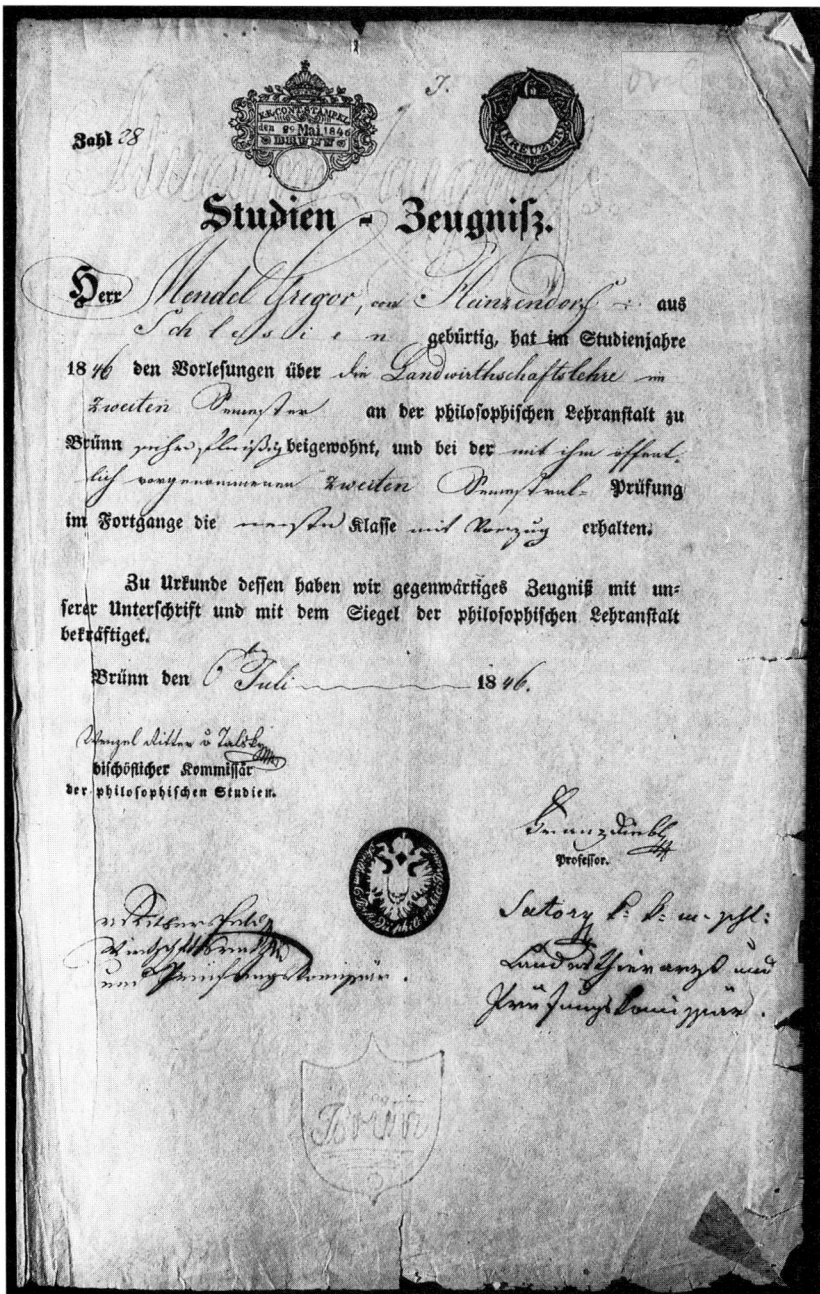

Fig. 3.2 Mendel's certificate from the lectures on agriculture which he attended at the Philosophical Institute in Brno in 1846.

1848—YEAR OF HOPE

In the mid-nineteenth century Brno was an important industrial and commercial centre, with a fast-growing population, especially in the suburbs.
Further expansion of industrial production was dependent on an increase
in the number of students with a new type of education. Up to 1848 the
feudal order still survived, and stood in the way of economic development.
Teachers of Czech extraction revived the ideas of the Enlightenment in combination with an effort to bring about a national revival, and political tensions in society steadily increased. The monastery was not immune from
this process: Abbot Napp held important positions in public and Church
institutions.

On 13 March 1848, revolutionary disquiet in Vienna came to a head,
and the absolutist government of Duke K.V.N. Metternich (1773–1859)
was forced to resign. This brought to an end a millenium of feudalism in
Central Europe, and a new, capitalist order opened the way to economic
and social development. The Emperor Ferdinand I (1835–48) promised to
abolish feudal labour, lift censorship, and summon a legislative assembly. A
wave of revolutionary fervour rolled across the provinces, and the
industrialized city of Brno was engulfed by it. On 17 March a mass meeting
was held in the city, at which a revolutionary poem written by Viennese
student H.J. Frankel called 'University' was read out. The Augustinian
P. Křížkovský promptly set it to music, and it was sung by the newly
formed revolutionary guard as it marched through Brno. The first verse of
the poem gives an idea of its content:

> Whose are the feet that tramp so proud?
> Their banner flutters, their weapons gleam,
> And with a sonorous roll of drums,
> The university is on the march.
>   (Eichler 1904.)

Frankel's poem sparked off a movement demanding social change in
Brno. Among the students at the head of the procession was the popular
teacher of agricultural science, F. Diebl. A wave of enthusiasm seized most
of the city's inhabitants, and the influential Abbot Napp was among those
who voiced their support. In a letter to Bratránek, Klácel explained how a
deputation of students had set out from Brno for Vienna, and was joined by
the Augustinians Bratránek and Gabriel. Later about twenty students came
to Brno, and were received in the monastery by Napp. On 18 March he
said mass for students who had fallen in the revolution, in the presence of a
delegation of students from Vienna, in the Minorite church in Brno

(Kříženecký 1965*b*, p. 16). The music was Cherubini's requiem, conducted by Křížkovský. Klácel gave the sermon, in which he demanded the abolition of feudal labour as being out of keeping with human dignity, and an improvement in the social conditions of the working people. He supported a Utopian ideal of socialism. Napp received the Viennese students in the Old Brno monastery. As an influential member of the provincial Diet the abbot immediately lent his support to calls for radical political reforms in the province. During the discussion of the new principles on which the Diet was to be elected he favoured proportional representation of rural areas. This aroused the anger of the city's captains of industry, who on 31 March organized a demonstration against the monastery. At this time of revolution Napp also helped Klácel to get appointed editor of a new Czech-language newspaper, the *Moravské noviny*, which defended Napp against accusations of 'Czechization' and 'pan-Slavism'. In fact all Napp was asking for was equal rights for the German and Czech languages.

The first provincial Diet in the Empire met in Brno on 31 May 1848. The best-represented group was the peasant farmers; this is why it came to be called the 'peasants' parliament'. It soon proposed a series of new measures, the most significant of which was, on 11 July, the abolition of feudal labour. There were other important proposals, but they were never to be implemented. The rest was decided in Vienna. In May the government convened an imperial constitutional parliament consisting of 380 delegates, only thirty-eight of whom were from Moravia and Silesia. The parliament met on 22 June, and it was not until 7 September that feudal labour was abolished, giving the peasants much broader civil rights. But the parliament's further deliberations proceeded at a snail's pace, leading to a new wave of unrest directed at the conservative delegates, and even at the dynasty itself. The Emperor left Vienna and took up residence in the Moravian city of Olomouc, and the parliament removed itself to the nearby town of Kroměříž, summer residence of the Archbishop of Olomouc. Shortly afterwards Emperor Ferdinand abdicated, and his place was taken by Franz Josef I (1848–1916), who first of all restricted the powers of the parliament, and then, on 7 March 1849, dissolved it. The incoming government reduced the new constitution to a mere shadow of its former self, and the revolution had *de facto* been crushed, and a new absolutism was to be gradually introduced. It was a period Mendel must have lived through with some anxiety: he was reaching the end of his theology studies, and his future was in the balance.

In the revolutionary year of 1848 many requests were sent to the imperial parliament and the provincial Diet, calling for reform not only of the system of government, but also of ecclesiastical institutions. The Brno provincial Diet alone received over three hundred such applications. One which was unique in its content was a petition by six of the Augustinian monks, dated 8 August 1884 (Huber 1978; Orel and Fantini 1983; Orel and Verbík 1984; Czihak and Sládek 1991; Zlámal 1991). Two days before that Mendel had received holy orders. But the bishop had already appointed him a curate of the parish attached to the monastery in July of that year. There was a lack of priests at the time, and the reasons given for his premature appointment were his strength of character and his piety. It was in this position that Mendel signed a petition stating that the petitioners considered it their duty to point out that 'in the frenzied call for liberty' no thought had been spared for the members of religious orders, who, on entering the monastic life, are provided with material support, but effectively deprived of civil rights. Their situation is compared with that of patients in a mental hospital. The laws so far enacted, the text of the petition continues, have, in contrast to those introduced in France, been in the interests of ecclesiastical bodies, but against those of their individual members. The Austrian monasteries are 'almshouses for poor and short-sighted youths' who, soon after entering the religious life, came to know 'enforced isolation' and a condition 'representing the nadir of degradation'. It would, it is said, be shameful for Austria if, in drawing up a new constitution, these prisons of citizenship and 'burial-places of constitutional liberty' were to be sanctioned. In conclusion, the signatories request that they be permitted to devote their entire efforts, according to their ability and past service, to public teaching and 'free, united, and indivisible citizenship' (Czihak 1984, pp. 11–12).

The petition was written by Klácel, who was also the first to sign it. He added to his signature the comment 'ex-teacher.' In the year of revolution Klácel hoped for a new opportunity to teach philosophy in Prague. His name was among the organizers of the Slavonic Congress which met at the beginning of June 1848. Representatives of the Slavonic peoples asked the Empire to allow them freedom of ethnic development, civil liberties, and the peaceful settlement of disputes between countries and nations. But the Congress was broken up with violence by Austrian soldiers. Klácel had to leave the capital to return to Brno. In the period of disappointment following the events in Prague, and in the hope that the constitutional parliament would enact a more liberal constitution, Klácel conceived the petition, whose language was very strong, in keeping with the radical thinking of the hour. Mendel and a further five members of the community appended their signatures.

Klácel, who was forty at the time and had fallen victim to persecution by the Bishop of Brno and the state authorities, had the best reason to express a desire for freedom. The second oldest, thirty-seven-year-old P. Gabriel, was a mathematics teacher at the Brno secondary school, and continued to teach up to the end of his life. Later, during Mendel's abbacy, he left the monastery to work as a teacher and lay priest. B. Fogler was thirty-six and taught French and Italian in Brno. J. Lindenthal was thirty-eight, a teacher of religion. Later, according to Iltis, he assisted Mendel in his plant-hybridizing experiments. In 1859 Mendel proposed him as a member of the natural science section within the Agricultural Society. A group photograph of 1862 shows Lindenthal standing next to Mendel, who holds a fuchsia flower in his hand. The youngest petitioner, C. Ciganek, was twenty-three years old and was also a curate at the parish church attached to the monastery. Like Mendel, he added to his signature the words 'candidate for teaching'. We do not know why no other members of the community signed the petition, nor how much Napp knew about it, especially its content.

The application was to be considered at the parliament in Kroměříž by a committee for questions of Church and State. When the parliament was suspended in March 1849, the petition, along with other documents, was transferred to Vienna, and it was only brought to light in 1955 in connection with the search for documents relating to Mendel (Huber 1978; Orel and Verbík 1984; Czihak and Sladek 1991).

After his theology studies were completed, on 20 July Mendel was made a curate at the monastery parish, and among the tasks he was entrusted with was the spiritual care of the sick at the nearby hospital. One of the reasons he signed Klácel's petition seems to have been that he was not happy in the post. He surely agreed with the proposal to give the members of religious orders more freedom to choose the teaching profession, and freedom to study science. He knew the history not only of Klácel, but also of Neděle, and he was probably acquainted with the end of Fr. Schreiber's career as a teacher in Kunín. Though possibly not fully appreciating the far-reaching significance of the petition, he must have been aware of political tension in the province. In any case he saw no future for studies of natural science in his job as a curate, and sought some way of working as a teacher. It is evident from the monastery accounts that he fell seriously ill in January of the following year, and that he was confined to bed for thirty-four days (Sajner 1968). The illness was surely a grave one, since he was attended by a nurse from the nearby hospital of the Brothers of Mercy, something which the abbot himself had to sanction: the cost of his treatment was considerable. It seems he was again suffering from exhaustion due to the stress he underwent during the period of major changes in public life, in which he placed great hopes of being able to continue his studies. Medical experts who have considered Mendel's illnesses at critical times in his life consider this ailment, too, to have been one of his specific nervous disorders.

Fig. 3.3 Photograph of Mendel from the group of Augustinians, taken in about 1848.

When Napp saw the situation Mendel was in, he lent a helping hand once more. The abbot had previously been in charge of *Gymnasia*, and his influence in them was still considerable. In September of that year representatives from the town of Znojmo (Znaim) in southern Moravia approached him and asked him to provide a teacher for the newly established seventh grade at their *Gynasium*, to teach classics and mathematics. On 4 October Napp informed the bishop that the provincial *praesidium* had appointed Fr. Gregor Mendel to the post. The reasons given were that 'this collegiate priest lives a very retired life, modest, virtuous and religious, thoroughly appropriate to his condition' and at the same time 'he is very diligent in the study of the sciences, but he is much less fitted for work as a parish priest, the reason being that he is seized with an unconquerable timidity when he has to visit the sick-bed or to see anyone ill and in pain. Indeed, this infirmity of his made him dangerously ill, and that was why I found it necessary to relieve him of service as a parish priest' (Iltis 1966, p. 38). Mendel later admitted he had 'followed this call with pleasure'.

The first years in the monastery brought the young scholar hope of fulfilling his ambition to become a natural history teacher. Mendel had first been motivated in this by the exemplary teaching of Fr. Schreiber at his village school. The notion formed more concretely when he attended the excellent lectures given at the Philosophy Institute by Franz, professor of physics at Olomouc University, and was reinforced by his contact with other members of the monastic community in Brno and with his teacher of agricultural science, the prominent plant breeder, F. Diebl.

The revolutionary year of 1848 was one of suspense for all the monks in Old Brno, but only a few of them actually signed Klácel's remarkable petition. Mendel's first experience of a curate's work certainly came as a great disappointment, and may well have been what made him seriously ill. Thanks to Abbot Napp's magnanimity, however, he was at last able to devote himself to teaching. He grasped the opportunity with enthusiasm, and his state of mind is perhaps best indicated by another quotation from his subsequent curriculum vitae: 'His sorrowful youth taught him early the serious aspects of life, and taught him also to work. Even while he enjoyed the fruits of a secure economic position, the wish remained alive within him to be permitted to earn his living.' But it soon became clear to him that many grave obstacles still lay before him if he was to achieve his goal.

How then did Mendel manage to study natural science, and where did he get his motivation for research into plant hybridization?

4

The making of a scientist

The views you have been good enough to express regarding Fr. Gregor
Mendel have decided me to send him to Vienna for a higher education.
—C.F. NAPP TO PROFESSOR A. BAUMGARTNER in 1851 (Iltis 1966, p. 75).

To his great satisfaction, Mendel took up the post of supply teacher at the
Znojmo *Gymnasium*, on 7 October 1849. He came to an ancient county
town lying on the southern Moravian plain, where the hard-working peasants grew mainly high-quality vegetables and grapes. The nearby
Premonstratensian monastery had in the previous century been home to
the notable physicist Prokop Diviš (1698–1765), who had, working independently, invented a better lightning conductor than that of Benjamin
Franklin (1706–90) in North America, though two years later than him.
The new political climate following the events of 1848 brought a new
interest in education in the Empire, and *Gymnasia* were being expanded
from six to eight grades, which is why the headmaster in Znojmo had asked
Napp to help him expand his staff. Such posts were often occupied by
priests without professional education, provided they had proven themselves capable of teaching. Mendel taught Latin, Greek, and mathematics,
with a teaching load of twenty lessons a week and a monthly salary of
360 guilders (Sajner 1974).

   Soon after arriving in Znojmo, Mendel wrote to his fellow monk
Rambousek asking him to send some money to help him pay his living
expenses, and indicated in the letter (Kříženecký, 1965b, p. 128) that he
would welcome an advance on the pocket money paid by the abbot. He also
admits to having already borrowed money from his fellows, saying he does
not want the abbot to know about it. Mendel assures Rambousek that he will
return the money at the earliest opportunity. It is clear from this letter that
Mendel sent his dirty linen to be washed in Brno. Up to the time he received
his first salary from the school, he lacked even the most basic essentials.

Mendel's arrival was welcomed by the other teachers. A surviving report on his teaching activity indicates that he soon learned his teaching material and mastered the basic teaching skills. In his spare time he visited a book club, and he and his colleagues went to the local theatre.

According to the new education law, teachers who did not have the required professional qualification had to take an examination of teaching competence at a university (Czihak 1984, p. 43). On 16 April of the following year the headmaster sent an application to Vienna asking for Mendel and two of his colleagues to be allowed to take the examination. In Mendel's case he stated that he was to teach natural history at all grades and physics in the lower grades. It was to this application that Mendel attached his curriculum vitae, an invaluable source of information on his life up to the age of twenty-eight. In it he states that from the beginning he had: 'made every effort to present his assigned subjects to the students in an easily comprehensible manner'. He hopes his endeavour has not been entirely unsuccessful, since 'during the private tutoring by means of which he earned his bread for four years, he found sufficient opportunity to collect experience relating to the possible accomplishments of students and the different levels of their mental capacity'. Mendel was here referring to his private teaching of grammar school pupils while he was studying in Opava and Olomouc.

An evaluation of Mendel's teaching by the headmaster of the Znojmo *Gymnasium* states that he was doing his duty with daily increasing accomplishment. It underlines in particular his 'vivid and lucid method of teaching', noting that Mendel had managed to keep up with the syllabus (Iltis 1966, p. 60). Such teachers were a welcome addition to the staff. The application for permission to take the teachers' examination also states that Mendel had devoted himself to the study of natural science without help from anyone else. All the contemporary evidence indicates that Mendel was anxious to gain the requisite knowledge for teaching. At the end of his curriculum vitae he writes that he would 'consider himself happy if he could conform to the expectations of the praiseworthy board of examiners and gain the fulfilment of his wish'. It was a wish he had cherished from the time he entered the monastery. Now the only thing that stood in the way of his ambition was this one examination, and he prepared himself for it thoroughly. The year before he had been faced with the prospect of having to work as a curate rather than a teacher, and everything had turned out all right in the end, a thought which must have encouraged him. His headmaster's favourable report must also have reassured him. Now he was studying for his examination while teaching an average of twenty periods a week, and he also had to prepare his lessons. The knowledge required of him by the examiners was such as full-time students would have spent several years acquiring.

FAILURE

Less than two months after the application for permission to take the examination had been filed, Mendel received the subjects of two written essays which he had eight weeks to complete. The zoology professor, R. Kner, well known for his publications in the fields of palaeontology and ichthyology, set him the following: 'The chief differences between rocks formed by water and those formed by fire, detailing the main varieties of the Neptunian strata in serial order according to their age and giving a short characterization thereof, and, in conclusion, giving in like manner a review of the igneous rocks, both Plutonian and Vulcanian'. Mendel's essay, published by Orel *et al.* (1983), was twenty-three foolscap pages long. He divided it into chapters, and concluded with a short list of literature. In his assessment Kner objected that Mendel had used outdated references and had devoted too much attention to describing minerals and rocks. He had touched on all the necessary questions, but not in enough detail. Kner does not, however, mention Mendel's having made any mistakes. It is almost strange that the professor does not refer to Mendel's original explanation of how plants and animals came into being, which eight years before the publication of Darwin's theory of evolution was surely something out of the ordinary.

On the third page Mendel writes: 'As soon as in the course of time the earth had achieved the necessary capability for the formation and maintenance of organic life, the first plants and animals of the lowest sorts appeared. The process of development was frequently interrupted by catastrophes, which were hostile to the life of organisms and often brought about their extinction.' Further on, Mendel writes in an even more evolutionary vein: 'Plant and animal life developed more and more abundantly; the oldest forms disappeared in part, to make room for new, more perfect ones.' He explains the appearance of new species of organisms without the action of a supernatural force. In the concluding chapter the main features of the Kant–Laplace theory of the origin of the Earth are described. This theory 'offers a simple and sufficient explanation of almost all spheres of phenomena in nature'. Mendel deduced that '... the Vulcanian and Neptunian formations are not yet completed, for the creative force of the Earth remains active. So long as its fires still burn and its atmosphere still moves, the history of creation is not finished.' This was an idea introduced into geological literature by C. Lyell (1797–1875). The actual mechanism of the evolution of species was not explained until 1859 by Darwin; he, too, set out from the ideas of Lyell.

Kner's final assessment was that the candidate's work did not come up to the required standard. He had not proved himself fit to teach natural history in the upper grades of a *Gymnasium*. Over a century later, Professor H. Wieseneder (1983) of Vienna University's Institute of Petrography, made a careful study of Mendel's essay in the light of the state of knowledge at the time, and concluded that Kner's assessment was unfair. Mendel referred to the necessary literature, and he arrived at a remarkable synthesis of the contemporary state of knowledge. Wieseneder admitted the possibility of Kner's being biased against Mendel as a member of a monastic order, as had already been suggested by Iltis (1966, p. 67).

The second essay subject Mendel received was set by Professor Baumgartner, then minister of trade. At the university he was chairman of the commission for examining teachers. The candidate was to explain the mechanical and chemical properties of air and the origin of wind (the original is preserved in the Mendelianum, sign. no. 34). The metereologist J. Munzar (1971) characterized this paper as Mendel's first contact with metereology. Baumgartner considered Mendel's concise and clear explanation satisfactory. The candidate mentioned the possibility of weather-forecasting according to the processing of meteorological observations undertaken at various points, which could be done using the telegraphic service which Baumgartner was at the time organizing in the Empire. Later, back in Brno, Mendel showed great interest in the subject of meteorology. His physics essay was also notable for its style and grammatical presentation. In the opinion of the examiner, T.G. Karajan, the manner of exposition was fairly fluent, natural, and clear, and he adds: 'The candidate says what he means to say, if not brilliantly, still scientifically, and satisfactorily on the whole (Iltis 1966, p. 65).' Because of the difference between the assessment of his two essays, Mendel was allowed to sit the oral examination, but at this juncture fate took a hand.

The university requested Mendel and two other candidates from the *Gymnasium* to come to Vienna on 15 July. The school year ended at the end of July. The headmaster was unable to spare his teachers at the end of the school year, so he asked the ministry of education to postpone the examinations until August. His request was granted, and the commission asked Mendel to attend on 12 August. The vacation was to begin on 15 August. But Mendel did not receive the invitation in time, and he arrived in Vienna on 13 August. His examination took place, but, as Iltis records, in an unfavourable atmosphere, since the lecturers were preparing to leave on vacation. According to custom, the candidate was first given the subject of his written examination. His task in physics was to describe how to make a permanent magnet from a steel bar. Mendel wrote five foolscap sheets on the subject, and Baumgartner considered his paper satisfactory; Professor Doppler agreed.

Professor Kner asked him to describe the systematics of mammals, and to characterize them. He was to give examples of animals useful to man. Mendel based his work on a study of the natural history textbook of J. Gistel (1848). He was not to know that Kner himself (1849) had published a text-book a year later, giving a more exact classification of animals, and Kner's assessment (published by Czihak 1984, p. 18) reflected this fact clearly. He accused Mendel of having an insufficient knowledge of systematics and of using unsuitable terminology. This criticism was fair. A critical reading of the essay suggests that Mendel was unable to concentrate, and, according to George (1967), wrote the first thing that came into his head. Iltis says that on setting the subject of the essay to be written at home Kner also set out the subjects for the subsequent oral examination, as if he wished to indicate what the candidate should prepare. The subjects were crossed out, a fact Mendel seems to have overlooked. Though Kner assessed the essay most unfavourably, Mendel was allowed to take the oral examination. He is said to have answered the questions, but not to have satisfied the exam-iners of his competence to teach even the lower grades. The commission therefore advised him to continue his studies and to retake the examination the following year. More detailed information about Mendel's examination at Vienna University has been published by Gickelhorn (1969, 1973).

With the benefit of hindsight we can see that the main reason for Mendel's failure was the fact that he had studied privately, without having enough time for proper study, and not knowing the manner in which uni-versity examinations were held. In addition, the twenty-eight-year-old can-didate was nervous, which contributed to his submitting a confused written paper in zoology. Disappointed with his failure, Mendel returned to Brno in dread of having to relinquish his dream of becoming a teacher.

The next information we have about Mendel dates from 1851. In April of that year J. Helcelet, the natural history teacher at the Brno Technical School, fell ill, and Mendel was asked to supply for him. It was written of him at the time that 'he is fully competent both in respect of his scientific knowledge, and as a teacher' (Iltis 1966, p. 73). At the end of term the headmaster thanked him, saying he had displayed great zeal, given most useful instruction, and been most considerate to his pupils. He added that Mendel had shown the most praiseworthy goodwill to all the other members of staff. Shortly after this Mendel was accepted as an extraordinary member of the Natural Science Section of the Agricultural Society, which may have been connected with his success as a natural history teacher. Clearly, his failure in the examinations in Vienna was not taken seriously in Brno.

Mendel's further road to learning depended largely on changes in the polit-ical and economic development of the country. The increasing importance of professionalism in industry and agriculture had even in the early nineteenth century led to proposals that technical subjects should be taught on a proper basis. Previously the only such teaching was at the aristocratic academy in Olomouc, founded in 1725 to educate the sons of privileged families. In 1811 a chair of agriculture had been set up there, and soon afterwards it was linked with the teaching of natural history and moved to the university. Manufacturing industry introduced new, more complex technology, calling for greater technical skill. Shortly after 1840 it was suggested in Brno that a Technical School should be established. In 1845 the idea was approved by the central authorities in Vienna, and in 1850 the project came to fruition. Abbot Napp was behind both the idea and its realization.

In 1849 secondary schools underwent reforms. The two-year period of study at institutes of philosophy was divided into two parts, the first being attached to the six-grade *Gymnasia* as the seventh and eighth grades, while the second, dealing with professional and technical skills, became the basis of teaching at the new Technical School. Natural history and agricultural science were taught for a further year by Diebl, who was now eighty, before being taken over by Helcelet, who came to Brno from Olomouc. In subsequent years teaching at the school was brought up to a high standard. In 1867 it was renamed the Institute of Technology, and in 1873 the Technical University.

The industrial city of Brno was also seeking to create a new type of secondary school suited to the needs of factory production. In 1851 J. Auspitz (1812–88) was commissioned to build a new type of school, the *Realschule*. Up to 1848 he had taught accounting at the Technical College in Vienna. In that revolutionary year he organized the students' legion, and consequently had to leave the city. He became a teacher at the Brno Technical School. He soon proved his organizational skill, which led to his being appointed headmaster of the *Realschule*. In 1852 the first pupils were enrolled, with three parallel classes, each comprising eighty to a hundred pupils. There was a major lack of teachers in Brno at the time, and Abbot Napp was still an influential figure in education. He was not informed of Mendel's examination failure, through the provincial governor's office, until almost a year later, on 9 August 1851.

The report stated that in his oral examination Mendel showed a more extensive knowledge than his written paper would suggest, adding that he lacked 'neither industry nor talent' (Czihak 1984, pp. 19–20). The chairman of the commission, Professor Baumgartner, recommended in a letter to Napp that Mendel be given the opportunity to study full-time at the University of Vienna in order to acquire a more detailed grasp of his subjects. On 3 October Napp, in a letter published by Iltis (1966, p. 75), replied that he had decided to send Mendel for 'higher scientific training', and that he would regret no expense devoted to his education. The content of the letter suggests that Napp had previously asked Professor Baumgartner for details of how the examination had progressed. His recommendation was decisive for Mendel's being sent to Vienna University. The day before he wrote to Baumgartner, Napp sent a letter to the Bishop of Brno, stating that 'Father Gregor Mendel has proved unsuitable for work as a parish priest, but has on the other hand shown evidence of exceptional intellectual capacity and remarkable industry in the study of natural sciences, and his praiseworthy knowledge in this field has been recognized by Count Baumgartner, but for the full practical development of his powers in this respect it would seem necessary and desirable to send him to Vienna, where he will still have full opportunities for study.'

Mendel's examination failure certainly came as a stunning disappointment to him. He did not return to Znojmo, though on 30 August 1850 the mayor of the town, A. Buchberger, turned to Napp with 'an urgent request' that Mendel should be permitted to teach there in the following year. He added that Mendel had, 'in view of his praiseworthy manners and especially in view of his universally appreciated teaching skills', won the respect of the local inhabitants. But on 7 September Napp replied that he was unable to grant the mayor's request, since he had 'already made other plans for Father Gregor Mendel' (Sajner 1967). He added that there was no one else at the monastery capable of filling the post. Napp had in mind his plans for Mendel to study at Vienna University; the abbot's decision must have been further reinforced by the mayor's praise for Father Gregor's talents, though Professor Baumgartner's recommendation was paramount.

At the age of twenty-nine Mendel now had the chance to study natural science at university, something he might never have dreamed possible. Had he passed his teachers' examination, he would have stayed on at the Znojmo *Gymnasium*, and several generations of secondary school pupils would have gained an excellent teacher. Science, on the other hand, would almost certainly have lost one of its leading discoverers.

AT UNIVERSITY

Mendel arrived at Vienna University shortly after 1848, a year of revolution, in whose aftermath natural science teaching underwent a process of reform (Weiling 1986). Education was now under a ministry instead of the previous 'Court Commission for Studies', and the university syllabus was radically affected by the change. Botany, zoology, and chemistry, previously taught at medical faculties and *Lycea* or philosophy institutes, had been transferred to the competence of the newly established faculty of philosophy. A new emphasis was being placed on a combination of teaching and experimental work. At first the newly constituted institutes suffered from a lack of space and instrumentation, but despite difficult conditions the teachers managed to inspire their students with enthusiasm for the nascent scientific disciplines and the research which accompanied their studies.

Napp sent Mendel to study exact physics at the newly established institute headed by Professor Christian Doppler (1803–53). What remains of Mendel's student records shows that from the second term he signed on for additional lectures in mathematics, chemistry, zoology, botany, the physiology of plants, and palaeontology. His syllabus was explained by Weiling (1967) and Czihak (1984, pp. 21–5). Two-thirds of his time was spent studying experimental subjects. In his second and third terms he had a total of thirty-two periods of lectures and practicals per week. He continued to pay most attention to physics, which was at the time a new and highly regarded course, intended to give secondary-school teachers the kind of preparation called for in the new industrial age.

The academic year began in October, but Mendel applied for permission to start on 5 November. The reason he stated was that he had to overcome obstacles that were not of his own making. This was a reference to Napp's negotiations with the bishop and his efforts to secure lodgings for Fr. Gregor in one of the Vienna monasteries. When this turned out to be impossible, Mendel had to take private accommodation a couple of kilometres' walk from the university buildings. On arriving in Vienna, he delivered to Professor Baumgartner a letter from Napp in which the abbot asked for Mendel to be given all assistance in starting his course. His late arrival also accounts for his somewhat exceptional timetable. On the recommendation of the influential Baumgartner Mendel was enrolled after the start of term.

Professor Doppler was appointed head of the new physics institute, whose foundation has been described by Dick (1986), in January 1850. Before that the subject had been taught by Professor A. Kunzek, in a course without any practicals. Doppler's name is associated with the Doppler effect, an optical phenomenon in astronomy. The new physics institute was situated in a building which had first to undergo reconstruction and was completed in 1851. By order of the ministry of education the institute was to teach secondary-school teachers; it was equipped for a maximum of twelve students, called *élèves*. Mendel was accepted even though the twelve places were already filled, which may have been why he was described as an extraordinary student. The *élèves'* course was three half-year terms, but

Mendel enrolled in physics for a fourth, possibly owing to his special interest in the subject, or simply because at the start of the second academic year Professor Doppler died, and his successor, A. Ettingshausen (1796–1878), did not take up his duties immediately. Thus Mendel may have applied for permission to extend his studies.

According to the ministerial order mentioned above, physics students were expected to acquire the necessary theoretical background and the skills required to perform experiments and undertake independent research. Mendel had already studied basic physics in Olomouc. Though he later studied other branches of science, he considered himself to be above all a physics teacher. Under both Doppler and Ettingshausen physics students had two periods of lectures daily, and the rest of the day was spent studying literature and doing the set work required by the professor. Before performing experiments, they had to submit a plan for approval. They wrote up the results, and some of these were later published. We do not know what practical work Mendel was set, or how his teachers evaluated his results, and none of his experiments were published.

In his guidelines for the physics course, Doppler emphasized that the students should seek the most appropriate solution to the problem, with the least possible expenditure on experimentation. Later, when Mendel was a teacher at the Brno *Realschule*, he stuck to this principle—a fact which is acknowledged in the headmaster's reports on his teaching. In the first term Mendel attended only lectures in experimental physics. The following year he also enrolled in lectures and practicals in zoology; however, these clashed with the physics timetable, and he later cancelled his enrolment. In the third term he was also enrolled for lectures in chemistry, palaeontology, mathematics, anatomy, and plant physiology, with practicals using a microscope. That was the most demanding term for Mendel, both from the point of view of the amount of time he spent attending lectures and practicals and from that of the number of subjects he studied. Mathematics lectures took place on Saturdays, and physiology practicals even on Sundays. In the summer term of his second year Mendel enrolled for three periods in the use of physics instruments, another three in higher mathematics, and five in organic chemistry.

The future researcher was greatly influenced by both his physics professors. Before coming to Vienna, Doppler had taught mathematics at secondary school and at the Technical College in Prague, later at the Mining and Forestry Academy at Banská Stiavnica, and finally at the Technical College in Vienna. The second edition of Doppler's (1851) arithmetic and algebra textbook was published during Mendel's studies in Vienna. In it Doppler outlined the principles of combinatorial theory and the theory of probability in relation to the needs of applied science. Later Mendel was able to put these principles to good use in his research on plant hybridization. Doppler's teaching of methods of experimental research and of the manner of solving scientific problems were also of great use to Mendel.

Ettingshausen had been teaching at Vienna University since 1821. In 1826 he published the textbook *Combinatorial analysis*, and a year later two volumes of lectures in higher mathematics (Ettingshausen 1826, 1827). From 1842 he cooperated with Professor Baumgartner on the seventh edition of the textbook *Basic physics* (Baumgartner and Ettingshausen 1842). Later Ettingshausen (1843) published his own physics textbook. He stressed the importance of the mathematical generalization of experimental results. He also mentioned the explanation of crystal forms as combinations of simple series. His pupil Mendel was later to apply this principle in expressing the conclusions of his research on plant hybridization. While Mendel was teaching at the *Realschule* twenty-nine crystallographic tables were sent from the school to the 1862 Great Exhibition in London. The tables were prepared by teachers in cooperation with students. Richter (1943, pp. 129–30) assumed that Mendel might have taken part in the project. Taking into consideration the attempt at a theoretical explanation of crystal forms, Matalová (1991) assumes that Mendel could have examined this idea in an attempt to explain the origin and development of hybrids.

While Mendel was studying in Vienna, a new specialization in the teaching of botany was introduced (Weiling 1986). On the death in 1849 of Endlicher, the botany professor, lectures were divided into two sets of courses. Morphology and plant systematics were taken over by the director of the botanical gardens, E. Fenzl, and the new subjects experimental plant anatomy and plant physiology were taught by F. Unger (1800–70), who had previously lectured in botany and zoology at the University of Graz. According to ministerial order, the newly established Institute of The Anatomy and Physiology of Plants was to undertake a microscopic study of the 'internal structure and independent life of plants'. In the academic year 1850–1 Unger continued to teach the development of plants from the viewpoint of palaeontology. Like Doppler, he introduced his students to methods of experimental research. Later, evaluations of his teaching stress how he replaced empirical botany with the new principles of inductive research (Orel 1972b).

Unger's scientific orientation was significantly influenced by J.M. Schleiden (1804–81), a name associated with the idea of cells in plants. In his most outstanding work, *Principles of scientific botany*, subtitled 'Botany as an inductive science', Schleiden (1849–50) stated that he was above all

offering a guide to the new manner of researching into plants. He was sure
that botanists lacked an essential knowledge of philosophy, chemistry, and
physics, without which the 'real development of the science of organisms'
was unimaginable. He considered himself a pupil of Professor F. Fries
(1773–1843), who taught at Jena University and propounded a new school
of thought in the natural sciences that he denominated the mathematical
philosophy of nature. Influenced by Fries, Schleiden (1849–50, p. 39) later
wrote 'Looking at the composition of our experience of nature from its
individual elements, we find facts to be subordinated to laws and determined through them'. On the next line Schleiden added: 'It follows from
this that a complete theoretical explanation, in which we explain the inter-
connection of facts that are subjected to laws in terms of the latter, is possible only on the basis of mathematics and only in so far as mathematical
treatment is feasible.' Schleiden himself developed not only the teaching of
Fries, but also that of his pupil E.F. Apelt, who, in his theory of induction,
stressed the connection of empirical and rational induction, which are principles *a priori*, to which the accidental facts of experience are subjected.
Kepler and Newton were considered models of this sort of procedure in
research (Orel 1979).

In his book, Apelt (1854, p. 46) describes empirical and rational induction. Setting out from Apelt's leading maxims, Schleiden (pp. 141–8)
formulated the main maxims of scientific botany. The first two are as
follows:

(1) the maxim of developmental history; and
(2) the maxim of the independence of the plant cell.

According to these maxims Schleiden drew two conclusions:

(1*a*) Every hypothesis and every induction in the science of botany must be
unconditionally rejected unless its orientation is guided by the developmental
history.
(1*b*) Every hypothesis and every induction in the science of botany must be unconditionally rejected if it does not have as its aim the explanation of any process
taking place in the plant as a function of the changes which take place in its
individual cells.

Schleiden's masterpiece was bought by Mendel, and it can be assumed
that he read the methodological part of it. His research was conducted in
accordance with Schleiden's maxims. Schleiden's emphasis on a new
methodological approach in botanical research also had a profound effect
on Professor Unger, who taught plant physiology at Vienna University from
1849 to 1866. Before that Unger had taught botany and zoology at Graz
University, during which period he had also shown an enthusiasm for the

study of geology and palaeobotany. He gradually moved from the investigation of plant forms of fossils to the influence of soils upon plants and to plant geography. At Vienna University he soon became well known as a cytologist, and proved that all cell multiplication is by division. Like his contemporaries in Central Europe, he based his ideas on *Naturphilosophie*. In 1843 he and Professor S. Endlicher published a botany textbook in which they spoke of the action of the 'vital force' (Endlicher and Unger 1843). Three years later Unger (1846, p. 62) explained in a textbook on the anatomy and physiology of plants that 'under the term *vital force* of plants we understand any complex of causes as yet unknown, resulting from the modification of molecular forces, which determine their origin, maintenance, and propagation as individuals.' Schleiden was already thinking in terms of the norm of development, and in referring to plant hybrids he emphasized that 'embryo formation determined from both sides (i.e. both parents) in a way represents the mean' (Schleiden 1848, p. 109).

Unger (1851) published in Vienna a book entitled *The primitive world in its various transitional periods.* In it he summarizes the conclusions he reached while teaching in Graz. He describes the production of numerous forms of plants and animals in former epochs, and hints at a development from lower forms to higher, until the time that man appeared on the scene. Soon afterwards Unger (1852) began publishing an anonymous series of seventeen 'Botanical letters' in the weekly supplements of the local newspaper *Wiener Zeitung,* between 28 May and 18 October. Olby (1967) pointed out that Unger's views attracted sharp criticism from the editor of the church journal *Wiener Kirchenzeitung,* Dr S. Brunner, at the time Mendel was in Vienna.

In the first of these 'Botanical letters', Unger (1852, p. 4) states: 'The lucky turn of botany is evident only in the most recent period and its success is secured to such an extent that it will come to be considered the physics of the plant organism.' This organism was, according to Unger, 'an artificial chemical laboratory, the most ingenious arrangement for the play of physical forces'. He rejected the notion of constant species, supposing a process of metamorphosis from one to another to occur. In his view 'It would be erroneous to assume that the diversity of species consisted in this process of metamorphosis; but who can deny that new combinations arise out of this permutation of vegetation, always reducible to certain law-combinations, which emancipate themselves from the preceding characteristics of the species and appear as a new species.' He was aware that the stability of species is limited, but he could not explain how species change. As an outstanding observer of nature, Unger knew that there was an unknown regularity in the formation of plant forms in the sequence of the individual generations, and was convinced that one day it would be elucidated through investigation.

The editor of the *Wiener Kirchenzeitung*, Dr Brunner, attacked Unger's palaeontological book in 1851 as scandalous, criticizing Unger in his capacity as dean of the philosophy faculty. In his eyes the Catholic University of Vienna ought not to be open to a man such as Unger. Brunner recognized the anonymous author of the 'Botanical letters' as Unger, and a week after the last of them was published launched an attack on the university itself for its part in spreading agnostic and socialist views. Without mentioning Unger by name, he spoke of the paganism which was taught at the university. Brunner's attacks intensified when the 'Botanical letters' were published in book form, this time with the author's name. Unger was in danger of being removed from his university post, but the minister of education, Graf Leo Thun, who had some respect for science, threw the weight of his authority behind Unger, and Brunner was forced to publish an apology in a number of newspapers. At the height of the affair, in 1852, Mendel was one of Unger's students.

While teaching in Graz, Unger had experimented on the relationship between the chemical constitution of the soil and the distribution of plants. He tried to explain the role of soil in producing local varieties. The results convinced him that the source of variability must be sought elsewhere than in the diversity of the chemical composition of soils. One of his pupils was Kerner von Marilaun (1831–98), a medical student who attended lectures in plant physiology in Vienna at the same time as Mendel. Later Kerner performed transplantation experiments in 1875–80, demonstrating that variations produced by growing plants under Alpine conditions are lost as soon as the progeny is transplanted to the lowlands. We may therefore assume that it was also Unger who prompted Mendel to carry out the transplanting experiments described in Chapter 5.

In 1842 Nägeli had studied with Schleiden in Jena. The result was Nägeli's discovery of dividing cells at the apex of shoots and roots. In his 'Botanical letters' Unger (1852, pp. 21–7) had also expressed his admiration for Nägeli, whose research would, in his view, penetrate 'deep into that labyrinth which no eye has yet seen'. Unger was not to know that it would not be Nägeli, but his own pupil Mendel, who in this labyrinth of speculation would explain with precision the laws governing the origin and development of hybrids, and thus at the same time the transfer of traits from parents to offspring.

Unger's writings on the hybridization of plants also drew attention to the use of the artificial pollination of plants to create new varieties. In 1860 he published his lectures to the Vienna Horticultural Society, mentioning in them the significance of artificial pollination for the production and improvement of all cultured plants. According to Unger (1860), 'Fruit-tree varieties not grown until this century, for example peach and many apple, pear, and vine bastards, the flower hybrids of pelargoniums, fuchsias, wind-flowers, tulips, lilies, etc., testify to this commercial practice'. Here Unger is mentioning plant species that were later to be used by Mendel.

Unger's lectures aroused the interest of his students not only in study, but also in experimentation. He also influenced a wider public with his extensive publications. It is not generally known that the Dutch botanist Hugo de Vries (1848–1935), regarded as the first 'rediscoverer' of Mendel's paper, placed beneath the title of his thesis the following quotation from Unger: 'It is the task of physiology to reduce the phenomena of life to known physical and chemical laws' (de Vries 1870).

Unger enjoyed an exceptionally high regard among the natural scientists in Brno, most of whom had studied in Vienna. When the Natural Science Society was founded, Unger was mentioned among the first twenty-four honorary members. Professor Niessl (1871) closed his obituary notice on Unger with the following words: 'Unger was altogether a man who saw far beyond his study walls. He was in the closest possible touch with both the practical and the moral worlds, a fact to which above all his widely used writings can testify. In short, he was a thorough man, and one of the most significant geniuses of our century.'

One of Unger's students who took a detailed interest in the new physiological problems of plants was Johann Nave (1829–64), who was to become a friend of Mendel's. Nave was studying law in Vienna, and out of interest he attended Unger's lectures and practicals in his free time. He was especially fascinated by algae. This group of plants was at the time becoming a model for microscopic and physiological research moving towards the elucidation of the hotly discussed question of fertilization. Later, back in Brno, Nave followed the career of a provincial civil servant, spending his spare time working with algae. Thus both these students of Unger's, Mendel and Nave, developed lines of research inspired by their professor. Nave paid particular attention to the question of the fertilization of algae. In his hybridization research in the sphere of higher plants, Mendel linked this enigma with that of heredity. We can assume that he later discussed the topic with Nave in Brno. Nave himself dealt with the question of fertilization in both his lectures and his papers.

Another experimental discipline Mendel studied was chemistry. His lecturer in the subject was Professor J. Redtenbacher (1810–70), who, after graduating from medical school in Vienna, studied chemistry at Giessen University under Professor J. Liebig (1803–73), with whom he cooperated on a publication relating to the atomic weight of carbon (Redtenbacher and Liebig 1841). Mendel attended Redtenbacher's lectures on general and pharmaceutical chemistry and methods of chemical analysis, and may have learned of Liebig's idea of chemical radicals as the hierarchically higher building-units of organic matter. Sturtevant (1967) expressed the view that in addition to the atomic interpretation of matter, Liebig's ideas on radicals as semi-permanent, substitutable building-blocks could have inspired Mendel's particular interpretation of heredity. Liebig was working on the principles of organic chemistry, and his work pointed to the possibility of applying recently acquired knowledge in practice.

Mendel's chemistry studies have in the past been rather neglected. M. Campbell (1985) has underlined their importance with regard to the methodological evaluation of Mendel's work. Chicago University Library has a copy of the title-page of F. Strumpf's (1853) textbook on the latest advances in chemistry in application to technology and pharmacy which bears the pencilled signature of Gregor Mendel. We might explain how it got there by supposing that a visitor to the Brno monastery (Fig. 4.1) after 1900 received either the whole book or the title-page, and took this document back with him to the United States.

Fig. 4.1 The library in the Augustinian monastery.

Later Mendel, cancelling his enrolment in zoology lectures, found a replacement course in the zoology department of Vienna Museum, whose head, Dr V. Kollar (1799–1860), was also from Moravian Silesia—one possible reason why his enthusiastic new natural history student might find favour with him. After carrying out research in the museum Mendel lectured (1853) on a pest of radishes (*Botys margaritalis*). At the January meeting of the Zoological Botanical Society he had been accepted as a member of the society. He maintained his membership subscription right up to the end of his life. Some natural science students published the results of their research. The only surviving record of Mendel's studies is the 1853 article on entomology already mentioned. On this subject Mendel investigated the occurrence of the pea pest *Bruchus pisi* after his return to Brno. Dr Kollar presented Mendel's (1854) findings at a meeting in Vienna. The pest in question was widespread in Moravia at the time, and Mendel seems to have become interested in it in connection with his research into peas. There was an article on it in the Agricultural Society journal in 1855, and a copy of it in the monastery library has marginalia by Mendel, proving his continuing interest in the subject (Anon. [V-g-1] 1854).

Little is known of Mendel's palaeontology lectures, given by F. Zeckeli. In his record of studies there is a note by Zeckeli that Mendel was 'extremely diligent, and most attentive right up to the end of term' (Mendelianum, sign. no. 46). In this course Mendel acquired detailed information on the world of fossils mentioned by Professor Unger, and saw in the context of geological time the picture of nature in flux painted by Klácel (1843).

At university Mendel came across the new scientific thinking in application to research into nature and into the position of man in nature. It was based mainly on the development of physics, which was the most advanced natural science at the time. Physics attributed everything in nature to the action of a small number of laws, the starting-point for which was the existence of certain indivisible particles of matter. The laws of nature were written in the language of mathematics, and the task of the researcher was to discover the existence and activity of these indivisible particles in various phenomena, and to find the pattern to which they conformed. A new theory should not simply be the result of random speculation, but should be based on planned experiments, the results of which were to confirm the theory in a mathematical generalization. Mendel learned these principles in the lectures he attended, and he was able to read them in Baumgartner's and Ettingshausen's (1843) textbook. These authors took physics to be a higher scientific discipline than chemistry and the other descriptive sciences such as botany and zoology, and in the context of this hierarchy pointed out that these descriptive sciences were also the least developed ones.

Physicists also underlined the importance of combinatorial mathematics and probability theory. A book which may have been of major importance to Mendel was that published by the Viennese astronomer L.L. Littrow (1835), who informed his readers of the probability theory developed by his fellow astronomer P.S. Laplace (1749–1827) and fellow mathematician J.F.K. Gauss (1777–1827). The book's introduction points to the importance of this theory in research. According to Littrow: 'All phenomena, even those which because of their triviality seem quite random and entirely independent of the great laws of nature, are without doubt a necessary consequence of the same material laws as those which regulate the movement of the sun and all the heavenly bodies. Only our ignorance of the dependence of these phenomena on those laws of the universe leaves these basic causes, as it were, dependent on sheer chance, since they do not seem to be subject to any visible environment or any order we can observe.' Littrow further arrives at the conclusion that 'All phenomena, even those which seem most dependent on pure chance, exhibit, when they are repeated often enough, a tendency to more and more constant relations, and are subject to some usually very simple law, from which, if it is once recognized with sufficient precision, the future conditions of the phenomenon can be predicted'. There is no evidence to show that Mendel did indeed read this; but, if that were the case, he would have found in it what almost amounted to a guide to experimentation with a large number of plants. The physics textbook written by his teachers refers to Littrow's work, and it was in the library of the Brno Museum. Another book on astronomy by Littrow (1825) can be found in the Augustinian library in Brno. It contains pencilled notes, most probably by Mendel. And thus Littrow's name at least was not unknown in the Brno monastery.

In the first half of the last century hybridization was also a problem of evolutionary theory, and in his 1849–50 masterpiece, which was very much written from that standpoint, Schleiden considered it one of the main lines of research in scientific botany. Gärtner, who based his work on that of Kölreuter, showed that it was possible to create new combinations of plant traits. Unger thought research in experimental hybridization could be carried out in connection with the cell theory and the problem of generation and fertilization.

During his university studies Mendel bought Gärtner's (1849) monograph on plant hybridization, which may have been brought to his attention by Professor Unger. In the introduction to his 1865 lecture Mendel drew attention to the views of Kölreuter, Gärtner, Herbert, and Lecoq. Gärtner described exhaustively the experiments of these other authors. He referred to Kölreuter on 164 pages, to Herbert on 85, and to Lecoq on 26.

We may suppose that Mendel gleaned his data on these experiments from Gärtner's monograph. W. Herbert, Dean of Manchester Cathedral, experimented with plant crossings to produce 'new and more valuable' varieties of horticultural and field plants. He paid particular attention to the debated question of the varying fertility of hybrids, which he put down to the effect of different climatic and soil conditions. Herbert rejected the views of his contemporaries that the fertility of hybrids might be a criterion in distinguishing varieties and species. In Gärtner's monograph Mendel underlined the reference to Knight's view, expressed in 1799, that hybrids between varieties are fertile and those between species infertile. He never cited Knight, however, despite the fact that Gärtner mentions him on eighteen different pages. H. Lecoq undertook experiments with plant hybridization while he was professor of natural sciences and director of the botanical garden. He referred with enthusiasm to the great opportunities there were for creating new species, and pointed out that the fertility of hybrids varied, though he did not proffer an explanation. These views were also underlined by Mendel in Gärtner's monograph.

Nor did Mendel fail to note the pea-crossing experiments of Goss (1824) and Seton (1824), published in England. In Gärtner's monograph he marked the reference to these authors on page 85. A note in the margin gives the name of the Bavarian gardeners' journal where a German translation appeared in 1837. Abbot Napp was among its subscribers, and one can assume with some confidence that Mendel read the article, where he would have learned of the advantages of peas for his crossing experiments to investigate the transfer of traits from parents to progeny. Another subscriber to the journal was Professor Zawadski (see pp. 79–80).

In the middle of the nineteenth century natural scientists were already studying specialized branches of science. They normally concentrated on either physics or chemistry, and on the study of either living or non-living nature. Mendel chose a combination of almost all branches of science with the exception of astronomy. In his work he investigated an entirely new problem, one which according to the classification of sciences then current might be regarded as a physiological one. Following the principles propagated by his physics teachers, he created a new method of research in order to investigate the question of the origin and development of hybrids.

Little is known of Mendel's private life at the time of his studies in Vienna. He lived in lodgings, and had the means to keep himself in a humble manner and to buy the scientific books required for his studies. Only a few of these have survived. He was ten years older than the rest of the students, and he must have been outstanding in the enthusiasm he had for study. He dutifully performed his obligations as a priest, saying and attending mass regularly. For him study was a means of assuring his future, and he spent all his free time on it.

Two letters have survived from this period. He asked his friend Rambousek to help him obtain some new shirts, since a parcel containing twelve of them had been lost on the way back from the laundry (Kříženecký 1965*b*, p. 128). The abbot had announced that during the summer the members of the monastery staff were to participate in spiritual exercises, and in his letter Mendel added with humour that on completing these he would not wish to be a 'new man' spiritually, yet wear a tattered shirt. The letter reveals that he was able to construe instructions from his superior according to the demands of his studies. He was supposed to return to Brno early, in view of the coming retreat, but since the academic year ended on 20 July, he announced that he would not be arriving at the monastery until 24 July. The letter refers to 'lucky' Klácel, whom the Abbot had permitted to make a trip to the surroundings of his native Česká Trebová to meet fellow patriots. Mendel liked to buy lottery tickets. He jokes in his letter that if he were to win the first prize of 25 000 guilders he would telegraph Mrs. Smejkalová, who did the monks' washing and whom they all knew. The other surviving letter is to his parents, dated 24 March 1853, where he writes with indignation of an attempt on the Emperor's life. He remarks that he is in good health and that nothing has changed in this respect, and that he is studying hard and 'hopes that time will tell' (Kříženecký 1965*b*, pp. 114–15). Did he mean his finals, or was he already thinking of the research in which he was to put his recently acquired knowledge to good use?

A TEACHING POST IN BRNO

On 21 July 1853, Mendel returned to the monastery from his studies in Vienna. He had completed the course, but did not take the examination necessary for him to qualify fully as a secondary school natural history teacher. Up to the end of May 1854 there is a gap in our records of what he was doing. Immediately on returning from Vienna he must have embarked on his investigation of pests of cultivated peas, since Kollar read his communication on the subject at the 5 April meeting of the Vienna Zoological and Botanical Society.

When the post of physics and natural history teacher at the *Realschule* fell vacant, Mendel was presented with a new opportunity. On his appointment to the post as supply teacher, Dr Auspitz stated that he had found a suitable teacher who had studied physics at the University of Vienna and zoology at The Vienna Museum. He added that he knew the appointee from the time of his supply teaching at the Brno Technical School (Iltis 1966, p. 85).

The *Realschule* had been built on the initiative of industrial interests in order to support progress in production methods by applying the latest findings of science. When the school was opened in 1851, the first-year intake amounted to over 300 pupils, divided into three parallel classes. In the next two years new students were no less numerous. To begin with the school was housed on temporary premises on the outskirts of the city, but in 1859 a new and extensive building was opened on Janská Street in the city centre. In 1965, on the occasion of the centenary of the publication of Mendel's plant-hybridization paper, a plaque was unveiled to commemorate his work as a teacher, and at the same time the place where the discoverer made his findings public.

At the *Realschule* Mendel joined a team of enthusiastic natural scientists who, in addition to their teaching work, did a great deal of research. Along with the staff of the Brno Technical School, which was an institute of higher education, they formed the core of the Natural Science Section that was set up within the Agricultural Society in 1849. Its chairman was Count von Mittrowski, and members consisted mainly of teachers from the schools mentioned and from similar schools throughout the province. In 1854 the head of the *Realschule*, Dr Auspitz, was able to enlist the services of an outstanding physicist, in the person of A. Zawadski (1798–1868), recently deprived of his professorial chair at Lemberg, now Lvov in the Ukraine (Orel 1972*a*). The authorities in Vienna held him responsible for the student unrest in 1848, and he was obliged to leave not only the university but also the city. He taught physics and natural history in Brno, and it was not long after he arrived there that Mendel joined the staff of the *Realschule*. Zawadski breathed the new spirit of organized scientific research into local natural scientists, who knew him from his publications before he came to Brno. He was one of the older generation of natural scientists, interested in research into botany, zoology, palaeontology, and meteorology.

Zawadski soon assumed a prominent position among natural scientists both at the *Realschule* and in the Natural Science Section of the Agricultural Society. He published in the annual journal of the *Realschule* a remarkable sixteen-page article entitled 'The requirements of present-day natural science research' (Zawadski 1854). In an endeavour to evoke interest in the natural sciences, he wrote as follows: '... learn to see sagaciously, think correctly and feel warmly and deeply'. In this connection Zawadski stressed the importance of great discoveries in natural science, first of all in physics, and then in geology. He cited examples of scientists who had 'made the dark hours of the world's evolution brighter'. In conclusion he turned to his pupils in the following vein: 'I have explained to you, dear pupils, the requirements made upon you by natural science. In them also resides the reason why you must not neglect this science, which along with mathematics and chemistry forms the basis of a realistic education, and which I, together with Linnaeus, call the amiable science.' Fifty years previously, André had encouraged interest in science in a similar tone. In his day the scientific community had consisted of a handful of enthusiasts who were just beginning to study and exploit natural science in the interests of industry and agriculture. Zawadski, early in the latter half of the nineteenth century, was addressing dozens of students, who were later to make a major contribution to progress in the natural sciences and their practical application. His efforts were described by T. Frey (1869) in an obituary; he says that Zawadski's lessons 'thrilled through the wealth of his subjects, his true enthusiasm for nature, and his ubiquitous, finely-tuned understanding of the relationship of nature to human life. He was a generally loved teacher of youth, worked here in many ways, and won over many to the study of natural science.' This prominent figure was among the people Mendel met daily when he started teaching at the *Realschule*.

In 1854, Zawadski was elected secretary to the committee of the Natural Science Section of the Agricultural Society, and thus he would have had a say in the choice of subject for the monthly meetings. According to a brief report, on 4 January 1855 he lectured on the subject of 'The Developmental History of Lower Animals, namely on the copulation or blending of two animals to a single one for the purpose of fertilization'. Zawadski offered examples to illustrate his subject. At the same meeting Mendel was accepted, on Zawadski's recommendation, as a full member of the Natural Science Section. Zawadski continued his lectures at a further two monthly meetings. They were not published, so we lack any detailed knowledge of their content (Orel 1970a). It was a time when the dispute among natural scientists regarding the manner in which the generation enigma was to be solved was coming to a head, and one may suppose that Zawadski was one of those who inclined towards the hypothesis that a conjunction of male and female germs led to the creation of a new individual. Zawadski may have based his views on a study by M. Barry dated 1839, where he first showed the penetration of spermatozoa into the mammalian egg. Wood (1989–90) has explained how this discovery, controversial at the time of publication, was accepted by Purkyně, who drew it to the attention of his friend Wagner in Göttingen. His interest in this key process may have rubbed off on Wagner's pupil N. Pringsheim (1855), who finally demonstrated the actual penetration of a single spermatozoon into an ovum of the freshwater alga *Vaucheria sessilis*. After taking up his post at the *Realschule*, Mendel took charge of the natural science collections, and in this capacity came into contact even more often with Zawadski, with whom he may well have discussed his research plans and the question of fertilization.

Mendel taught physics and natural history in the second and third grades. His classes were large, between 62 and 109 pupils. He had a minimum of eighteen to twenty-seven lessons in a week, along with responsibility for the natural science collections. This led Richter (1943, pp. 129–35) to suppose that Mendel also helped to prepare crystallographic wall displays sent from the *Realschule* to the Great Exhibition in London, which Mendel subsequently visited (see Chapter 7). The authorities at the *Realschule* soon remarked upon Mendel's teaching skill. The headmaster, Dr Auspitz, noted in his 1855 report (Iltis 1966, p. 95) that Mendel's teaching was 'clear, logical and well suited to the needs of his hearers'. Other documents which refer to the young man's abilities confirm the picture of a teacher who conveyed to his pupils the import of natural history and physics, and kept his own knowledge up to date by studying literature and carrying out research. These qualities were none other than those which had already been outlined as the attributes of a good natural history teacher by Professor Diebl.

The first details of Mendel's teaching activity only started to come to light after the turn of the century, when there were still many former pupils of the *Realschule* alive. They now saw their teacher hailed as a great discoverer, and began publishing reminiscences in the newspapers, or sending them to A. Matoušek at the monastery. The material he started to collect is in the archives of the Brno Mendelianum museum. The part played by Matoušek in all this was pointed out by Orel and Marvanová (1966). Most of it was published in the Iltis biography (1966, pp. 88–95), usually with the names of his sources. These recollections harked back several decades, and we should be justified in assuming many of them to have been embellished under the influence of the kudos now generally afforded to the rediscovered propounder of the genetic laws. They are, however, the memories of his contemporaries, and as such help us to build up the personality jigsaw of this historic scientist, whose physical likeness is widely known from the painting which Iltis called a 'handsome portrait of Mendel'. The picture which emerges is that of a striking figure striding purposefully towards the town centre along Pekařská Street, to begin his day's work at the school on Janská Street. 'A man of medium height, broad-shouldered and already a little corpulent, with a big head and high forehead, his blue eyes twinkling in the friendliest fashion through his gold-rimmed glasses. Almost always he was dressed, not in a priest's robe, but in the plain clothes worn by members of the Augustinian order who worked as teachers: frock-coat, usually rather too big for him; short trousers tucked into top-boots.'

These early twentieth-century memories of Mendel also include an image of him 'standing in front of his pupils, looking at them in a most friendly fashion' and 'his cordial tone of voice, and his smile'. Another informant recalls: 'All his pupils agreed in extolling his method of instruction, his justice and conscientiousness, his gentleness and kindness. There was no need for him to have recourse to terror as a supplement to instruction. His clear and luminous method of exposition, reinforced in case of need by subsequent friendly elucidations, enabled all his hearers to understand what they wanted to understand. He himself delighted so much in his work as a teacher, and was able to present every topic so agreeably and invitingly, that we always looked forward to our lessons.' He presented the rudiments of zoology, botany, and physics to his pupils in a pleasant fashion. His manner of explaining the material, accompanied by good-humoured comments, sometimes engendered laughter, which he was always able to control, redirecting his pupils' attention to the subject-matter. He always kept his pupils well in hand, and was respected, as well as being popular.

Mendel the teacher was always considerate and patient. 'It pleased him much more that his pupils should have a lively interest in their work than that they should know a great many facts. Hardly ever did he allow any pupil to have a setback.' At the end of the school year he asked the pupils who already knew their grades whether they wished to be re-examined. Then he would tell them to set each other questions. Pupils were also permitted to visit Mendel in the monastery garden, where he offered additional explanations or special tuition. Mendel's enthusiasm for numerical evaluation of the phenomena he observed was also evident from the fact that ' ... each boy had his own number, determined by his progress in the class. Mendel, fluttering the pages of his notebook, would select one of the numbers.' He would choose a number, twelve for example, and then say 'Twice twelve is twenty-four, and twenty-four plus twelve is thirty-six'. Then he would examine the pupil whose number that was.

Some of Mendel's pupils recalled visiting him at the monastery, where he would receive them with a friendly smile. One Langer, who attended the *Realschule* in 1856–7, states that at the monastery their teacher showed them 'the plants with which he was doing crossing experiments, and his beehives'. In fact it is still not known for certain whether there *were* any beehives at the monastery during the period that Mendel taught at the *Realschule*. Mendel's own beehives did not appear until 1870, when as abbot he gave instructions for them to be built. According to Langer, when teaching in class Mendel would sometimes 'give formal demonstrations of the way in which crosses of plants were effected, showing how the flowers were protected from disturbing influence by paper caps'. Yet he did not as a rule explain that he himself was engaged in such experiments. When speaking about breeding experiments, Mendel is said to have used plain terms to describe reproduction, and if any pupil tittered he would say: 'Don't be stupid! These are natural things.'

Mendel and his pupils undertook botanical excursions in the surroundings of Brno, and he impressed them with his knowledge. J. Líznar (1902), later a professor of meteorology, recalls: 'I had the good fortune to have Gregor Mendel among my teachers. It was he who aroused in me a love for natural sciences.' Líznar also took part in Mendel's trips to southern Moravia, to the area around Lake Čejč, famous for its specialized flora. There Mendel dug up some plants for the monastery garden, especially hawkweeds. The period of Mendel's teaching at the *Realschule* is closely connected with his plant-hybridizing experiments, which form the subject of the next chapter.

The surviving letters to his parents and relatives show that Mendel kept in touch with his birthplace. Soon after the death of his father, on 7 June, probably 1857, he apologizes to his mother for not having been able to come as he had promised, and assures her he will come during the school holidays (Kříženecký 1965*b*, pp. 115–16). He mentions the damage caused to fruit trees by frost, and lists the prices of various types of grain at market in Brno, something his family was interested to know. Later, on 25 June 1859, another letter criticizes the Emperor Napoleon for the suffering inflicted upon the armies and the civilian population in Italy. He writes again of grain prices (Kříženecký 1965*b*, pp. 115–16). On 11 January 1862, he wrote to the husband of his younger sister, Theresa, following his mother's prolonged illness, and enclosed thirty guilders, requesting that they use the money to buy anything his mother might need, adding: 'Even if I did not know the Fourth Commandment, I should be anxious to relieve the burden of her old age (Kříženecký 1965*a*, pp. 119–20).' Another letter to his brother-in-law, dated 14 July the same year, is of particular interest (Kříženecký 1965*b*, pp. 120–1). In it he expresses gratitude for a gift from home, which according to the contents of the letter would appear to have been some home-made butter. Mendel mentions that on Thursday, 24 July he is leaving on a lengthy trip abroad, taking in Vienna, Salzburg, Munich, Stuttgart, Karlsruhe, Strasburg, and Paris, and thence on to London to visit the Great Exhibition. A photograph has survived showing the Moravian participants in the trip outside the Grand Hotel in Paris, close to the site where the Opera was to be built. The party was to spend a week in Paris and a week in London. They expected to return in mid-August. Mendel looked forward to visiting his home when he got back, and recounting his experiences. At the end of this letter Mendel notes the bad weather, which was unfavourable for the village harvest.

During his teaching years Mendel continued meticulously to comply with his priestly duties at the somewhat unusual monastery. The rest of his time was spent on study and scientific work, which brought him much satisfaction. We might now reflect that it was a period when his youthful ambitions were being fulfilled; but it was also a period which placed new and unforeseen obstacles in his path.

A SECOND CHANCE

No one has yet found the answer to the question why Mendel did not take his teachers' examination a second time immediately after completing his university studies, in 1853. We also do not know why he did not try to obtain a doctorate, as in the case of his fellow monks Alt, Bratránek, and Gabriel. It might at first sight seem that nothing stood in his way.

After Mendel's return from Vienna the situation at the monastery was extraordinary, affecting all the members of the community. Following the suppression of the revolution in 1849 a new, absolutist government took power in the Empire, and the status of the Catholic Church in society began to be restored. In 1855 a concordat was initiated between Church and State. The ecclesiastical authorities began a campaign to renew the spiritual life of monasteries. The bishops of the Habsburg Empire had undertaken visitations of monasteries even before that. On orders from Rome, the Archbishop of Prague asked the Bishop of Brno, A.E. Schaffgotsche, to check up on the spiritual life of the Augustinian monasteries in the provinces of Bohemia and Moravia. It transpired that he had not forgotten past disputes between the monastery and the bishop's office, especially the rebellious petition of the six Augustinian monks to the Constitutional Assembly in Vienna in 1848.

In February 1853 the monasteries the bishop was to visit received a list of questions the visitation was intended to answer. Abbot Napp quickly sent a letter (9 April) to the archbishop describing the privileges enjoyed by the monastery and drawing attention to the order of the Emperor for members of his community to devote themselves to the study of scientific subjects with a view to their working as teachers. Napp defended the scientific inclination of the monastery as a whole, and of each member of the community separately. He wrote of Mendel that he was studying physics in

Vienna at Professor Doppler's institute, in order to acquire a higher scientific education. The documents relating to the visitation were published by K.A. Huber (1978) in the original Latin with a commentary in German. The Latin documents have been published in German (Czihak and Sládek 1991; Zlámal 1991).

The bishop made his visitation on 7 and 8 June 1854. He spoke individually to all the members of the community, asking about their attitudes to the study of science and especially about the attitude of Abbot Napp in relation to the spiritual calling of the monastery. The bishop made a written report on his visitation, dated 7 September, to the archbishop. In it he stated that Abbot Napp held so many public offices that he was unable to devote sufficient attention to his main responsibility, that of running the monastery. As a result, 'the last ray of spiritual life' had faded there. The bishop saw the scientific and teaching activities of the monks as being in contradiction to their spiritual calling, and he cited as examples of the ill effects such wayward pursuits might produce the cases of those members of the community who had been forced to leave their teaching posts over the last thirty years. At the top of his list was Neděle, who, he said, had been deprived of his post because of his 'explanation of the bible in terms of the errors of the rationalists'. In second place is Klácel, who, it is said, had to leave because his philosophy teaching was permeated by 'pantheist fantasies'. Křížkovský is described as an outstanding musician, able to conduct a large orchestra, which 'wins him applause'. Fr. Rambousek, on the other hand, had to be admonished for bathing 'almost naked' in a public place. Mendel is said to be studying science at a secular institution in order to teach at a *Realschule* (Fig. 4.2).

The conclusions contained in the report were potentially catastrophic for the monastery. The bishop complained that not a single member of the community was willing to admit the error of his ways, and that they stood behind their misguided superior to a man. They even had the nerve to ask for a change in the rules of the monastery which would allow them to devote even more time to science and to teaching. In the bishop's view the monastery was irreformable, and the best thing would be to dissolve it by papal decree. He proposed that Napp be given a life pension which would allow him to continue to devote himself to public service, while the fate of the other monks would be decided individually after the dissolution of the monastery.

In this critical year, Mendel took the same resolute attitude to the study of science and to teaching as the rest of the monks. They appear to have had confidence in the protection of the influential Napp, who did indeed find a way out of the difficult situation. On the very first day of the visitation he recognized which way the wind was blowing, and immediately sent

Fig. 4.2 The teaching staff of the *Realschule*. Mendel is seated second from the right; Alexander Makowsky is standing fifth from the left; Joseph Auspitz is seated fifth from the left; and the sixth from the left is Alexander Zawadski.

a memorandum signed by all the members of the community. It mentioned briefly the privileges the monastery had enjoyed in the past and the fulfilment of its new duties. It proposed changing the monastery into one belonging to the order of *Chorherren* Augustinians, which would allow its community to study and teach science. The memorandum recalls the exceptions from the medieval Augustinian *regula* granted by decree of Pope Sixtus IV in 1484.

In July 1855, the archbishop sent letters to the monasteries which had received visitations, drawing attention to their shortcomings and requiring that their obligations be maintained both in and out of the monastery. The decree sent to Brno says nothing about dissolving the monastery. After receiving a reminder, Napp admitted to the faults mentioned, and promised to fulfil the obligations of his community. Later, on 27 December 1855, the archbishop sent to Rome the bishop's original findings from his visitation, including the radical demand that the monastery be dissolved. All these documents can be found in the Vatican archives, but no one knows what reply was sent if any. One way or another, Napp successfully defended the existence of the monastery and its scholarly aspirations, including, in effect, the experiments Mendel was beginning.

The monastery accounts show that on 9 April 1855, Mendel went to Vienna to arrange for his examination to be postponed, and he went again on 2 May to take the examination. Iltis (1966, p. 95) states on the basis of documents he found at the monastery that Mendel went to sit his examination on 5 May. Nowotny, a school inspector who had taught with Mendel at the *Realschule*, told Iltis that when Mendel returned from Vienna he was ill. The reason was supposed to be that he had got into a dispute with E. Fenzl, one of the examiners, and thus had failed his examination. Early this century the procurator of the monastery, A. Matoušek, put in the archives a letter from Klácel to Bratránek dated 8 May 1856, according to which Mendel was unhappy when he came back from Vienna (Křiženecký 1963*b*). The letter states that Mendel was given easy questions for his work at home, and it was not difficult for him to answer them. But when he was required to do his written examination *in persona*, he lost his nerve and could not write anything. Klácel adds that Mendel had been subject to fits of nerves since his youth. He must have been quite seriously ill after this second examination failure, since his father and brother came to see him on 31 May. R. Gickelhorn (1969) found in the Vienna University archives Mendel's examination application, dated 3 July, the date of the examination being given as 3 August. Mendel seems to have been in Vienna in May, when there may have been some sort of conflict which led to his illness. We do not know the circumstances under which the July application was submitted, nor whether Mendel attended the examination again.

Today one can only surmise that in Vienna Mendel met Professor Fenzl, who had set his written work at home. A conflict of opinion may have occurred, due to Mendel's explanation of the fertilization of plants. According to Wunderlich (1982), E. von Tschermak said that his grandfather, Fenzl, held right up to the end of his life that the plant embryo arose from a pollen tube and a germ cell responsible for its nutrition. Mendel's research was based on the idea that both parental forms were involved in the creation of the embryo. Thus he may have fallen victim to this dispute among botanists, which was just reaching its climax at the time, and in which Mendel was in favour of what turned out to be the correct view.

After returning from Vienna, Mendel went through a critical period, fearing the effect his examination failure might have on his position at the school. For Abbot Napp, too, it must have come as a disappointment, for he considered the successes of his monks to be those of the community as a whole. No evidence has survived to indicate what his reaction was. The fact remains, however, that this new failure had no practical effect on Mendel's teaching career. The headmaster continued to give him good reports. On the other hand, he had to be satisfied with a supply teacher's half pay.

In spite of his illness, Mendel continued with his *Pisum* experiments after his return from Vienna, artificially pollinating selected varieties of peas. His disappointment in the academic sphere was counterbalanced by his enthusiasm for his research, which was producing its first successes. The physics and natural history teacher was slowly turning into an eager researcher, starting to acquire the self-assurance he needed to produce hypotheses and to verify them experimentally. In doing so he was helping to unravel the very problem which may have led to his examination failure.

THE BRNO SCIENTIFIC COMMUNITY

Professor Zawadski at the *Realschule* nurtured his students' interest in the natural sciences, and in the learned societies he belonged to he encouraged members to perform research. In 1854, the year he came to Brno, he proposed to the Natural Science Section of the Agricultural Society that his young colleague, physics teacher Fr. Gregor Mendel, should be confirmed as a full member. Zawadski's series of three lectures that year on the enigma of procreation must have fascinated Mendel, since he had attended Professor Unger's lectures on the same topic, and it was closely connected with the subject of his own research. Zawadski was interested in the application of hybridization to fruit-tree improvement. He was also a subscriber to the journal of the Bavarian Horticultural Society, which carried a German translation of Seton's and Goss's accounts of their pea-crossing experiments. In 1856 Zawadski attended a congress of German natural scientists and physicians in Vienna, and subsequently published in Brno a report (Zawadski 1857) on the part of the proceedings dealing with physiological problems in plants, mentioning Professor E. Regel's lecture on the crossing of goatgrass with wheat in order to verify the origin of cultural forms of wheat. He pointed to the significance of hybridization in plants from both the evolutionary and the breeder's points of view—an indication which cannot have escaped Mendel's attention.

N. Pringsheim (1855) proved by experiments with the fresh water alga *Oedogonium* the fusion of spermatozoa, the male agent, with the oogonium, the female agent. This was the first experimental evidence of fertilization involving the fusion of two germ cells. Physiologists based on it their further research on the fertilization of higher plants. J. Nave (1858) lectured in Brno on Pringsheim's research in connection with the generation enigma at a meeting of the Natural Science Section of the Agricultural Society. His lecture was published in the section's yearbook. He gave a summary of various views of the problem of generation, placing particular emphasis on Pringsheim's research with algae.

Among Mendel's friends at that time was A. Makowsky, with whom he had taught at the *Realschule* in 1858–9, and who later taught geology at

the Technical School. He published significant works on botanical, zoological, and geological subjects, and shared with Mendel an interest in the
latest findings in the natural sciences. At the beginning of 1865 Makowsky
gave an enthusiastic lecture on Darwin's evolutionary theory (for details
see Chapter 5, p. 192). He later visited Abbot Mendel a number of times at
the monastery and was interested to know about his latest research.
Another active natural scientist was J. Kalmus, a medical practitioner who
had studied medicine in Prague, and one of whose teachers had been the
noted physiologist Purkyně. As well as being a practising physician, he did
research on mosses and, according to Iltis (1966, p. 101), was in close
scientific contact with Mendel. A no less important contact for Mendel was
the Brno pharmacist C. Theimer, who had studied in Vienna at the same
time as Mendel (Čižmář 1979). Apart from his profession, he was active in
the field of floristics, and paid great attention to the occurrence of hybrid
forms of plants in nature. He had already lectured on hybrids in 1862. As
vice-president of the Natural Science Society, in 1865 and 1869 Theimer
presided over meetings at which Mendel lectured on his experiments with
the crossing of peas and hawkweeds respectively.

In 1856 Mendel came across the subject of meteorology at a meeting of
the Natural Science Section. The city medical officer, P. Olexík (1801–76),
informed the committee of the section of the results of ten years' observations
in Brno, and according to the minutes Mendel 'was kind enough to offer to
draw graphs' of these data. This gave rise to Mendel's first meteorological
study (Mendel 1863). When in 1857 the committee of the section nominated
experts as consultants on various natural phenomena, Mendel and Olexík
were named as meteorology experts (Weiling and Orel 1967). The two later
became friends, and cooperated in the breeding of ornamental plants.

The Natural Science Section of the Agricultural Society was set up in
1849, but its members did not enjoy all the privileges of its governing body.
The committee of the Agricultural Society for the most part consisted of the
representatives of landowners, and they sometimes interfered in the affairs
of the Natural Science Section in a manner its members considered
unjustified. The discontent of the latter became apparent in 1859 with the
fall of the absolutist government in Vienna, which led to a certain relief of
the political tension in the Empire. The committee of the Natural Science
Section exploited this situation to propose a change in the Agricultural
Society's constitution. When the negotiations were dragged out, they
decided to form an independent Natural Science Society in order to devote
themselves to 'pure science' (Orel 1970b). The proposal also states that the
foundation of the new society did not in itself guarantee progress in science,
but that the decisive role would be played by the spirit in which the mem-

bers conducted their affairs. The organizers therefore placed the greatest
emphasis on encouraging their members to carry out research.

Prominent places on the committee were occupied by teachers from the
Realschule and the Technical School. Mendel's friend Nave was in the forefront; at the time he was conducting experiments with algae in his spare
time. In the background of the move to independence was Auspitz, headmaster of the *Realschule*. At the beginning of December Professor Zawadski
called a meeting of the members of the Natural Science Section, where
Nave read the proposal to set up a separate society. Unlike the Natural
Science Section, the new society was no longer to be a 'mere servant-girl' of
the Agricultural Society. Another meeting, on 15 December, was attended
by eighty-seven members, who approved the proposed constitution. The
minutes of the meeting show that Mendel and one of his fellow monks, Dr
A. Alt, were present. On 21 December the constitutive assembly took place
in the building of the *Realschule*. In his introductory speech Auspitz recalled
that the organizers of the independent society had to overcome opposition
to their scheme, and he implied that there had been a conflict between the
proponents of the new concept of science and the conservative committee
members of the Agricultural Society. The minutes of meetings show that
the Agricultural Society committee member Napp took a more circumspect
attitude. He formally welcomed the splitting off of a society for pure science,
but he expressed the hope that the Natural Science Section would continue
its work to develop scientific knowledge for application in agriculture. This
hope was not, however, to be fulfilled: in fact, all the members of the section joined the new society.

The chairman of the Natural Science Section, Count von Mittrowsky,
became chairman of the Natural Science Society, and the deputy chairman
was Professor Zawadski. The secretary was Dr Schwippel, who taught
natural history at the Brno *Gymnasium*, and Nave was made treasurer.
Mendel was among the 142 founder members. In the following year the
number of members rose to 171. The committee proposed twenty-four
honorary members, among them leading European natural scientists such
as R.W. Bunsen, F. Unger, R. Virchow, F. Wöhler, and J.E. Purkyně. The
Section had also had twenty-four honorary members. Most of them had,
however, been representatives of State and Church institutions, among
them Abbot Napp. In its choice of honorary members, therefore, the new
society clearly indicated the path it wished to tread. At the end of the constitutive assembly of the new society the physician C. Alle, who did
research into the medicinal effects of mineral waters, recited a poem in
which he expressed enthusiasm for modern natural scientific research. It
was printed on a single loose leaf inserted in the society's *Proceedings* and

has been republished by Orel (1970*b*). The tone of the poem is clear from the following excerpt:

> If matter acts upon matter, a force develops,
> To eavesdrop on the laws of nature and their explanation
> Is what science is aiming for.

Alle expressed the eagerness and ferment of new ideas among the natural scientists with whom Mendel came into contact in Brno. This was a reaction to the great discoveries which had been made in organic natural science, expressed in 1863 by R. Virchow (1821–1902): 'In all branches of the organic natural sciences there obtains a state such as that in a country which has undergone deep political upheaval, where all is at stake which previously seemed a closed chapter, where the authorities are losing their strength and each is left to his own devices (Uschmann 1974).' During this period Mendel was working on a problem his contemporaries considered unsolvable. He was forced to rely on himself in producing his original theory. But he had the moral support of his superior, and the backing of the authority of the members of the new Natural Science Society, led by a senior staff member from the *Realschule*, Professor Zawadski.

The new society was founded at the end of 1861, and at the very first working meeting, in January 1862, the problem of plant hybridization appeared on the agenda. At the same time plant hybridization was being discussed by the members of the Horticultural Association, which had come into being within the Agricultural Society in 1849 to replace the original Pomological Association. In December 1862 they discussed the use of hybridization in breeding new varieties of flowers. No details of this discussion are known. In October 1864, the matter was discussed again, and according to the published minutes Napp pointed out that all this was 'still more a question for science than for practice' (Orel 1975*a*).

In the 1830s Napp had explained to sheep-breeders that the problem of heredity was a subject for physiological research. Not quite thirty years later, he was now explaining to plant-breeders that the use of hybridization was still a theoretical problem. But he must already have been aware that Mendel was doing experiments on it in the monastery's experimental garden. The brief minutes of the Horticultural Section meeting give no details of Napp's contribution. But we can ask ourselves the question: how would Mendel have thought, given his cultural background, when he conceived his experiments, and, above all, how did he arrive at his innovative theory?

5

The researcher

Genetics, an important branch of biological science, has grown out of the humble peas planted by Mendel in a monastery garden

—DOBZHANSKY (1964).

After the year 1900, biologists for the most part wrote only of Mendel's *Pisum* experiments, from which they generalized Mendel's laws of heredity. Exceptionally, his second publication on experiments with *Hieracium* was mentioned. Only when C. Correns (1905) published Mendel's letters to Nägeli in German did both biologists and the new geneticists learn of the no- less-extensive experiments he carried out on other plant species.

In describing the laws of heredity most authors supposed Mendel to have performed his experiments at the monastery in his spare time, and merely for amusement. Only when he made his final evaluation, they assumed, had he happened upon the phenomenon of dominance, and the uniformity and segregation of traits in the hybrid progeny. On the occasion of the centenary of Mendel's birth Correns (1922) commented that 'Mendel's laws of heredity' as set out in the textbooks of the day were not in fact formulated by Mendel himself, but merely derived from the facts contained in his *Pisum* paper. Historians of science began seeking some forerunner of the discoverer, chiefly among the ranks of botanists who had performed plant-crossing experiments. Examples are a fine English monograph by H.P. Roberts (1929), or the book published in Russian by A.E. Gaissinovitch (1936). At that time Nilsson (1930) presented Mendel's achievements along with a short biography of Linné and Darwin.

The first to break the stereotyped image of Mendel was R.A. Fisher (1936). He had been fascinated by the originality of Mendel's research even as an undergraduate at Cambridge in 1911, when he began to consider the statistical significance of Mendel's numerical findings from the viewpoint of probability theory (Norton and Pearson 1976). After a critical examination of the *Pisum* paper, he (Fisher 1936) wrote of the 'tale' of Mendel's discovery of the laws of heredity and their rediscovery in 1900. A reconstruction of the individual *Pisum* experiments convinced him that Mendel had not published all his data, and that as 'an experienced and successful teacher' he might well have 'adopted a style of presentation suitable for the lecture-room without feeling under any obligation to complicate his story by unnecessary details'. Fisher pointed out that 'when the history of science is taken seriously, the number of enquiries which such a story suggests is somewhat formidable'. He wanted to know: What did Mendel discover? How did he discover it? And what did he think he had discovered? These questions had never before been asked: none the less, they remained ignored for a long time by both geneticists and historians of science. Fisher (1955a) added to his critical remarks about Mendel that it was essential to seek an answer to the question: What was Mendel's purpose in the series of experiments? Fisher was certain that the *Pisum* paper was more 'a contribution to the methodology of research into plant inheritance'.

In 1950 geneticists held a Golden Jubilee of Genetics at Columbus University, Ohio, in honour of the fiftieth anniversary of the rediscovery (Dunn 1951a). C.D. Darlington (1951) noted that the rediscoverers understood only the explanation of trait segregation in Mendel's paper, overlooking the far more important idea of determinants of heredity in the cells. Later, in 1965, geneticists commemorating the centenary of the publication of the *Pisum* paper referred to Mendel as the founder of genetics and the discoverer of the concept of the gene. At the time historians of science were starting to pay increased attention to Mendel. R.C. Olby (1966) explained the origin of Mendelism in the context of preceding research by botanists into plant hybridization and recent investigations into sexuality in plants and the physiology of fertilization. Around the same time A.E. Gaissincvitch (1965b, 1967) undertook a new analysis of Mendel's theory in connection with his predecessors in plant hybrid experiments, making use of all the data available, including Mendel's research into other plant species.

In the period since 1965, historians and philosophers of science have paid much attention to the way in which the great scientific discoveries were arrived at, and new angles on Mendel's work have been found. Criticisms were raised regarding the previous glorification of Mendel, and a reconsideration of the origin and essence of Mendel's discovery was called for. So before we make an examination of Mendel's scientific achievements, attention will be paid to a reconstruction of his experiments, not only with *Pisum*, but also with other plant species.

When Mendel arrived back from university in 1853, he apparently already had a research programme in mind, based on his recently acquired

knowledge of plant hybridization and the transmission of traits from generation to generation, and also on his previous experience in plant breeding. He must have pondered which aspect of the problem it was in his
power to investigate in the monastery. He may have acquired a more precise idea of what his research programme should involve when he studied
more literature on the subject, and during discussions with the natural
scientists and plant-breeders he came into contact with in learned societies
and at school. As a trained physicist he was able to put together well-
founded hypotheses, which he then tested gradually and proved in his
experiments. An analysis of this approach promises to explain the thought
processes of a researcher who supposed that it seemed 'the one correct way
of finally achieving the resolution of a question whose significance for the
developmental history of organic forms must not be underestimated'
(Mendel 1866a, p. 4).

THE PREMISSES

Mendel's 1866 paper differed from the publications of his predecessors in
plant hybridization in both form and content. It is divided into eleven parts.
After the introduction, Mendel devotes two sections to his choice of experimental plants and the organization of individual experiments. The next five
sections deal with the data obtained from the experiments and a detailed
evaluation of the nature of hybrids sharing one, two, and three pairs of
traits, together with their occurrence in further generations. All this formed
the subject of his first lecture, on 8 February 1865. In a subsequent lecture
on 8 March he described experiments to investigate the transfer of the traits
of parents to their offspring through determinants in the germ cells. In the
penultimate part he makes a brief mention of experiments with other plant
species, especially *Phaseolus*, and it is not until the final section that he sets
out his theoretical ideas on the heredity of traits, in connection with the
mechanism of fertilization, which was still not understood.

Scholars in the early part of the twentieth century failed to notice this
peculiar organization of Mendel's paper, which was not at all typical of the
biological publications of its day. They were mainly interested in why
Mendel chose *Pisum* as his model plant, and the manner in which he
crossed plants differing in various pairs of alternative traits. It was this
which became the point of departure when laws or principles of heredity
associated with Mendel's name came to be derived.

In the introduction to his paper, Mendel first recapitulated the experience
of those who bred new plant varieties to obtain new colours. In his second

sentence he recalls the 'striking regularity with which the same hybrid forms reappeared whenever fertilization took place between the same species'. He had in mind the experiments of botanists who had crossed various forms of plants. His introduction also points to the significance of the research for the developmental history of organic forms, including the enigma of generation, heredity, and fertilization. All these three sources provided the *motivation* to Mendel's research.

An important feature of his research was the careful *planning of experiments*. In his first letter to Nägeli on 31 December 1867, after describing the *Pisum* experiments, Mendel writes: 'I plan to use the same procedure next summer with *Hieracium* as well (Correns 1905, p. 197).' A year later he informed Nägeli of his intention to perform well-planned experiments with *Hieracium*, after two years spent acquiring experience. Mendel must have thought along similar lines before setting out on his *Pisum* experiments. He says: 'The selection of the plant group which is to serve for experiments of this kind must be made with all possible care if it be desired to avoid from the outset every risk of questionable results (Mendel 1866a, p. 5).'

At the same time Mendel emphasized that the experimental plants must possess constant differentiating characters: hybrid plants must be protected during the flowering period from the influence of all foreign pollen; and all hybrids and their offspring should suffer no marked disturbance in their fertility. From the beginning he tried to find the most suitable model among the *Leguminosae*, and preliminary experiments with several members of this family led him to believe that the genus *Pisum* had the necessary qualifications. He started out from a study of the literature on experiments with plant hybridization, and from his own initial experiments. Mendel mentions only briefly the unsuitability of the plant species *Phaseolus* and *Lathyrus*, because of the decreased fertility of their hybrids. *Pisum* was thus selected as being the best model plant for Mendel's research. He states as the reasons for this that its seed and plant have some quite suitable characteristics that are easily and reliably distinguishable. He notes particularly the fact that *Pisum* yields fertile hybrids and that the pollinated flower can easily be protected from cross-pollination, adding that the plant is easy to grow in both experimental gardens and greenhouses.

Mendel's plant-hybridization experiments formed a long-term programme, in which the *Pisum* experiments were only the first stage. The results he obtained allowed him to investigate ever more complex forms of the transmission of parental traits to the offspring in experiments with other plant species. He was limited in his research both in time and space, a fact which is reflected in his choice of further model plants and the extent of individual experiments.

THE *PISUM* EXPERIMENTS

From the *Pisum* paper it is not clear exactly when individual experiments were begun or how they were arranged in individual years. In his second letter to Nägeli, dated 18 April 1867, Mendel explains that: 'The paper which was submitted to you is an unaltered reprint of the draft of the lecture mentioned; hence the brevity of the exposition, as is essential for a public lecture.' After the second lecture, the secretary of the Natural Science Society, Niessl, asked him to publish the text. Mendel read it carefully, and 'after having re-examined my records for the various years of experimentation, and not having been able to find a source of errors', he handed the manuscript over for publication. He was aware of the fact that the paper in the form in which it was published might not be sufficiently clear, a fact he drew to Nägeli's attention (Correns 1905, p. 200).

When Fisher wrote his critical paper in 1936, he was not acquainted with Mendel's letters to Nägeli, and merely guessed that the first experimental crossing might have been in 1857. The *Pisum* paper states that the experiments were completed in 1863, saying that Mendel wished to gain time and space for growing other experimental plants. Mendel also draws Nägeli's attention to the fact that he had been obliged to abandon the *Pisum* experiments in 1864 for another reason, the devastation caused by the beetle *Bruchus pisi*. In previous years this insect had been found only rarely on the plants, and had not interfered with the experiments.

It was a mystery to Fisher how Mendel could have conducted such extensive experiments in such a small experimental garden (35 m × 7 m). No matter how close he sowed them, he could not have grown more than 5000 plants there in one year. Fisher overlooks the mention Mendel makes in the *Pisum* paper of having been able to place a number of potted plants in a greenhouse during the flowering period, considering this to be a 'control on the main experiment in the garden against interference by insects' (Mendel 1866a, p. 9). At the start of Mendel's experiments Abbot Napp had a greenhouse measuring 22.7 m × 4.5 m built (Fig. 5.1)(Orel 1975b). This was at the time of the bishop's visitation, which is described in Chapter 4. The greenhouse was essential for Mendel's research.

In 1855 the headmaster of the *Realschule* wrote that Mendel 'is a good experimentalist and gives excellent demonstrations with rather scanty equipment, both in physics and in natural history', adding that his exposition was 'plain, logical and well-adapted to the capacity of his listeners' (Iltis 1966, p. 95). It was with these faculties that Mendel designed his *Pisum* experiments and adopted the style of presenting both experimental results and a theoretical explanation to the audience at the Natural Science Society meeting.

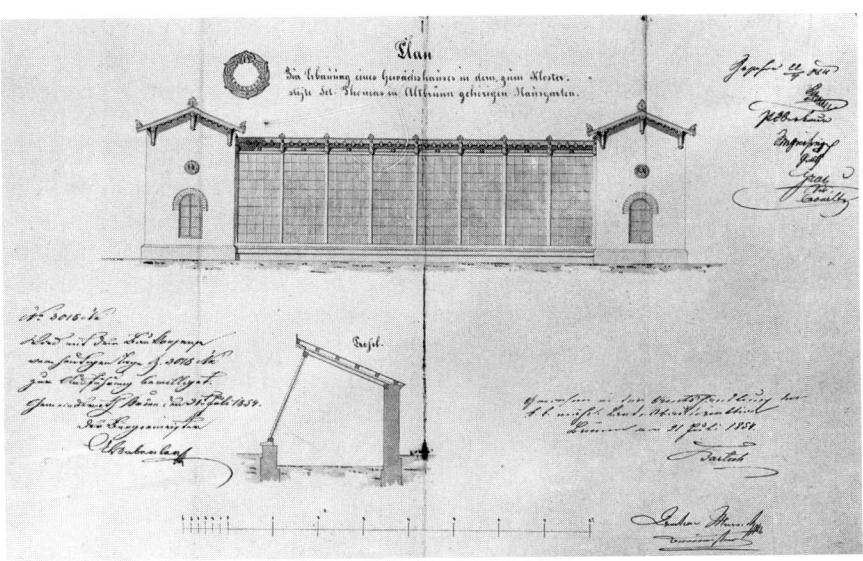

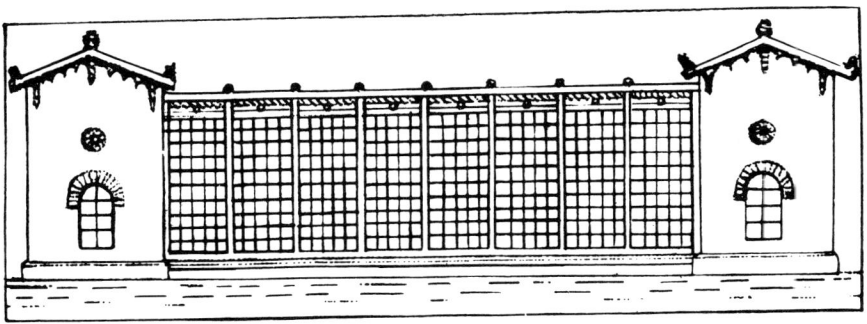

Fig. 5.1 The greenhouse built for Mendel's experiments.

Mendel obtained 34, more or less distinct varieties of peas for his experiments, and spent two years testing the constancy of the traits to be investigated. He chose 22 of them which had fertile hybrids. These he then planted annually throughout the period of the experiments. They remained stable without exception. Mendel notes that the systematic classification of the selected forms is a difficult and uncertain business. He supposed the majority to belong to the species *Pisum sativum*, and the remainder to be subspecies of *P. sativum* or separate species such as *P. quandratum*, *P. saccharatum*, and *P. umbellatum*. He makes the critical comment that: 'In any event, the ranks assigned to them in a classification system is com-

pletely immaterial to the experiments in question. Just as it is impossible to
draw a sharp line between species and varieties, it has been equally impossible so far to establish a fundamental difference between the hybrids of
species and those of varieties (Mendel 1866a, p. 7).'

In describing his choice of pea traits for the experiments, Mendel makes a
brief mention of experiments with ornamental plants undertaken in previous
years, in which he also observed intermediate forms of traits in hybrids, such
as shape and size of leaves and the pubescence of individual parts in *Dianthus*.
In other cases, however, he noticed that one of the two parental traits was so
preponderant that it was difficult or quite impossible to detect the other in the
hybrid. When he had undertaken a study of the suitability of 14 alternative
traits of seeds and plants, he found that some of the traits did not permit a
definite and sharp separation, since the difference rested on 'a more or less'
which was difficult to define. He did not consider such traits suitable for his
experiments, and selected only seven pairs of traits which allowed concordant
behaviour in the hybrid union to be demonstrated, and where a classification
of the segregated traits in the hybrid progeny was possible.

As a novice in the field of research, Mendel most often quoted Gärtner,
considering him an authority on plant hybridization, though he admitted
that Gärtner did not 'diagnose his hybrid types sufficiently' (Correns 1905,
p. 195). Later, in December 1866, Mendel wrote to Nägeli that he was not
able to follow Gärtner's experiments completely, 'not in a single case'. This
was the critical voice of a physicist who wished to describe a research
method which required a repetition of experiments to obtain the same
results. Gärtner (1849, pp. 276–94) characterized plant hybrids according
to the whole plant habitus into three types: (1) *gemischte* (mixed) forms,
representing blending of parental types; (2) *gemengte* (comingled) forms,
exhibiting some traits from the pollen and some from the seed plant: and
(3) *decidirte* (biased) forms, with a prevailing habitus of one or other of the
parental forms. Mendel lent a new meaning to the term hybrid. Reducing
the problem to be investigated to discrete traits of seeds or plants which
existed in alternative forms and were reliably distinguishable, he replaced
the then current holistic image of the species as an indivisible entity with
the atomistic or mosaic concept. It is therefore unnecessary to examine the
question of how Mendel developed his 'original true-breeding varieties' for
the mono-, di-, and trihybrid experiments which Corcos and Monaghan
(1984) attempted to subject to analysis. At the start of his experiments into
plant hybridization he set out from another of Gärtner's data sets, distinguishing two types of hybrids: (1) variable hybrids, in the progeny of which
parental traits appeared anew; and (2) constant hybrids, which reproduced
themselves for generations without any change in the progeny.

In the *Pisum* experiments he was investigating the laws of the origin and development of the variable hybrids as regards the following seven pairs of traits:

(1) the shape of the ripe seeds: either round or nearly round, with depressions, if any occur on the surfaces, always very shallow, or the peas are irregularly angular and deeply wrinkled;
(2) coloration of the albumin of the ripe seeds: either pale yellow, or bright yellow and orange, or more or less intense green;
(3) coloration of the seed coat: either white, associated with white flower colour, or grey, grey-brown, or leathery, with or without violet spotting;
(4) shape of the ripe pod: either smoothly arched and never constricted anywhere, or deeply constricted between seeds and more or less wrinkled;
(5) colour of the unripe rod: either coloured light to dark green, or vivid yellow, which is also the coloration of the stalks, leaf-veins, and calyx;
(6) position of flowers: either axillary, i.e. distributed along the main stem or terminal, or bunched at the end of the stem and arranged in what is almost a short cyme; and
(7) difference in stem length: 1.9–2.2 m or 0.24–0.46 m.

In describing the pattern and sequence of his experiments, Mendel states briefly, referring to numerous experiments by others without mentioning their names, that when two plants which differ constantly in one or more traits are crossed, the traits they have in common are transmitted unchanged to the hybrids and their progeny. He adds that on the other hand, when traits are united in the hybrid to form a new trait, this is usually subject to changes in the hybrid progeny. It was upon this fact that his experiments were based. They were intended to 'observe these changes for each pair of differing traits, and to deduce the law according to which they appear in successive generations'.

A further methodological novelty in Mendel's experiments was his *two-year prior testing of the constancy of the traits investigated*. His conclusion was that 'during the two years no essential change could be noted' (Mendel 1866*a*, p. 6). His precursors among botanists who performed hybridizing experiments failed to test the constancy of traits in the experimental plants. Right from the start Mendel was aware that he could arrive at a new theory *only by using a large number of experimental plants, to eliminate 'a mere chance effect'*. Here we find a further methodological element introduced by Mendel; he may have been inspired by Gärtner's monograph, in the margin of which he noted that one should employ 'a large number of observations' to arrive at a standard and to explain leaf shape in hybrid plants.

A hybrid connecting a single pair of traits

To begin with, Mendel crossed varieties of *Pisum* differing in shape and colour of seeds. These were the experiments he dealt with in greatest detail, noting that they 'lead most easily and surely to the goal' (Mendel 1866*a*. p. 19). In the same year as he performed artificial pollination he was able to observe traits of the hybrid generation in the ripe pods of the resulting seed. He also experimented on a further five pairs of differing traits of plants, which gave the same results. In the case of traits of the plant itself the results did not become apparent until a year later. All seven experiments were very simple, but Mendel's contemporaries were unable to grasp their significance, which did not become clear to biologists until after the year 1900.

An examination of the results Mendel obtained for one of the seven trait pairs serves to illustrate the whole experiment (Figs. 5.2 and 5.3). He took 15 plants from a pure-bred variety with round seeds, and crossed them with a pure-bred variety with angular seeds. The resulting hybrid seeds were all round, i.e. they acquired their shape from the round-seeded parent.

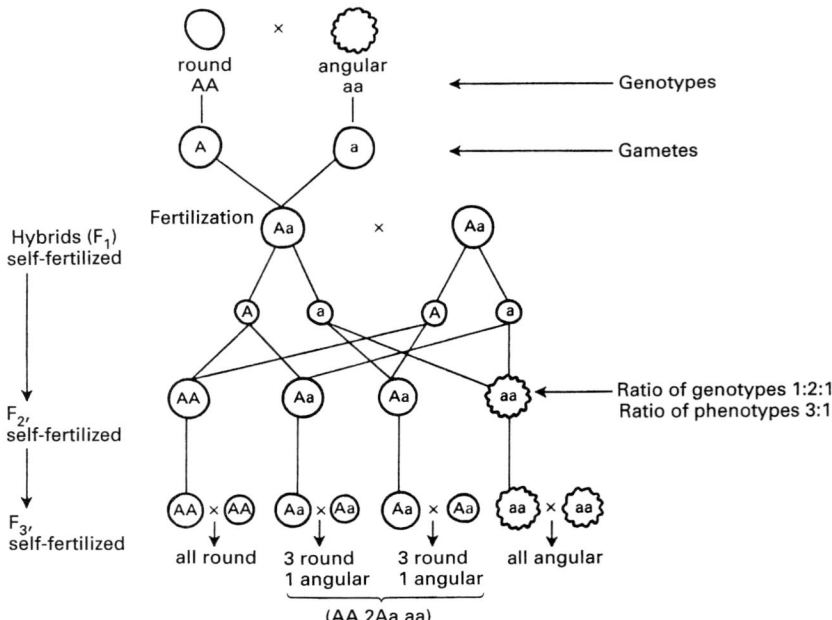

Fig. 5.2 Cross of two different true-breeding parental lines, 'round' and 'angular'. The hybrids are all round, but when crossed together do not breed true: round and angular peas segregate in the ratio 3:1.

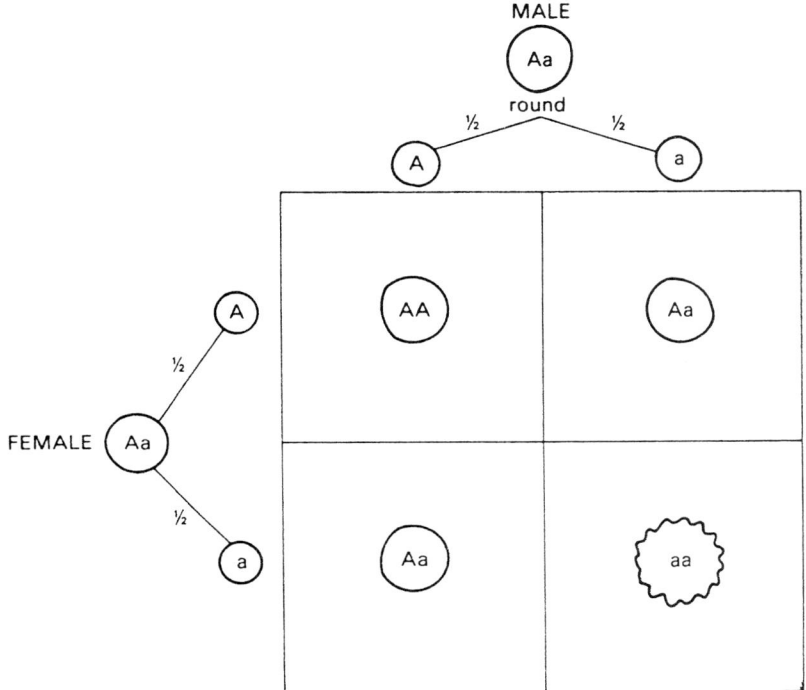

Fig. 5.3 Cross of two hybrid round peas, showing probabilities of producing different kinds of progeny.

The result was the same regardless of which parental variety provided the ovules and which the pollen. To support his conclusion, Mendel (1866a, p. 11) referred to the monograph of Gärtner, according to whom 'even the most practised expert is unable to determine from a hybrid which of the two species was the seed and which the pollen plant'.

In the next year Mendel planted the round hybrid seeds, and by self-fertilization obtained from them a generation of plants which yielded 5474 round seeds and 1850 angular ones. By introducing a numerical analysis of individual traits, he was able *to evaluate qualitative traits quantitatively*. Similar proportions of parental traits were recorded in the other six experiments, demonstrating that these traits segregate in the hybrid progeny *in the ratio of 3:1*. The seeds of the hybrid generation were not visibly distinguishable from those of the round-seeded parental plants, but in the next generation (F_2) the angular shape of the seeds reappeared. Mendel concluded that it had been present in some latent form in the hybrid, and appeared in the progeny with-

out any contamination by the round. Because the round shape had prevailed over the angular, Mendel called the former *dominating* (from about
1900 geneticists began to refer to *dominant* traits), and the angular shape,
which had temporarily receded from view, he denoted *recessive*.

Table 1 gives the numbers of artificial fertilizations performed in the
seven crossing experiments with plants differing in one pair of traits, and
Table 2 the numbers of experimental plants in individual experiments.
Using this exceptionally large number of experimental plants, Mendel
proved that different plant traits do not blend in hybrids, but are
transmitted discretely to further generations.

In generalizing that the segregation ratio was 3:1, Mendel, an experienced
lecturer, pointed out that this figure was only apparent when a large number
of observations was involved. Where the number of observations was small,
quite different results might be obtained; by way of an example he stated that
in one plant he found 43 round seeds and only two angular ones. The other
extreme of random occurrence was a plant which yielded 20 seeds with the
dominant yellow colour and 19 with the recessive green colour.

In the next season Mendel planted both round and angular seeds produced in the F_2 generation, and the resulting plants, when self-fertilized,
yielded the seeds of the next generation (F_3). An analysis of these seeds
showed that those from the angular seeds were exclusively angular. But
one-third of the round seeds (F_2) produced exclusively round seeds, while
the other two-thirds gave both round and angular seeds, again in the ratio
3:1. Mendel's conclusion was that 'the ratio of 3:1 in which the distribution of the dominating and recessive traits take place in the first generation
therefore resolves itself into the ratio of 2:1:1 in all experiments if one differentiates between the meaning of the dominating trait as a hybrid and as
a parental trait' (Mendel 1866*a*, p. 16). He had already pointed out that

Table 1   Artificial fertilization experiments with one trait pair

Experiment no.	Number of fertilizations	Number of plants
1	60	15
2	58	10
3	35	10
4	40	10
5	23	5
6	34	10
7	37	10
Total	287	70

Table 2 Experimental crossing of peas differing in one trait pair

Trait	Dominant		Recessive		Total	Ratio (dominant : recessive)
	Number	Per cent	Number	Per cent		
Form of seed	5474	74.74	1850	25.26	7324	2.96 : 1
Colour of seed	6022	75.06	2001	24.94	8023	3.01 : 1
Colour of seed coat	705	75.89	224	24.11	929	3.15 : 1
Form of pod	882	74.68	299	25.32	1181	2.95 : 1
Colour of pod	428	73.79	152	26.21	580	2.82 : 1
Position of the flowers	651	75.87	207	24.13	858	3.14 : 1
Length of stem	787	73.96	277	26.04	1064	2.84 : 1
Total seeds	11496		3851		15347	
Total plants	3453		1159		4612	

the dominating trait can have *double significance*, namely that of a parental trait or that of a hybrid trait. Here he was explaining the difference between the appearance of the dominating trait, which after 1900 came to be called the *phenotype*, and its hereditary basis, later called the *genotype*. This simple type of experiment is today called a *monofactorial* experiment.

Mendel investigated the segregation of the traits of plant forms in simple experiments up to the fourth or even seventh generation, and in each generation the offspring of the hybrids split into the hybrid and constant forms in the ratio 2:1:1. He introduced alphabetic symbols to denote dominant and recessive traits. Dominant traits were denoted using upper-case letters, and recessive traits using lower-case letters. Thus the designation Aa indicated a combination of differing traits in the hybrid form. Constant traits were denoted by a single letter, and variable (i.e. hybrid) traits by two letters. For the progeny of hybrids connecting one trait pair of differing traits Mendel derived the mathematical expression A + 2Aa + a, denoting it a *simple series*. It implies both the segregation ratio 3:1 of phenotypes, and that of 2:1:1 of genotypes.

Mendel used the results of his experiments with crossing plants differing in a single trait pair to clarify the views of his predecessors Kölreuter and Gärtner that hybrids have a tendency to revert to the parental forms. He tabulated the model of the progeny of self-fertilized hybrid plants assuming equal fertility and the production of four seeds in each plant in all generations (see Table 3). In the progeny of hybrid plants he always illustrated half hybrids and half constant forms in equal proportions. It is clear that the number of hybrids derived from self-fertilization decreases significantly from generation to generation as compared with the number of constant forms. Yet hybrid forms can never disappear entirely. Mendel showed clearly that in the tenth

Table 3   The decreasing prevalence of hybrid forms in the progeny of a self-fertilizing hybrid

Generation	A	Progeny[a] Aa	a	Ratio A: Aa: a
1	1	2	1	1: 2: 1
2	6	4	6	3: 2: 3
3	28	8	28	7: 2: 7
4	120	16	120	15: 2: 15
5	496	32	496	31: 2: 31
n				$2^n - 1$: 2: $2^n - 1$

[a] Assuming equal fertility and the production of four seeds by each plant in each generation.

generation, out of every 2048 plants there are 1023 with the constant dominant trait, 1023 with the recessive trait, and only two are hybrids.

A hybrid connecting several pairs of traits

In his next experiment, Mendel investigated whether the law deduced for peas differing in one trait pair would hold good in the case where plants differed in two or more trait pairs. First of all he crossed pure-bred pea varieties whose seeds differed in both shape and colour, denoting the different alternative traits A, a and B, b. This experiment is set out in Fig. 5.4. Textbooks often use the scheme Fig. 5.5. Again Mendel stressed that experiments with seed traits led most easily and assuredly to success. Growing peas from the 15 hybrid plants he obtained 556 seeds, comprising not only both parental types, but also combinations of them. Some pods even contained all four kinds of seeds. The seeds he obtained were as follows:

> 315 round and yellow
> 101 angular and yellow
> 108 round and green
> 32 angular and green.

Mendel was not surprised to find the re-combination of plant traits, since he knew about the combination of traits from the breeding of new varieties of flowers through artificial fertilization. According to the results of his experiments with plants crossed on the basis of a single trait pair, he could expect the progeny of hybrids to develop as a combination of two simple series for each two trait pairs, as follows:

	B	**b**	**2Bb**
A	AB	Ab	2ABb
a	aB	ab	2aBb
2Aa	2AaB	2Aab	4AaBb.

In his paper Mendel denoted the combination of the two simple series in a simplified form as :

$$A + 2Aa + a$$
$$B + 2Bb + b.$$

To this he added the expansion of the expressions of the *combination series*:

$$AB + Ab + aB + ab + 2ABb + 2aBb + 2AaB + 2Aab + 4AaBb.$$

Mendel planted all 556 seeds from the self-fertilized F_2 generation the next year, and demonstrated that the 629 plants produced nine different forms of progeny, in very unequal numbers, as follows:

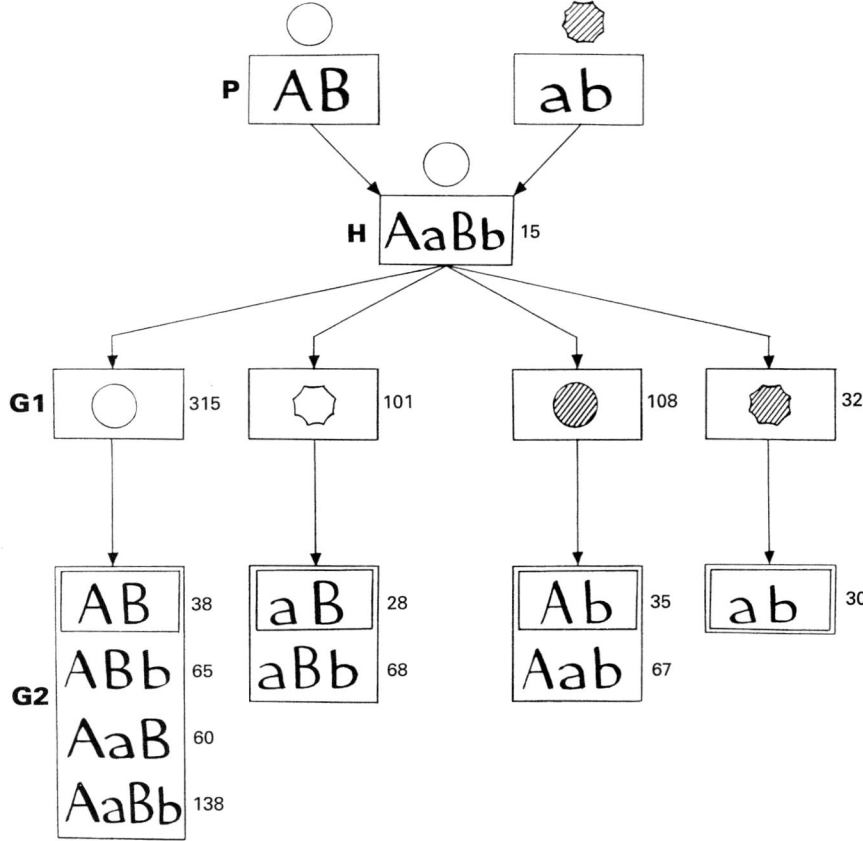

Fig. 5.4 Mendel's bifactorial experiment. The two factors are colour (which can be either yellow or green) and shape (round or angular). Shading indicates green peas. The numbers of plants Mendel obtained are shown to the right of each box. **P**, parents; **H**, hybrids; **G1**, filial generation 1 (commonly written as 'F₁'); **G2**, filial generation 2 (commonly written as 'F₂').

   38 plants with the designation AB
   35 plants with the designation Ab
   28 plants with the designation aB
   30 plants with the designation ab
   65 plants with the designation ABb
   68 plants with the designation aBb
   60 plants with the designation AaB
   67 plants with the designation Aab
138 plants with the designation AaBb.

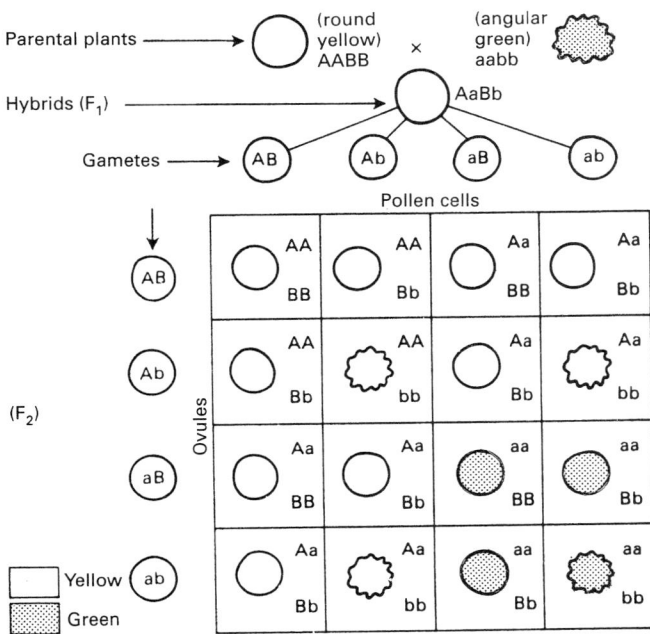

Fig. 5.5 Chart showing the results of a dihybrid cross. Expected numerical ratios of progeny can be easily read off the chart: they are 9 round yellow, 3 angular yellow, 3 round green, and 1 angular green.

Fisher (1936) graphically expressed the numerical data in this experiment in the form shown in Table 4.

Mendel placed these forms in three essentially different groups. The first comprises the groups with designations AB, Ab, aB, ab, possessing only constant traits which do not change in following generations. Each one of these forms is represented 33 times on average. The second group contains the forms ABb, aBb, AaB, Aab. They are constant for one trait and hybrid for the other. In the next generation they vary only with respect to the hybrid

Table 4 Classification of plants grown in the bifactorial experiment (according to R.A. Fisher, 1936)

	AA	Aa	aa	Total
BB	38	60	28	126
Bb	65	138	68	271
bb	35	67	30	132
Total	138	265	126	529

trait and each of them appears 65 times on average. The form AaBb occurs 138 times, is hybrid for both traits, and behaves exactly like the hybrid from which it is descended. All these forms can be summarized as follows:

> 33 plants with forms AB Ab aB ab
> 65 plants with forms ABb aBb AaB
> 138 plants with forms AaBb.

Examining the numbers in which the forms in these groups occurred, Mendel generalized the average proportions as 1:2:4. After 1900 geneticists also wrote in their textbooks, as the ratio 3:1 was used to explain the segregation of traits in the progeny of a hybrid with one trait per pair, the ratio 9:3:3:1 was used to explain the segregation of four different trait combinations in the progeny of hybrids. Mendel did not explicitly state such a ratio in his paper, leading some historians of science to suppose that he did not know the ratio 9:3:3:1. But probably he simply did not see any point in expounding on this logical result of the application of combinatories to the segregation of traits in this form.

Figure 5.5 shows 16 combinations of the traits of the parental forms in the progeny of a hybrid combining two pairs of different traits. Of these the forms AB and ab are identical to the parental types. The forms Ab and aB are a combination of the parental forms. These four forms are pure-bred types. After the year 1900 geneticists called such forms *homozygotes*. These four combinations occur in the diagonal top-left–bottom-right. The other forms are hybrid in one or the other of the traits or in both of the traits, and these were called *heterozygotes*. Of the 16 combinations, nine gave round and yellow seeds, three round and green seeds, three angular and green seeds, and only one gave angular and green seeds. This clearly indicates the segregation ratio 9:3:3:1, which Mendel had in mind even though he does not state it explicitly in the *Pisum* paper. After 1900, experiments with the crossing of organisms differing in two trait pairs came to be called *bifactorial* experiments.

The logical step from the experimental crossing of peas differing in two trait pairs, explaining the operation of the principle of segregation of parental traits in hybrid progeny with two different trait pairs, was to examine the segregation ratio in the progeny of hybrids with three different trait pairs. According to Mendel this was the most laborious experiment, confirming the possibility of using mathematics in research. He crossed varieties differing in three trait pairs. The initial plants were designated by Mendel as follows:

ABC seed plants	**abc pollen plants**
A seed shape:round	**a** seed shape:angular

B colour: yellow   **b** colour: green
C seed-coat: grey-brown   **c** seed-coat:white

In his experiment Mendel crossed plants differing in two trait pairs of the seed and a further trait pair of the plant. He was aware that *the grey-brown seed-coat was constantly associated with violet-red flowers and the white seed-coat with white flowers.* During the preparations for the experiment he must have predicted the free combination of traits. From 24 hybrid plants he obtained 687 seeds, all of which were round or angular and yellow or green. The seed-coat was coloured grey-brown or grey-green. In the next generation he obtained from these seeds 639 plants, and according to the way the traits appeared in the progeny classified the plants he obtained in terms of the ancestral traits they manifested as follows (number of plants showing each combination of characters):

8—ABC	22—ABCc	45—ABbCc	78—AaBbCc
14—ABc	17—AbCc	36—aBbCc	
9—AbC	25—aBCc	38—AaBCc	
11—Abc	20—abCc	40—AabCc	
8—aBC	15—ABbC	49—AaBbC	
10—aBc	18—ABbc	48—AaBbc	
10—abC	19—aBbC		
7—abc	24—aBbc		
	14—AaBC		
	18—AaBc		
	20—AabC		
	16—Aabc		

Of these 27 differing categories, eight were constant for all three traits, and each occurred ten times on average; 12 were constant for two traits and hybrid for the third, and of these each appeared on average 19 times; six were constant for one trait and hybrid for the other two, and each of these turned up 43 times; and one form occurred 78 times, and was hybrid for all three traits (Fig. 5.7). The observed ratio 10:19:43:78 approaches the ratio 10:20:40:80, or 1:2:4:8, so closely that the latter doubtless represents the correct values.

Mendel described the *combination of three simple series* as an experienced physics teacher, masterfully applying combinatorial theory in explaining the development of the progeny of hybrids AaBbCc in accordance with the expression:

ABC + ABc + AbC + Abc + aBC + aBc + abC + abc + 2ABCc + 2AbCc + 2aBCc + 2abCc + 2ABbC + 2ABbc + 2aBbC + 2aBbc + 2AaBC + 2AaBc + 2AabC + 2Aabc + 4ABbCc + 4aBbCc + 4AaBCc + 4AabCc + 4AaBbC + 4AaBbc + 8AaBbCc.

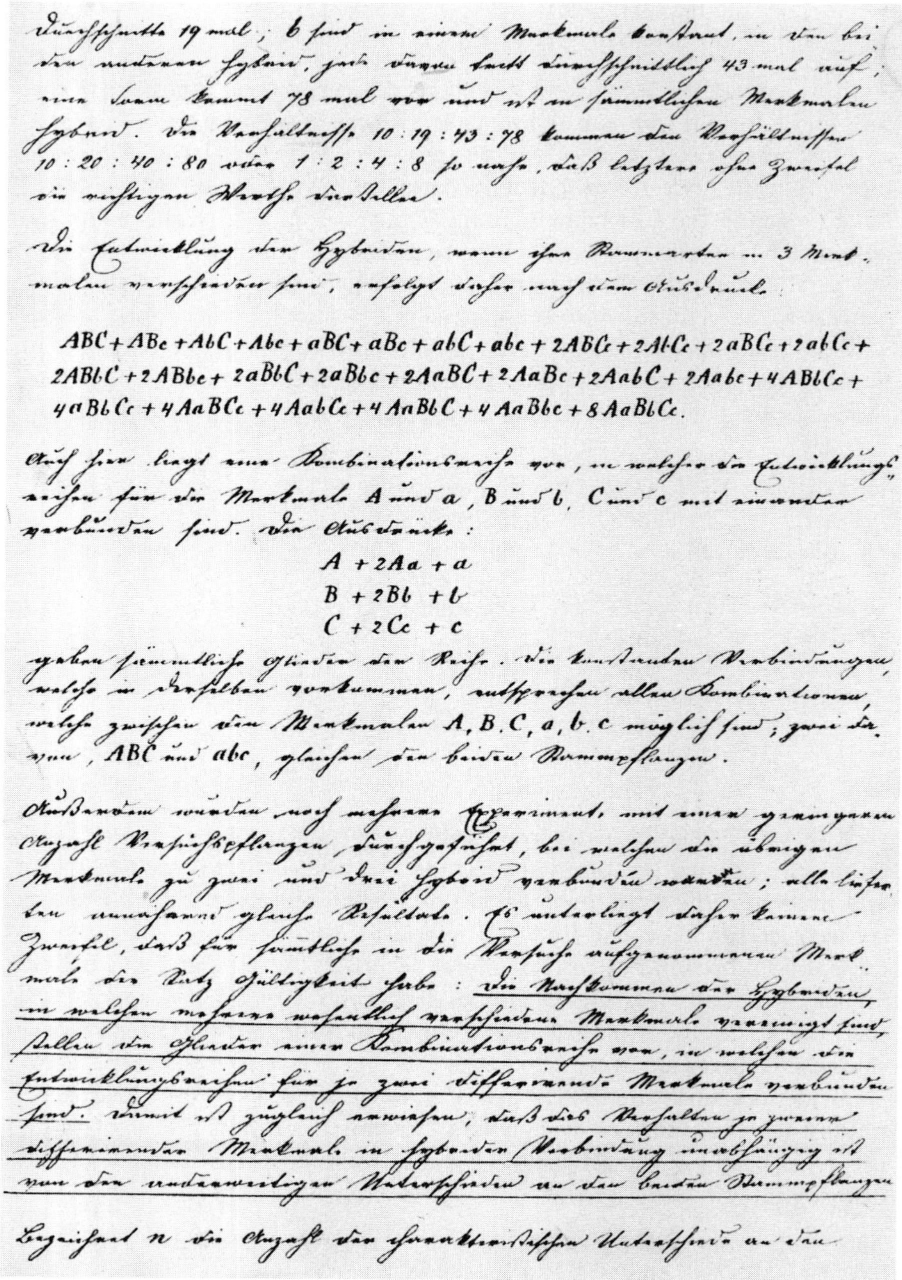

Durchschnitte 19 mal; 6 sind in einem Merkmale constant, in dem andern hybrid, jede dieser Arten kommt durchschnittlich 43 mal vor; eine Form kommt 78 mal vor und ist in sämmtlichen Merkmalen hybrid. Die Verhältnisse 10 : 19 : 43 : 78 kommen den Verhältnissen 10 : 20 : 40 : 80 oder 1 : 2 : 4 : 8 so nahe, daß letztere ohne Zweifel die richtigen Werthe darstellen.

Die Entwicklung der Hybriden, wenn ihre Stammarten in 3 Merkmalen verschieden sind, erfolgt daher nach dem Ausdrucke:

$$ABC+ABc+AbC+Abc+aBC+aBc+abC+abc+2ABCc+2AbCc+2aBCc+2abCc+2ABbC+2ABbc+2aBbC+2aBbc+2AaBC+2AaBc+2AabC+2Aabc+4ABbCc+4aBbCc+4AaBCc+4AabCc+4AaBbC+4AaBbc+8AaBbCc.$$

Auch hier liegt eine Kombinationsreihe vor, in welcher die Entwicklungsreihen für die Merkmale A und a, B und b, C und c mit einander verbunden sind. Die Ausdrücke:

$$A+2Aa+a$$
$$B+2Bb+b$$
$$C+2Cc+c$$

geben sämmtliche Glieder der Reihe. Die constanten Verbindungen, welche in derselben vorkommen, entsprechen allen Kombinationen, welche zwischen den Merkmalen A, B, C, a, b, c möglich sind; zwei derselben, ABC und abc, gleichen den beiden Stammpflanzen.

Außerdem wurden noch mehrere Experimente mit einer geringeren Anzahl Versuchspflanzen durchgeführt, bei welchen die übrigen Merkmale zu zwei und drei hybrid verbunden wurden; alle lieferten annähernd gleiche Resultate. Es unterliegt daher keinem Zweifel, daß für sämmtliche in die Versuche aufgenommenen Merkmale der Satz Gültigkeit habe: die Nachkommen der Hybriden, in welchen mehrere wesentlich verschiedene Merkmale vereinigt sind, stellen die Glieder einer Kombinationsreihe vor, in welcher die Entwicklungsreihen für je zwei differirende Merkmale verbunden sind. Damit ist zugleich erwiesen, daß das Verhalten je zweier differirender Merkmale in hybrider Verbindung unabhängig ist von den anderweitigen Unterschieden an den beiden Stammpflanzen.

Bezeichnet n die Anzahl der charakteristischen Unterschiede an den

Fig. 5.6 Page 21 of Mendel's manuscript 'Experiments into plant hybridization', showing his derivation of combination series.

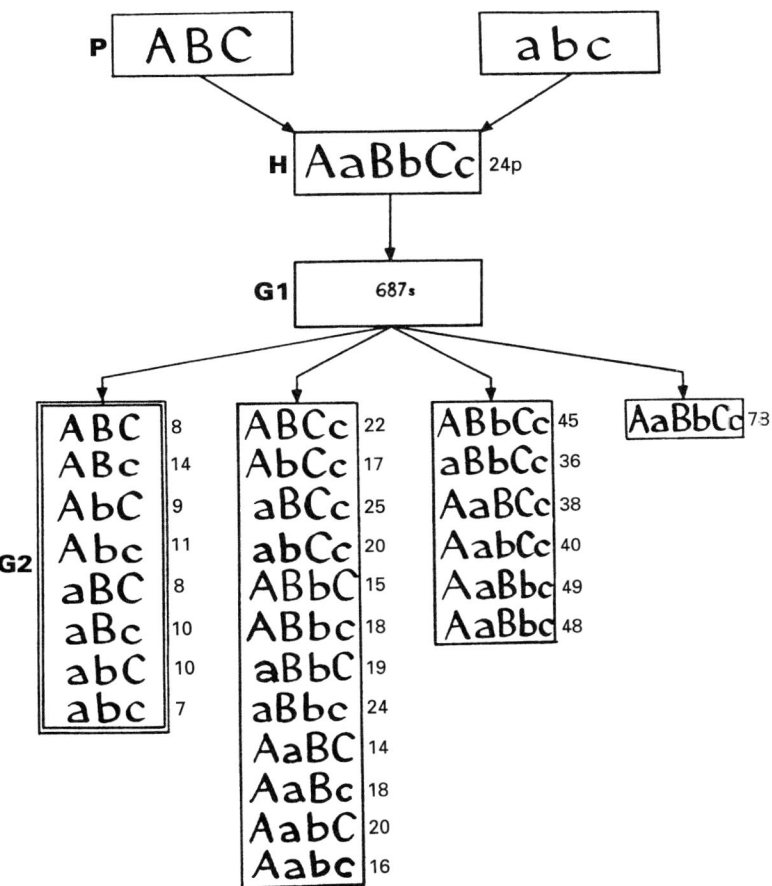

Fig. 5.7 Mendel's trifactorial experiment.

Mendel's conclusion was that 'constant traits occurring in different forms of a plant kindred can, by means of repeated fertilization, enter into all the associations possible within the rules of combination' (Mendel 1866a, p. 23). In addition he mentioned 'several more experiments carried out with a small number of experimental plants' in which the remaining traits were combined in twos or threes in hybrid fashion, all giving identical results. His generalization was that 'the behaviour of each pair of differing traits in a hybrid association is independent of all other differences in the two parental plants'. Mendel added that he had actually obtained all the constant associations theoretically possible in *Pisum* through a combination of all seven traits which he studied, and there were 128 of them, or 2^7. These combinations achieved by Mendel have been analysed by Benedik (1971).

Mendel's idea was to present his theory in terms of mathematics. His conclusion was therefore: 'If n designates the number of characteristic differences in the two parental plants, then 3^n is the number of terms in combination series, 4^n the number of individuals that belong to the series, and 2^n the number of combinations that remain constant.' (Mendel 1866a, pp. 22–3.) In conclusion to his lecture Mendel states that the results he had obtained justified the assumption that the same behaviour can be attributed to traits exhibited less distinctly in the plants and not included in the experiments described. He gives as an example his experiment on flower stems of different lengths, where the distinction and classification of the forms could not be accomplished with the certainty 'that is indispensable to correct experiments' (Mendel 1866a, p. 24).

Mendel's mathematical expressions were after 1900 incorporated into the teaching of genetics, to become the standard manner of planning genetic experiments and predicting the appearance of new combination of traits. His explanation of the segregation of parental traits in the hybrid progeny and random combination of any number of traits came to be known as *Mendel's Law (or Principle) of Segregation and the Law (or Principle) of Independent Assortment of Traits.*

Reproductive cells of hybrids

In the introduction to his second lecture, Mendel states that the results obtained by crossing peas with differing pairs of traits suggested further experiments 'whose outcome would throw light on the conclusion as regards the composition of the seed and pollen cells of hybrids' (Mendel 1866a, p. 24). The point of departure for this new series of experiments was the observation in the *Pisum* experiments that constant forms appeared in the progeny of hybrids in all combinations of associated traits. In applying the theory of probability Mendel concluded: 'Since the different constant forms are produced in a single plant, even in just a single flower, it seems logical to conclude that in the ovaries of hybrids as many kinds of germinal cells and in the anthers as many kinds of pollen cells are formed as there are possibilities for constant combination forms, and that these germinal and pollen cells correspond in their normal make-up to the individual forms.' He added that: 'Indeed, it can be shown theoretically that this assumption should be entirely adequate to explain the development of hybrids in separate generations if one could assume at the same time that the different kinds of germinal and pollen cells of a hybrid are produced on average in equal numbers.'

Mendel's new hypothesis regarding heredity and fertilization was verified with a relatively small number of plants in experiments which seem at first sight to be very simple. They are illustrated in Figs. 5.8 and 5.9. For clarity's sake they are divided into four groups. First Mendel crossed *AaBb* hybrid plants combining different shapes and colours of seeds with plants with both the dominant traits, *AB* constant. In the next experiment he

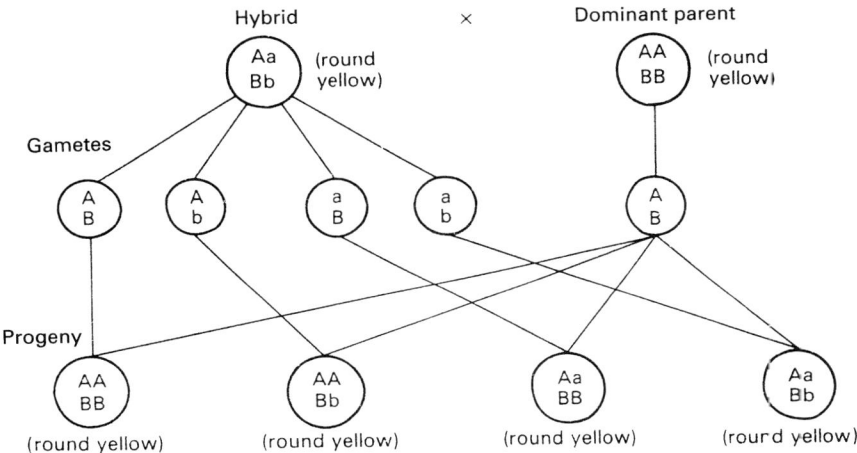

Fig. 5.8 Backcross of the hybrid with the dominant parent.

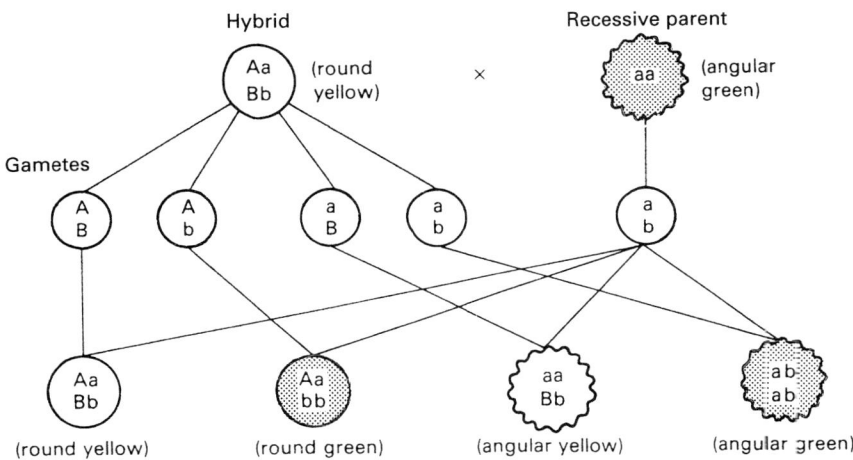

Fig. 5.9 Backcross of the hybrid with the recessive parent.

crossed the same hybrid plants with constant ab recessive forms. In addition he used the hybrid plant as the maternal and the paternal parent alternately. He denoted these experiments as follows:

(1) the hybrid crossed with the pollen from AB;
(2) the hybrid crossed with the pollen from ab;
(3) the AB form crossed with the pollen from the hybrid plant; and
(4) the ab form crossed with the pollen from the hybrid plant.

This form of cross is today called reciprocal back-crossing. In each group Mendel fertilized all flowers on three plants, and expected the following combinations of trait determinants.

(1) AB, ABb, AaB, AaBb;
(2) AaBb, Aab, aBb, ab;
(3) AB, ABb, AaB, AaBb; and
(4) AaBb, Aab, aBb, ab.

According to Mendel (1866a, p. 26), 'if the premises were correct' then the progeny in the first and third groups and in the second and fourth 'had to give identical combinations'. The results which confirmed this assumption were set out as follows:

experiment 1, 98 seeds, all round and yellow;
experiment 3, 94 seeds, all round and yellow;
experiment 2, 31 round yellow, 26 round green, 27 angular yellow and 26 angular green seeds; and
experiment 4, 24 round yellow, 25 round green, 22 angular yellow, 27 angular green seeds.

Mendel's conclusion was that 'a favourable result could hardly be doubted any longer.' None the less, he performed a check on the traits decision. He summarized the results of the experiment using the symbols he had introduced. His conclusion was that: 'In all experiments, therefore, all forms postulated by the preceding hypothesis appeared, and did so in nearly equal numbers' (Mendel 1866a, p. 27).

Mendel exhibited his most brilliant reasoning in planning further experiments with mutual reciprocal back-crossing of forms differing in traits of flower coloration (*A* and *a*) and stem length (*B* and *b*). According to his theoretical knowledge he produced the initial hybrid forms *Aab* and *aBb*, expecting in the progeny the following four combinations:

$$AaBb + aBb + Aab + ab.$$

Out of 45 fertilizations he obtained 187 seeds, and from these 166 plants reached the flowering stage in the following year. Among them the individual combinations appeared in the following numbers:

Combination	Flower colour	Stem length	Occurrences
1	violet-red	long	47
2	white	long	40
3	violet-red	short	38
4	white	short	41

In order to explain more clearly the role of probability, Mendel adds that in the progeny from the cross, half the plants should have had violet-red flowers (*Aa*), half white flowers (*a*), half long stems (*Bb*), and half short stems (*b*). Mendel (1866a, p. 29) states that 'in this experiment too, the proposed hypothesis finds adequate conformation'.

He briefly mentions another experiment undertaken 'on a small scale' on traits of pod shape, pod colour, and flower position and asserts: 'The results obtained were in full agreement; all combinations possible through union of the different traits appeared when expected and in nearly equal numbers.' In this case he does not even mention the number of experimental plants, already convinced that 'pea hybrids form germinal and pollen cells that in their composition correspond in equal numbers to all the constant forms resulting from the combination of traits united through fertilization'. At this point Mendel recapitulates the conclusion of his first lecture, where he had explained segregation ratios. Evoking the simple series $A + 2Aa + a$, containing four individuals in three different terms, he shows the germinal and pollen cells in fertilization in accordance with the law of probability, as follows:

Mendel visualized the result of fertilization by writing the designations for associated germinal and pollen cells in the form of fractions: pollen cells above the line and germinal cells below. The striking phenomenon that 'hybrids are able to produce, in addition to the two parental types, progeny that resemble themselves' he saw as being explained by the combination of terms and 'both giving the same association, regardless of which of the two traits belonged to the pollen and which to the germinal cell'. He placed the average course of fertilization of hybrids in the context of the simple series, as follows:

$$\frac{A}{A} + \frac{A}{a} + \frac{a}{A} + \frac{a}{a} = A + 2Aa + a.$$

In this connection Mendel (1866a, p. 31) emphasized the statistical character of his theory. 'The true ratio can be given only by the mean calculated from the sum of as many separate values as possible; the larger their number, the more likely it is that a mere chance effect will be eliminated.' He gives a theoretical account of this conclusion using the example of hybrids in which two and three different kinds of differing traits were associated, concluding this part of the lecture with the generalization: 'The law of the combination of different traits according to which hybrid development proceeds thus finds its basis and explanation in the proven statement that hybrids produce germinal and pollen cells that correspond in equal numbers to all the constant forms resulting from the combination of traits united through fertilization.' He thus explained lucidly the process of fertilization proved in the segregation ratio: 1:1 in agreement with the segregation ratio 3:1 and the expression for simple series.

In his concluding remarks, Mendel (1866a, p. 43) compares the results of his experiments with the observations published by Kölreuter and Gärtner, 'the two authorities in this field'. Both of them state that in external appearance hybrids either maintain a form intermediate between the parental strains, or they approach the type of one or the other, sometimes being barely distinguishable from them. According to these two authors, the majority of individuals produced from self-fertilization of hybrids maintain the form of the hybrid, while a few become more like the seed plant, and the occasional individual very nearly matches the pollen plant. This, however, they did not consider valid for all hybrids without exception.

Mendel drew attention to Gärtner's view that the precise determination of 'whether a form bears a greater resemblance to one or other of the two parental types often presented great difficulty, since much depended on the subjective viewpoint of the observer'. According to Mendel, there is another circumstance that could contribute to making the results variable and uncertain in spite of the most careful observation. When one is dealing in a general way with a degree of similarity, account must be taken not only of the traits that stand out sharply, but also of those that are often difficult to put into words, but yet suffice, as every specialist knows, to give such forms a peculiar appearance. He added that according to the law valid for *Pisum* the series obtained in each separate experiment must comprise very many forms, because the number of terms increases with the number of differentiating characters as the power of three. With a relatively small number of experimental plants the results could be only approximately correct, and occasionally could deviate not inconsiderably. As a striking example, Mendel (1866a, p. 39) refers to an experiment in which the two original stocks differed in seven traits, and if 100 to 200 plants were raised from the seeds of their hybrids for evaluation of the degree of relationship of the progeny, one can easily understand how uncertain such a judgement must be, since the series for seven differing traits contains 16 384 individuals appearing in 2187 different forms. Sometimes one relationship, and sometimes another, would assert itself more strongly, depending on whether the observer found, by chance, a larger number of this or that form.

Mendel also replied to Gärtner's objection that the two parental types themselves were not represented among the offspring of the hybrids, and that he had observed only occasional individuals closely approximating to them. Using the same example of a crossing of plants differing in seven traits, Mendel showed that each parental form occurs only once in more than 16 000 offspring of the self-fertilized hybrid. For this reason there is not much likelihood of finding them among a small number of experimental plants. Yet with a reasonable degree of probability, one may expect the appearance of a few forms that approximate to those in the series.

Mendel (1866a, p. 42) was convinced that his *Pisum* experiments had proved his theory of the origin and development of *Pisum* hybrids. But he states in the paper that 'the principles proposed for *Pisum* need confirmation, and repetition of at least the more important experiments is therefore desirable; for instance, the one on the comparison of hybrid fertilization cells'. In his second letter to Nägeli he writes that he had attempted to inspire some control experiments, but 'no one undertook them'. He again stresses the importance of experiments on germ cells, believing that their results might 'furnish an explanation for the development of hybrids as observed in *Pisum*', and adding that 'these experiments should be repeated and verified' (Correns 1905, p. 199). Mendel indicated furthermore that he expected his new theory to apply to other plants besides *Pisum*, but he added critically: 'Whether variable hybrids of other plant species show complete agreement in behaviour also remains to be decided experimentally; one might assume, however, that no basic difference could exist in important matters, since unity in the plan of development of organic life is beyond doubt (Mendel 1866a, p. 43).' Here we see the view of a physicist experimentally investigating the general laws of nature. This methodological part of Mendel's paper is discussed further below; see pp. 163–80.

Transformation experiments

The last three pages of the *Pisum* paper are devoted to a comparison of the new theory with the idea of the transformation of one species into another by artificial fertilization, which Gärtner considered 'the most difficult thing in hybrid production'. Mendel quoted Gärtner's statement that 'the characteristic force toward change and transmutation of the maternal type that a species exerts in reproduction is very different in various plants, and consequently the length of time required for one species to become transformed into another, and the number of generations it takes, must also be different; transformation is accomplished after more generations in some species and after fewer in others'. This idea was based on Kölreuter's experiments in which, by repeated pollination of hybrid plants with the pollen of the paternal plant, he explained reversion of the hybrid to the original paternal form. In Gärtner's monograph Mendel marked page 446, where the author reaches the conclusion that the inclination of hybrids to revert to the parental forms refutes the existence of constant hybrids. Mendel offered an explanation of transformation using the model of a pair of discrete traits. In three pairs of differing traits the hybrid form produces the following eight kinds of germinal cells:

ABC, ABc, AbC, aBC, Abc, aBc, abC, abc.

When the hybrid plant is fertilized with the pollen cells abc progeny is obtained as follows:

AaBbCc + AaBbc + AabCc + aBbCc + Aabc + aBbc + abCc + abc.

Mendel used this example to demonstrate that the form abc occurs once in the series of eight terms, and therefore there is little likelihood that it would be missing from the experimental plants, even if only a fairly small number were raised. Accordingly, for a single trait it would be possible to attain a transformation of that trait even in the second generation. From this Mendel deduced that 'the smaller the number of experimental plants and the larger the number of differing traits in the two parental species, the longer an experiment of this kind will last, and, furthermore, a delay of one or even two generations could easily occur with these same species, which is what Gärtner observed' (Mendel 1866*a*, p. 45). Gärtner stated that for a transformation of the species A into species B a different number of generations was needed to that required to transform species B into species A. He therefore disagreed with the idea of Kölreuter that 'the two natures in hybrids are in perfect equilibrium'. According to Mendel (1866*a*, p. 45), Kölreuter does not deserve this reproach, since Gärtner had overlooked an important point, that it 'depends on which individual is chosen for further transformation'. Here Mendel very briefly compared Kölreuter's species hybrids with his model hybrids reduced to the level of the recombination of different trait pairs. He also marked the margin of page 473 of Gärtner's monograph explaining that the deviation in the transformation process consists in a particular *Anlage* of the egg. (He therefore recommended 'further investigation'.) This may have led Mendel to undertake his experiment designed to explain transformation. He chose two plants differing in five traits; those of species A were all dominant, those of species B were all recessive. To effect mutual transformation Mendel fertilized plant A with pollen from B and plant B with pollen from A. The same procedure was repeated in both hybrids in the following year. In the first experiment, B/A, 87 plants in the 32 possible forms were available in the third experimental year from which to choose individuals for further fertilization; in the second experiment A/B, the external appearance of all 73 plants obtained coincided completely with that of the pollen plant, although their internal constitution must have been just as varied as the forms from the other experiment. Intentional selection of forms was possible only in the first experiment. In the second experiment only a few plants could be selected at random. Only a few flowers of the latter were fertilized with pollen from A, the rest being allowed to self-fertilize.

Out of every five plants used for fertilization in the two experiments, the next year's culture showed the following agreement with the pollen plant:

No. of plants showing agreement	Experiment 1	Experiment 2
In all 5 traits	2	—
In 4 traits	3	—
In 3 traits	—	2
In 2 traits	—	2
In 1 traits	—	1

In the first experiment, therefore, transformation was completed, while in the second, which was not continued, two more fertilizations would probably have been necessary. This was Mendel's explanation of the process of transformation of hybrid forms to the paternal form according to his theory of variable hybrids.

In the complete transformation of one species to another Gärtner saw the proof that a species has fixed limits beyond which it cannot change. Although, in Mendel's view, this opinion cannot be considered unconditionally valid, 'considerable confirmation of the previously expressed conjecture on the variability of cultivated plants is to be found in the experiments performed by Gärtner'. Mendel (1866a, p. 46) gives the examples of three hybrid forms of ornamental plants which, according to Gärtner, lost none of their stability after four or five repetitions of hybrid association. Using the model of variable hybrids, Mendel explained the mystery of the transformation of one species into another. He also mentioned Gärtner's views on constant hybrid forms, which he considered to be 'of particular importance to the developmental history of plants, since constant hybrids are able to attain the status of a new species'. But Mendel considered this a mere hypothesis, which he tried to verify in later experiments using other plant genera. Here one can see Gärtner's typological concept of species and Mendel's idea of the variation of discrete traits.

The number of experimental *Pisum* plants

In his attempt at a statistical analysis of the numerical data from Mendel's experiments with *Pisum*, Fisher (1936) soon realized that Mendel was an experienced and successful teacher, and had selected for his lectures only those data which he considered important for explaining his theory 'without feeling under any obligation to supplement his story with unessential details'. In 1867 Mendel gave Nägeli more detailed information regarding the way in which individual experiments were performed. An important piece of information for any reconstruction is Mendel's statement that 'For two generations all experiments were conducted with a fairly large number of plants. Starting from the third generation, it became necessary to limit the numbers because of a lack of space, so that, in each of the seven experiments, only a sample of those plants of the second generation could be observed further. The observations were extended over four to six generations. Of the varieties which bred true, some plants were observed for four generations' (Correns 1905, p. 200).

In describing his experiments with plant traits (the third to seventh) Mendel does not give the number of hybrid plants. Fisher estimated approximately 250 plants, the same number given by Mendel in the first two experiments with seed traits. In Table 5 the reduced number of 150 plants is given, because in these experiments Mendel also reduced the number of artificial fertilizations from the average of 59 for two experiments to 33.6 for the remaining ones. Even from 150 plants Mendel could have obtained three times more seed than he needed for seeding in the succeeding generations. In the experiments with plant traits Mendel (1866*a*, p. 16) tested the constancy of the dominating trait in such a way that '100 plants that possessed the dominating trait in the first generation were selected, and in order to test this trait's significance 10 seeds from each plant were sown.' Therefore the table gives a figure of 100 test plants. One must note that not all seeds germinate. Mendel gives a germination rate in his experiments involving two trait pairs of 95.2 per cent. It can be assumed that he obtained at least 950 plants for testing. In the fifth experiment with the lowest number of fertilizations and of plants germinated from the hybrid seeds, Mendel repeated the test for the constancy of the dominant trait, since the original result 'showed the greatest deviation'.

It can be seen from Table 5 that in his experiments with the crossing of plants differing in one trait pair Mendel must have grown up to 17 000

Table 5 Estimated numbers of *Pisum* plants differing in one pair of traits

Year	Seed shape	Seed colour	Colour of seed coat	Shape of pod	Colour of unripe pod	Position of flowers	Length of flowers	Total number
1856	30	30	—	—	—	—	—	60
1857	253	258	20	—	—	—	20	551
1858	565	519	150	20	10	20	150	1434
1859	60	60	929	150	150	150	1064	2563
1860	60	60	1000	1181	580	858	1000	4739
1861	60	60	300	1000	1000	1000	300	3720
1862	60	60	300	300	1000	300	300	2320
1863	—	—	300	300	300	300	300	1500
Total	1088	1037	2999	2951	3040	2628	3134	16887

plants. The start of this experiment is estimated to have been in 1856. In view of the fact that there was limited space in the experimental garden, the beginning of the experiments with plants differing in several trait pairs is placed in 1859. In the *Pisum* paper Mendel (1866a, p. 9) states that the experimental plants were grown 'in the garden beds, except for a few in pots, and were maintained in their natural upright position by means of sticks, twigs, and taut strings'. According to Fisher (1936), judging by the advice of the Swedish plant-breeder Rasmussen, Mendel could have grown a maximum of 5000 plants on a plot measuring 35 m × 7 m. With rows sown 25 cm apart and plants 15 cm apart it was not possible to fit more in.

Table 6 gives an estimated total number of experimental plants of 24 034. This does not include plants grown annually for testing the varieties used in the experiment, or those for experiments relating to the germ cells of hybrids, transformation experiments, and others of which Mendel makes only passing mention in the *Pisum* paper. One might also consider the statement by Mendel (1866a, p. 23) that 'all constant associations were actually obtained through repeated crossing'. Another example would be the experiment with crossing peas differing in four different trait pairs, which he mentions in his second letter to Nägeli (Correns 1905, p. 201). In connection with the limited area of the experimental garden, one should bear in mind that in parallel with his *Pisum* experiments Mendel was also performing experiments with other plant species.

But the *Pisum* paper does not state how exactly Mendel tested the constancy of experimental plants, nor how he chose the plants for individual experiments. Corcos and Monaghan (1984), who examined Mendel's manner of setting up individual experiments when he crossed peas differing in one, two, and three trait pairs, came to the conclusion that monohybrid

Table 6 Estimate of experimental *Pisum* plant numbers 1856–63

Year	Monohybrid	Dihybrid	Trihybrid	Total number	Carefully observed plants
1856	50	—	—	50	—
1857	551	—	—	551	551
1858	1434	30	30	1494	1184
1859	2563	15	24	2602	1993
1860	4739	529	639	5907	2619
1861	3720	180	4730	8630	4730
1862	2320	180	400	2900	—
1863	1500	—	400	1900	—
Total	16877	934	6223	24034	11077

experiments were performed with varieties differing in several traits, but in each case Mendel concentrated his attention on one trait tested for constancy. The authors also assumed that Mendel had parental varieties for his dihybrid and trihybrid experiments before he started them. The enigma of the origin of Mendel's experimental plants has recently been addressed again by Di Trocchio (1989, 1991), working from Bateson's (1909, p. 332) emphasis on Mendel's formulation that he used for monohybrid experiments plants which 'differed in only one essential trait'. Were we to take this literally, Mendel must have carried out monohybrid experiments with fourteen varieties differing from each other in this way. Di Trocchio restates Tschermak's (1940, p. 66) view that Mendel knew in advance the inheritance of separate traits, and could by crossing have formed plants for all his experiments. Di Trocchio pointed out an article written by T.H. Morgan in 1932 for the *Encyclopedia italiana di scienze, lettere ed arti*, in which he illustrated how Mendel might have created experimental lines for his dihybrid experiment by hybridization. Di Trocchio (1991, pp. 486–7) concludes that Mendel's account was 'not to be taken too literally' and that 'most of the experiments described in Versuche are to be considered fictitious'. He therefore proposes a new reconstruction of the experiments, which he calls a 'forked-line system', which takes no account of Mendel's description of individual experiments. He considers the monohybrid experiments as fictitious in the sense of having been performed 'on paper, by disaggregating the data from many polyhybrid experiments' (Di Trocchio 1991, p. 511).

The data in Table 6 take no account of Di Trocchio's reconstruction of the pea experiments. However, it is clear that Mendel grew more experimental plants than are mentioned in his paper, and some of them must surely have been grown elsewhere than in the experimental garden. We are unlikely ever to know in more detail. The tabulated data also show that Mendel used different numbers of experimental plants for different experiments, a fact which is connected with the creation of various hypotheses which he proved in his experiments, followed up by further hypotheses. He paid greatest attention to the simplest experiments, dealing with plants differing in a single trait pair, above all seed shape and colour. This was also the point of departure for his long-term research programme, whose theoretical basis is discussed below, see pp. 163–80.

EXPERIMENTS WITH OTHER PLANT SPECIES

Following Mendel's second lecture in 1865, the Brno daily *Neuigkeiten* mentioned the fact that Mendel had in recent years carried out artificial fertilizations of many named, related species of plants in order to obtain hybrids, and that his successful results had encouraged him 'not only to study hybridization further, but to give an account of these results'. The *Neuigkeiten* reports were published in English by Sir Gavin de Beer (1966). In his second letter to Nägeli of 18 April 1867, Mendel also mentions a number of hybridizations undertaken in 1863 and 1864, which convinced him of 'the difficulties of finding suitable plants for an extended series of experiments'. He had, on New Year's Eve in 1867, stated with sadness that 'it is very regrettable that the worthy Gärtner did not publish a detailed description of his individual experiments with different species' nor 'diagnose his hybrids sufficiently, especially those resulting from like fertilizations' (Correns 1905, p. 195).

When Mendel tried to repeat Gärtner's experiments as far as possible and re-examined them carefully, he found 'agreement with the laws of development which he found to be true for his experimental plants'. In conclusion, Mendel adds that 'in most cases, it can at least be recognized that the possibility of agreement with *Pisum* is not excluded'. At the end of 1866 he wrote to Nägeli that 'a decision can be reached only when new experiments are undertaken in which the degree of kinship between the hybrid forms and their parental species is precisely determined, rather than simply estimated from general impressions'. He simplified the method of verifying the applicability of his theory to the evaluation of the hybrid progeny as follows: 'If, for two differing traits, the same ratio and series which exist in *Pisum* can be found, the whole matter would be decided.' Here Mendel indicates the direct connection between *Pisum* and his further experiments.

In his second letter of April 1867 to Nägeli (Fig. 5.10) Mendel enclosed pea-seed samples designated symbolically according to their traits, inviting the learned man to verify the correctness of his results with the germ cells of hybrids. In his reply Nägeli promised to carry out such experiments, but he would appear subsequently to have forgotten about this. He was more enthusiastic about Mendel's remark in his first letter that he had chosen the genera *Cirsium*, *Hieracium*, and *Geum* for his further experiments. Mendel wrote that with *Cirsium* and *Hieracium* he was entering a field in which Nägeli possessed 'the most extensive knowledge, knowledge that can be gained only through many years of zealous study, observation and comparison of the manifold forms of these genera in their natural habitat' (Correns 1905, p. 201). He asked Nägeli for his advice. The professor, obviously flattered, enclosed with his reply 12 reprints of his papers dealing with plant hybridization, which he had published mainly in 1853–66 in the *Proceedings of the Bavarian Academy of Science* (Nägeli 1853, 1865a–d, 1866a–f). Incidentally, these *Proceedings* were in any case being

Fig. 5.10 Carl Wilhelm Nägeli (1817–91), with whom Mendel was in correspondence for a number of years, exchanging experiences and scientific papers.

supplied to the library of the Agricultural Society in Brno, where they were available to Mendel.

In his reply, dated 25 February 1867, Nägeli writes: 'It seems to me that your experiments with *Pisum* are far from being completed and are indeed only just beginning. A shortcoming of all the more recent experiments is that they have shown so much less perseverance than Kölreuter and Gärtner. I am pleased to note that you are not making this mistake and are following in the footsteps of your celebrated predecessors. However, you should try to excel them. In my view this will be possible—and this is the only way to make advances in the theory of hybridization—if exhaustive experiments are conducted upon a single object in every possible direction. Such a complete series of experiments giving irrefutable proofs of the most important conclusions has by no means yet been undertaken.' In another part of the letter Nägeli expresses an interest in receiving seeds from Mendel's experiments for cultivation in the university garden 'so as to see whether they remain constant under different conditions. Those I should particularly wish to receive are A & a (offspring of Aa), AB, ab, Ab, aB (offspring of AaBb).' Mendel must have been pleased that the famous professor was interested in his experiments.

Fragments of three surviving letters from Nägeli to Mendel are preserved in the Mendelianum archives, and were published by Hoppe (1971). In fact, Nägeli was repeating in this letter the criticism of contemporaries who undertook plant-hybridization experiments which he had previously published in his speculative paper on the formation of bastards in the plant kingdom (Nägeli 1866*b*). We might ask with what degree of care he read Mendel's work, since he failed to understand that in it Mendel was offering an answer to many of the questions he was asking.

In his second letter to Nägeli, Mendel writes that he had not been surprised when the professor expressed his view of his experiments 'with mistrustful caution', adding: 'I would not do otherwise in a similar case (Correns 1905, p. 200).' Nägeli did not believe that there would be constant parental forms as well as hybrids in the hybrid progeny. Mendel explained to him that his statement that 'some of the progeny of hybrids breed true to type thus includes only those generations during which observations were made; it does not extend beyond them'. He was thus inclined 'to regard the separation of parental traits in the progeny of the hybrids in *Pisum* as complete, and thus permanent'. But he emphasized that the progeny of hybrids 'carries one or other of the parental traits, or the hybrid form of the two'. He never observed a gradual transition between the parental traits or a progressive approach toward one of them. His closing statement is apt: 'The course of the development consists simply in this: that in each generation the two parental traits appear, separated and unchanged, and there is nothing to indicate that one of them has either inherited or taken over anything from the other.'

Mendel thought it important to assume a standpoint regarding Nägeli's criticism that 'you should regard the numerical expressions as being only empirical' (Correns 1905, p. 202). First he repeated what he had already said about simple series and combination series in the *Pisum* paper. To this he added: 'Up to this point I do not believe I can be accused of having left the realm of experimentation. If then I extend this combination of simple series to any number of differences between the two parental plants, I have indeed entered the rational domain. This seems permissible, however, because I have proved by previous experiments that the development of a pair of differing traits proceeds independently of any other differences.' In the same place he points out to Nägeli that his statement has also been confirmed in experiments with the germinal pollen cells of hybrids whose results 'furnish the explanation for the development of hybrids as observed in *Pisum*'. The conclusion is that 'these experiments should be repeated and verified'. Here the beginner in plant-hybridization research gives a method-ological lesson to an acknowledged scholar, and thus tries to provoke him into at least repeating the experiments to convince himself of the justification of the conclusions.

From the content of Mendel's first letters Nägeli realized he was corre-sponding with a researcher of exceptional quality. In the first five letters Mendel's salutation is 'Highly Esteemed Sir'. From the fifth letter, dated 12 June 1868, Mendel switched to the salutation 'Highly Esteemed Friend' (Correns 1905, p. 222). Mendel's letters also end differently (see Fig. 5.14). He ended the first of them with the words 'Monastery Capitular and teacher at the *Oberrealschule*' after his signature. The second letter ends 'Your devoted', the third 'Your sincere admirer'. The fifth refers to 'Your devoted friend'. Only three fragments of Nägeli's letters have survived, and we do not know how Nägeli addressed Mendel. However, the former must have taken the initiative for the latter to change the form of salutation. In a letter dated 27 April 1870, we see how Nägeli expressed his pleasure at having found in Mendel 'so skilful and successful a colleague' (Hoppe 1971, p. 138). This must have constituted high praise for Mendel at the time.

The exchange of views with Nägeli regarding research on further plant species was of prime importance to Mendel, a fact which commentators have explained in various ways. He had established contact with an acknowledged scholar, and that must have been a source of satisfaction to him. But Mendel's contacts with natural scientists in Brno were also of great importance. At the time they were interested in the occurrence of hybrids in nature and were also seeking a theoretical explanation of the role of hybrids in connection with the appearance of new species; this is a fact which has largely been neglected in the past.

Plant hybridization and the Brno naturalists

The Natural Science Society was established in Brno in December 1861, and at the first working meeting, on 8 January 1862, Professor Niessl (1863*a*) reported on a paper by A. Zeising (1854) regarding the new pro-portion theory, suggesting the possibility of its being used in research into the appearance of new plant forms. Mendel must have known about this paper soon after it was published, since in December 1855 he pencilled notes in the weekly journal of the Agricultural Society, *Mittheilungen*, which constituted views on the theorem of the golden section (Orel 1978*a*). Analysing these notes, H. Kalmus (1983) reached the conclusion that Mendel was seeking a point of departure for his research, and initially was as bewildered by his results as his lecture audience was. According to Iltis

(1966, p. 177) the latter might have thought of his mathematical botany as a sort of number mysticism, not unlike some Pythagorean speculations. Kalmus was convinced that Mendel had succeeded in creating order out of bewilderment, and that in this respect he resembled Kepler, who, two hundred years earlier, had also produced in his notes a bizzare mixture of fundamental insights into celestial mechanics and strange ideas and calculations.

At the February meeting of the Natural Science Society the pharmacist C. Theimer (1863) spoke of bastard-formation in the plant kingdom, and showed hybrids from the Brno area of *Cirsium praemorsum* (Michl. *Cirsium oleraceo-rivale* Dc) and the form *Cirsium cano-oleraceum* Koch, which had already been found in Moravia. He gave a detailed description of leaf, flower, and stem traits in the parental and the hybrid forms. Theimer pointed out that it was a mistake to consider hybrids infertile, stating that this had been disproved in many new experiments, though he was no more specific than that. He added that according to recent repeated experiments, hybrids produce seeds capable of germinating. In the same year A. Makowsky (1863), considered in the society to be an expert in phanerogams, published in the *Proceedings* a monograph on the flora of the Brno region, and in his conclusion stated that the 1264 specimens described came from the systematic list of plant species, though bastards 'as chance phenomena' were not included in the plants enumerated. Among the twelve forms of *Cirsium* the following four hybrids are included: *Cirsium palustri-oleraceum* Näg., *Cirsium cano-oleraceum* Reichb., *Cirsium oleraceo-rivale* Dc, and *Cirsium cano-palustre* Wim. Makowsky found these hybrid forms among the stem-parents. He attributed the finding of the second and third hybrid forms to C. Theimer and that of the fourth to C. Römer, a clerk at a textile factory in Náměšť, a small town to the north-west of Brno. Among the eleven forms of *Hieracium* mentioned is the hybrid *Hieracium pilosella-praealtum* Neil, which Theimer found in a meadow to the south-west of Brno. Among the nine forms of *Verbascum* mentioned there is also the hybrid *Verbascum thapsiformi-nigrum* L., with a note by its discoverer Römer, and the remark that it occurs only rarely among the stem-parents.

On 10 June 1863, Makowsky (1864a) reported on an excursion to southern Moravia, and the first thing he mentioned was finding the hybrids *Hieracium pilosella* and *auricula*, and the no less common occurrence of the hybrid *Hieracium pilosella-auricula* Nl (*H. bifurcum* M.B.). At a meeting on 8 July, Niessl *et al.* (1863b) had presented a written report on a botanical excursion to the north-west part of Moravia, where they also reported finding various species of *Cirsium* and *Hieracium*. In one locality they found the hybrid *Cirsium palustri-rivale*. They also described the species *Cirsium palustre* Scop., *C. canum* M.B., and *C. rivale* Link. Among these three species Niessl sought a hybrid form, and was pleased that in one place where all three forms were present he found the hybrid *Cirsium palustri-rivale*. According to Reichenbach's '*Icones*' it was somewhat different, and he therefore presented this hybrid form for critical evaluation. With this report Niessl submitted a three-page appendix with a detailed evaluation of this hybrid form, including the shape of the roots, stems, leaves and blossoms, and the colour of the flowers. There is a reference in the text to the German literature of the day, which the participants in the excursion would seem to have been familiar with.

At a meeting on 14 December 1864 Niessel (1865) handed over to the committee floristic notes on phanerogams, and the second item was the finding of the hybrid *Hieracium auricula-pilosella* Fries. The description of the hybrid also points to the traits of the two parental forms; Niessl adds that he had 'in almost all parts observed a *Mittelbildung* between the parental forms'. Third, he mentions the hybrid *Verbascum lychnitidi-phlomoides* Reissek, with a question-mark. Obviously he was not entirely sure of his identification. He describes in detail individual traits of flower and plant, and adds that the form has not previously been described in Moravia. At the next meeting, on 14 June 1865, in the *Proceedings* (p. 52) it is stated that 'Mendel showed in the fresh state two hybrid forms of *Verbascum phoeniceum* × *V. blattaria* and *Campanula media* × *C. pyramidalis* which he had bred'.

On 13 December 1865 Niessl (1866) reported on an excursion to southern Moravia, and in the list of plant species found he mentions the hybrid *Hieracium bifurcum* and the forms *Verbascum phoeniceum* and *rubiginosum* W. Kit. In a further part of his report he explains that *Verbascum rubiginosum* W. Kit is a hybrid of *Verbascum phoeniceum* × *V. orientale* L., and not, as Reichhardt wrongly stated in 1861, *Verbascum phoeniceum* × *V. migrum*. Niessl found this form among the two stem-parental forms. Here he emphasizes that the hybrid plant forms found in various places 'must constantly be fully described, so that from the fluctuating numerical data certain more constant ones can be found'. For this reason Niessl added a description to the specimen he presented to the society's herbarium.

A meeting with an interesting bearing on Mendel's experiments was the lecture given by Niessl (1867a) on 11 April 1866, in which he demonstrated a specimen of *Asplenium heufleri* Reichn. as the hybrid *Asplenium trichomanes* × *A. germanicum*. In a further paper Niessl (1867b) states that Baron von Heufler in Südtirol had described the form as a species. Niessl explains that it was in fact a hybrid form. He mentions in his lecture hybrid forms of the fern family, and says he expects more hybrid forms to be discovered, in which context he states that 'an understanding of the sexual reproduction of cryptogams will bring great progress in every division of the plant kingdom'. Subsequently, Niessl outlines the latest findings regarding the sexual reproduction of plants. Among these was Hedwig's discovery of fertilizing bodies in mosses and Valentin's of *Archegonium*, later shown to be a female flower. Niessl pays great attention to the investigation of fertilization in higher crytopgams conducted by Hofmeister (1849, 1851). He also mentions Nägeli's discovery of sexual reproduction in fungi, 'where fertilization takes place analogously with that of phanerogams'. Referring to a publication of de Barry's of 1863, without quoting the paper, Niessl describes a sort of copulation between each of the crossing threads of mycellae. In conclusion, he draws attention to the fact that hybridity is very difficult to determine in cryptogams, with a lower degree of organization and a smaller number of traits. Niessl's detailed description of the history of research into the generation in lower plants runs to eight pages, and he makes comparisons with higher plants, phanerogams. Mention is made of discoverers; no sources are cited, but it is clear from the report that Niessl knew the latest publications in the field.

At a meeting on 12 December 1866, C. Römer (1867) handed over to the society a list of twenty species of plants found in the area of the town of Náměšť. Among these he mentions *Hieracium virescens* Sonder *(Hieracium boteli-racemosus, Hieracium echioidi-pilosella,* and *Hieracium bifurcum W.K. pilosella echioides Schultz)*. Later Niessl (1867c) presented a report on botanical research undertaken on cryptogams. Mention is made of a large number of hybrids of *Hieracium pilosella-praealtum* observed 'among the parental forms'. Five pages are devoted to the hybrid forms *Cirsium palustre × C. rivale, Cirsium canum × C. palustre,* and *Cirsium rivale × C. oleraceum*. In conclusion he mentions *Verbascum phlomoides × orientale*. Niessl gives a similar description of the stem, leaves, and flowers. Of the first form, *Cirsium palustri-rivale*, he adds: 'I found this form most often, and gave one specimen to Professor Mendel for culture. It yielded a rich crop of seeds, and these produced flowers in the same year.' Here one can see how Niessl and Mendel cooperated.

At a meeting on 8 May 1867, Professor Makowsky (1868) demonstrated 'a specimen of *Geum urbano-rivale/intermedium* Ehr artificially pollinated by Professor Mendel, which is quite in accord with the wild form'. The published report contains a description of six traits of this hybrid form. For the first time Brno natural scientists were able to see an artificially obtained hybrid form; previously they only knew those occurring in nature. Attached to the published report of the society meeting of 11 December 1867, is Niessl's report on the further occurrence of the *Hieracium hybrids praealto-pilosella* and *pilosella-praealtum* (Niessl 1868*a*). In the same year Niessl (1868*b*, 1869) reported *Mittelformen* of *Asplenium* and *Cirsium*.

The published reports from meetings of learned societies are always very brief. Nevertheless, the reports of the meetings of the Natural Science Society in Brno for the period when Mendel was carrying out his plant-hybridization experiments show the connection between the interests of natural scientists, who observed and evaluated the occurrence of plant hybrids in nature, and Mendel's choice of plant species for his further experiments once he had finished investigating *Pisum*. Like Mendel, the natural scientists evaluated alternative traits in the parental and hybrid forms. In the context of Mendel's research it is also interesting to note Niessl's efforts to explain the latest findings relating to the fertilization of lower and higher plants. In this scientific community there was an exchange of views, and there are repeated records in the *Proceedings* of Mendel's evaluation of experimental hybrid forms which the natural scientists could only observe. Mendel was the only one who was presented here as an experimenter. In this scientific community he might have received the stimulus to undertake experiments with *Hieracium* even before he approached Nägeli in 1866 asking for advice on the methods of artificial fertilization he should use in his further, more demanding experiments. But the literature on the subject still asserts that it was Nägeli who instigated Mendel to carry out experiments with *Hieracium* (Oldroyd 1980; Bowler, 1989, p. 84).

Experiments with *Phaseolus*

In his second lecture in 1865 Mendel (1866*a*, p. 32) mentions having recently begun several experiments, of which two 'fairly small ones with species of *Phaseolus*' had been completed, so that he was able to report on them. He crossed *Phaseolus vulgaris*, with a stem 10–12 cm long and yellow pods constricted at maturity, with *Ph. nanus*, with a dwarf-like stem and green, smoothly-arched pods. Mendel gives eight constant trait combinations in the hybrid progeny of this crossing, corresponding to the 2^3 ratio derived from the *Pisum* experiments, with plants differing in three pairs of traits. According to him, the green colour, arched pod-shape, and tall stem are dominating traits, again 'as in *Pisum*'. He adds laconically that the constant association 'also proceeds according to the law of the simple combination of traits, exactly as in *Pisum*'.

In a second experiment Mendel crossed 'very different *Phaseolus* species', denoting the results as 'only partly successful'. The seed plant was *Phaseolus nanus* L., with white blossoms in short racemes and small white seeds in straight, and arched, smooth pods. The pollen plant was *Phaseolus multiflorus* W., with a tall, winding stem, crimson blossoms in very long racemes, rough sickle-like crooked pods, and large seeds with black flecks and splashes on a peach-blossom-red background. The hybrid bore the greatest resemblance to the pollen plant, but the blossoms were less intensively coloured. In this case Mendel was crossing different species, and met with the problem of the reduced fertility of hybrids. From 17 plants, which developed a total of many hundreds of blossoms, only 49 seeds were harvested. These were of medium size, and bore a pattern similar to that of *Phaseolus multiforus*. The background colour also did not differ basically. In the following year he obtained 44 plants, of which only 31 reached the flowering stage. The traits of *Ph. nanus*, which became latent in the hybrid, reappeared in various combinations. Mendel (1866a, p. 33) was able to see broad variability of traits. For some traits only, such as stem and pod-shape, he found a proportionality 'as in *Pisum*, almost exactly 1:3'. Numerical data are not given here.

Mendel was fascinated by the remarkable colour change in the blossoms and seeds. In hybrid progeny in *Pisum* he had observed only one or the other of the parental traits. In this experiment with *Phaseolus* the white flower and seed colour of *Ph. nanus* appeared, but the remaining 30 plants developed flower colours that represented several gradations of colour from crimson to pale violet. Again there was a high rate of infertility, especially in the red-flowering forms. Well-formed seeds were harvested from 15 plants only. In the next generation Mendel obtained seed from only four plants, and three of these had seeds similar to those of the paternal plant, though with more or less pale background coloration, while the fourth plant yielded only one seed, which was plain brown. Forms with predominantly violet flower-colour had dark-brown, black-brown, and totally black seeds. He continued the experiment for two more generations, and again found reduced fertility or infertility. The forms which exhibited recessive traits in the hybrid progeny remained constant in all cases. Individual plants with violet flowers and brown or black seeds were already constant in these traits. Most, however, gave the same progeny, some with white flowers and similarly-coloured seed-coats. Red-flowering plants remained so poorly fertile that nothing could be said about their future development.

In spite of all the obstacles he came across in this experiment with *Phaseolus*, Mendel (1866a, p. 34) concluded that 'development of hybrids follows the same law as in *Pisum* with respect to those traits concerned with the shape of the plant'. But in the case of coloration he could not find sufficient agreement. Apart from the colour of the parental forms he observed a whole range of colours, from purple to pale violet and white.

Out of 31 flowering plants only one received the recessive white colour, while in *Pisum* this was true of every fourth plant on average. Here we can see Mendel's readiness to seek a new theoretical explanation based on the principles discovered in *Pisum*, which attracted the attention of Mueller (1951). The new premiss was that 'in *Phaseolus multiforus* the colour of flowers and seeds is composed of two or more totally independent colours that behave individually exactly like any other constant traits of the plant'. According to this idea the blossom colour A would be composed of independent traits $A_1 + A_2 + \ldots$ which produce the overall impression of crimson coloration. Through crossing with the white-flowering plants the hybrid association $A_1 + A_2 + \ldots$ would have to be formed. The coloration of the seed coat might be similar. This was the basis of Mendel's idea of the *combination of two simple series*:

$$A_1 + 2A_1 a + a; \text{ and}$$
$$A_2 + 2A_2 a + a.$$

Accordingly the terms of these series can enter into nine different combinations, each of which represents the designation for another colour:

1	A_1, A_2	2	A_1a, A_2	1	A_2, a
2	A_1, A_2a	4	A_1a, A_2a	2	A_2a, a
1	A_1, a	2	A_1a, a	1	$a, a.$

Mendel's conclusion (p. 35) was that 'if colour development really occurred in this manner, then the above-mentioned case of white blossom and seed-coat colour appearing only once among 31 plants of the first generation would have an explanation'. He added that this coloration in his experiments occurred only once in the series, and therefore could be expressed only in every 16 plants on average. He did, however, concede that plant colour could be composed of three colour traits, in which case the white blossom would only be expressed in one in 64 plants. It must, however, be stressed that Mendel considered this explanation 'mere supposition', recommending a study of colour development in hybrids further by similar experiments, 'because it is probable that through this approach we can learn to understand the extraordinary diversity in the coloration of our ornamental flowers'.

Following this conclusion, Mendel passed on to the then topical problem of the effect of the environment on the variability of traits. He rejected the view that species stability was to a large extent or even entirely destroyed by cultivation. Mendel (1866*a*, pp. 37–8) knew the history of cultivated plants, and was able to argue that 'during over 1000 years of cultivation under the most diversified conditions, numerous varieties have arisen, yet these maintain stability under constant living conditions, just as do species growing wild'. His conclusion was that 'our cultivated plants, with few exceptions, are members of different hybrid series whose development along regular lines is altered and retarded by frequent intraspecific crosses'. The probability of this view is supported by the fact that among the great arrays of variable forms solitary examples are always found which in one character or another remain constant if only foreign influences are carefully excluded. These forms, according to Mendel, develop 'like certain members of the *composite hybrid series*', whose composition was intimated in the example of the coloration of *Phaseolus* flowers. Cultivated *Dianthus* species described by Gärtner are mentioned by Mendel as being instructive examples for such a conclusion.

From Gärtner's monograph Mendel obtained information regarding previous experiments with the crossing of various members of the bean family. He began his own experiments in 1859, when the *Pisum* experiments proved the segregation of the traits of the parental forms in the hybrid progeny. In his lectures in 1865 he was able to explain not only simple and combination series found in the *Pisum* experiments, but also the composite series found in *Phaseolus*. After the year 1865 he turned his attention to other plant species, which was a continuation of the long-term research programme begun with the *Pisum* experiments.

Experiments with other plants

After the *Pisum* experiments ended in 1863, Mendel set out to test his theory of variable hybrids in other plant species.

Cetl (1973) outlines the individual experiments in chronological order, stating the number of successive generations involved (Table 7). While with *Pisum* Mendel crossed varieties inside the species, in the *Phaseolus* experiments he was crossing various species of plants. Here he came across the problem of decreased fertility, and he also had difficulty with the technique of artificial pollination. He also had to take into account the limited space he was able to devote to growing a large number of experimental plants. For this reason he brought most of the experiments to an end after evaluating the hybrid generation and their offspring. In exceptional cases he went as far as the fourth generation.

Table 7 shows how Mendel experimented with the largest number of plant species in the years 1866–7. Among them were also species which Gärtner had described as constant hybrid forms. First of all Mendel became interested in the hybrid *Geum urbanum* × *G. rivale*. He was not convinced that this hybrid, which according to Gärtner produced non-variable progeny when self-pollinated, was actually *Geum intermedium* Ehrh. In his first

Table 7 Mendel's experiments with 14 plant genera. (After Cetl 1973)

Combination/Year	1864	1865	1866	1867	1868	1869	1870
Verbascum phoeniceum × V. Blattaria	C	F_1					
Tropaeolum majus × T. minus	C	F_1	F_2				
Campanula medium × C. pyramidalis	C	F_1					
Cirsium arvense × C. oleraceum			C	F_1			
C. arvense × C. canum			C	F_1	F_2		
C. canum × C. oleraceum			C	F_1	F_2		
C. lanceolatum × C. oleraceum			C	F_1			
C. premorsum (= *C. oleraceum × C. rivale*)			N	F_1			
C. premorsum × C. palustre			C	F_1			
C. premorsum × C. canum			C	F_1	F_2		
				C			
C. canum × C. palustre			N	F_1			
C. rivulare × C. palustre			N	F_1			
Geum urbanum × G. rivale			C	F_1	F_2		
				BC			
Linaria vulgaris × L. striata			C	F_1	F_2		
				BC			
L. vulgaris × L. genistaefolia				C	F_1		
L. vulgaris × L. triphylla				C			
Aquilegia canadensis × A. vulgaris			C	F_1			
A. canadensis × A. atropurpurea			C	F_1			
A. canadensis × A. Wittmaniana			C	F_1			
Matthiola annua (varieties)			C	F_1	F_2	F_3	F_4
M. annua × M. glabra (?)				C	F_1	F_2	
Ipomoea purpurea (varieties)			C				
Antirrhinum majus (varieties)			C				
A. majus × A. rupestre (?)					C	F_1	
Calceolaria salicifolia × C. rugosa			C	F_1			
Zea Mays major × Z. M. minor			C	F_1			
Z. M. major × Z. Cuzko (?)			C	F_1	F_2	F_3	
Mirabilis jalapa × M. longiflora				C	F_1	F_2	
M. jalapa (varieties)				C	F_1	F_2	F_3
Lychnis diurna × L. vespertina (= *Melandrium rubrum × M. album*)				C	F_1	F_2	
No. of individual operations	3	3	21	26	13	7	3

C, Crossing; F_1 to F_4, Hybrid generation; BC, Backcrossing; N, A natural hybrid

letter to Nägeli he had indicated that he would be able to obtain an artificially-bred specimen of the hybrid in the following year (Correns 1905, p. 197). In his next letter, dated 17 April 1867, he stated that according to

Gärtner the hybrid *Geum urbano-rivale* had mostly large flowers, like *rivale*, and only a few specimens had small flowers like *urbanum* (Correns 1905, p. 208). Mendel, on the contrary, found about half the flowers to be the size of *Geum rivale*, while the other traits corresponded, as far as could be judged, to those of *Geum intermedium* Ehrh. The explanation, in Mendel's view, was to be sought in the fact that he, unlike Gärtner, had crossed pure species. Shortly after this, on 8 May 1867, Professor Makowsky (1868) showed Mendel's artificially produced hybrid form at a meeting of the Natural Science Society; it was designated *Geum urbano-rivale, intermedium* Ehrh., which in Makowsky's view corresponded to the form that occurred naturally. The published report gives a detailed description of the traits of the hybrid form as follows: 'a fork-shaped ramified state of blossoms of *Geum urbanum*, the lower secondary leaves are divided, the upper ones not, the persistence of leaves is in between that of the two parents; the same is true of the dimensions of the flower petals. The blossoms are wrapped round, first yellow, later cup-like with a reddish bloom.'

   On the flyleaf at the back of Gärtner's monograph Mendel gave six traits of *G. urbanum* and *G. rivale*. The traits of *G. rivale* are denoted recessive. The following symbols are used,

$$ABcDEe$$
$$ABcdEe,$$

representing two hybrid forms differing in the traits *D* and *d*, which according to Mendel's description on the flyleaf of the monograph were forms with erect or nodding flowers. The fifth trait, Ee, is given as the hybrid form. The symbols used, with the description of the alternative traits, are written in ink.

   There is an additional note by Mendel in pencil which says 'Hoffmann *Geum* 421'. This is a reference to the book by Hoffmann (1869) on the definition of the significance of species and varieties. That author had cited Mendel's *Pisum* paper, and also mentioned that on Gärtner's p. 698 he could not find anything to substantiate Mendel's statement that Gärtner had raised in the progeny of the hybrid of *Geum urbanum* and *G. rivale* exceedingly fertile and constant hybrids. Mendel now noted on the cover of Gärtner's monograph a reference to page 421, where Gärtner lists *Geum* hybrids among constant hybrid forms, stating that 'several simple exceptionally fertile hybrids propagate themselves without change of type, like pure species', and that 'several botanists are inclined to admit that these hybrids are stable species', thus again corroborating, though with a different reference, his original statement. Callender (1988) suggests that Mendel's main aim was to prove the existence of constant hybrids, in agreement with the results of his research into *Pisum*. He concludes that Mendel had misrepresented Gärtner's views in that, when he listed the types of hybrids, he omitted Gärtner's statement that there 'was always a steady loss of fertility and a general breaking up of the species'. Mendel later noted in pencil on the cover of Gärtner's book a reference to pages 430 and 431, with the comment 'very important'. The pages in question describe various forms of hybrid progeny, among them those which do not change. Mendel also made a note in ink of page 553, using the same words 'very important'. In this case the passage referred to contains a description of Gärtner's experience with more constant forms of hybrids of *Verbascum, Geum,* and *Dianthus*. There is also a reference to Wiegemann's similar findings. In the case of the *armeria–deltoides* hybrids of *Dianthus*, Gärtner observed constancy of hybrid traits over ten generations. This was cited in the *Pisum* paper. Mendel seems to have known about the criticism raised in Hoffmann's book relating to his explanation of constant hybrids, and the note in ink on page 553 might have been written after he had read Hoffmann's book. He was pointing to the fact that Gärtner considered the possibility of constant hybrids. Callender's view that he deliberately misrepresented Gärtner's views is not, however, justified.

In his third letter to Nägeli, dated 6 November 1867, Mendel briefly mentions reciprocal back-crossing experiments with *Geum* as follows (Correns 1905, p. 216):

> *G. urbanum* with the hybrid;
> *G. rivale* with the hybrid;
> the hybrid with *G. urbanum*; and
> the hybrid with *G. rivale*.

The plants obtained from these fertilizations, and those produced by self-fertilization from the hybrid, were planted in the garden at the beginning of August. Two years later, in his seventh letter to Nägeli dated 15 April 1869, Mendel states that he is sending, according to Nägeli's wishes, hybrids of *Geum* with the parental forms (Correns 1905, pp. 226–7). He wished to convince his correspondent that the hybrid of *Geum* was variable, since he had demonstrated segregation of parental traits in the progeny. In his letter Mendel quotes Gärtner, according to whom 'the progeny of these hybrids show no variation'.

In his first letter to Nägeli, Mendel also mentions experiments with *Cirsium*, and describes an improvement in his method of artificial fertilization, adding that he intended to use it the following year in the artificial fertilization of *Hieracium* (Correns 1905, pp. 196–7). In his second letter he mentions the hybrids *Cirsium arvense* × *C. oleraceum, Cirsium arvense* × *C. canum, Cirsium praemorsum* × *C. palustre, Cirsium praemorsum* × *C. Canum, Cirsium canum* × *C. palustre,* and *Cirsium rivale* × *C. palustre* (Correns 1905, pp. 208–9). Some of these forms did not survive the winter. Mendel found that the flower-heads which developed first and last on the stems did not form pollen, and were thus completely sterile. This led him to investigate the physiology of the capacity for fertilization in connection with the hybridization of plants.

Later, on 6 November 1867, Mendel wrote of the further hybrid *Cirsium praemosum* × *C. canum,* from which he obtained only two plants with differing traits (Correns 1905, p. 215). These gave no seed. From the two parental forms he obtained about 50 offspring. He expected that their improvement the next summer would 'show whether or not variants do appear following self-fertilization, and what kinds, and what relation they bear to the differences between the two hybrids'. In his seventh letter to Nägeli, dated 15 April 1869, he describes another two hybrid plants, from which he had obtained progeny with segregation of traits. His conclusion was that '*Cirsium* would be an excellent experimental plant for the study of variable hybrids if it required less space'. Because of this he stopped experimenting with *Cirsium,* though he had already proved hybrids of the genus to be variable, like *Geum* hybrids (Correns 1905, p. 226).

In his second letter Mendel mentions to Nägeli the luxuriant form *Linaria vulgaris* × *L. purpurea* (Correns 1905, p. 209). In the spring of the same year he altered this designation to *Linaria striata* C. (Correns 1905, p. 216). As in the case of the hybrid of *Geum,* he carried out back-crossing experiments, and was able to inform Nägeli that 'the seeds of the hybrid show considerable variation'. He also sent him, as he had with the *Geum* hybrid, seed from the hybrid and the parental forms, offering him the possibility of verifying whether or not it was a variable hybrid. Mendel also mentions luxuriant hybrid forms of *Calceolaria salicifolia* and *C. rugosa,* and *Zea major* × *Z. minor* and *Zea major* × *Z. cuzko* (Correns 1905, pp. 216–7). To study colour development in the flowers of hybrids he also made a cross-fertilization between varieties of *Ipomea purpurea, Cheiranthus annuus,* and *Antirrhinum majus,* adding that an experiment with hybrids of *Tropaeolum majus* × *T. minus* (first generation) must also be mentioned. For the current year he planned further exploratory experiments with *Veronica, Viola, Potentilla,* and *Carex.* He added that unfortunately he had only a small number of species. On this occasion Mendel mentioned the largest number of plant species considered for his experiments.

Owing to the lack of space, Mendel began the experiments with only a small number of plants. From the viewpoint of the fertility of the hybrids and the flowering period, he decided to undertake an extensive study of the hybrids *Tropaeolum majus* × *T. minus*, though in this form he had observed only partial fertility. He had also planted 'the hybrids *Aquilegia canadensis* × *A. vulgaris, Aquilegia canadensis* × *A. atropurpurea*, and *Aquilegia canadensis* × *Aquilegia* wittmaniana, the seedlings of which survived during the flowering period'. On the flyleaf of Gärtner's monograph he pencilled beside the hybrid *Aquilegia vulgaris* × *A. canadensis* the words '423 Hybr. I. Gener. 1:2:1'. Gärtner placed this hybrid form among the constant hybrids. Mendel in this connection gave his explanation of the segregation of traits in the progeny of hybrids, which was proof of the variable hybrid.

In spring 1867 Mendel underwent a period of suspense waiting for the results of his experiments. He was aware, however, that further experiments would be slow, and thus in the conclusion to his letter to Nägeli dated 18 April 1867, he writes: 'In the beginning some patience is required, but later, when several experiments are progressing concurrently, matters are improved. Every day, from spring to autumn, one's interest is refreshed daily, and the care which must be given to one's wards is thus amply rewarded. In addition, if I should, by my experiments, succeed in hastening the solution of these problems, I should be doubly happy' (Correns 1905, p. 210).

On 6 November 1867, Mendel devoted most attention to the experiments with *Hieracium*. He adds information relating to *Geum, Cirsium*, and *Linaria*, and in conclusion reports an observation he had made the previous year in a *Verbascum* hybrid (Correns 1905, pp. 211–15). In 1864 he had undertaken hybridization between several species of *Verbascum*, and the hybrids raised in the garden were completely sterile. Accidentally, one plant from the cross *Verbascum phoenicum* × *V. blattaria* was left in the seed bowl, and remained there in a corner of the garden without any care throughout the summer. He discovered this very stunted plant in the autumn, and planted it out in the open, along with its luxuriantly developing siblings. Although it became fairly vigorous during the following year, it did not bloom, and overwintered for a second time, whereas its siblings died as two-year-old plants after blooming. In the summer of 1867 the plant flowered from June to September, producing 'more than 100 well-formed seeds'. Mendel impatiently awaited the coming summer 'in the hope that several hybrids will exhibit their progeny in flower for the first time'. He adds that 'Care has been taken that they may appear in large numbers, and I only hope that they will reward my anticipation with much information about their life history'. He expected to test segregation of traits with the large number of progeny. There is no obvious explanation why we find no mention of this hybrid in later letters. Previously, on 14 June 1865, the minutes of a Natural Science Society meeting briefly state that 'Professor Mendel showed in the fresh state two artificially created hybrids of *Verbascum phoenicum* blossoming in white and *V. Blattaria*' (Mendel 1866*b*).

On 3 July 1870, Mendel mentions the first completed experiments with the hybrids *Matthiola annua* and *glabra, Zea,* and *Mirabilis.* He notes briefly that 'their hybrids behave exactly like those of *Pisum*' (Correns 1905, pp. 236–7). Mendel adds that 'Darwin's statements concerning hybrids of the genera mentioned in *The variation of animals and plants under domestication* based on the reports of others need to be corrected in many respects'. More detailed descriptions are to be found only of the colour experiments with *Matthiola,* which had been going on for several years. He considered himself to be getting to the bottom of the problem. He was aware that he was confronted with very complex plant coloration, and noted that lack of a reliable colour chart had greatly hindered his experiments. Though he had ordered from Erfurt 36 named colours, this proved insufficient for his purpose. He was especially interested in this experiment now, and promised Nägeli he would inform him of the results 'as soon as the inspection of the 1500 specimens of this year's culture has been completed'. He began these experiments at the time he was finishing the *Pisum* experiments, and at that time came across complex coloration in *Phaseolus,* for which he found the explanation in composite series in the hybrid progeny.

Three months later, writing mostly of his *Hieracium* experiments, Mendel states that 'the colour experiments with *Matthiola annua* have made only minor progress this year in spite of the great number of experimental plants' (Correns 1905, p. 240). Nevertheless, he believed that 'an agreement with *Pisum* appears probable'. Instead of the expected colour there frequently appeared in the whole series of specimens 'a higher or lower octave of colour or both appearing jointly'. Sorting became very unreliable, since it was easy to put together what should be separated, or make the opposite error. The numbers thus obtained for the frequency of the different colour variants are used for the deviation of the series. He therefore mentions a new group of plants for which he expected that he might succeed 'in obtaining a simple series with them'. This was the last information on the experiments with *Matthiola.* It also expresses his conviction that he would succeed in developing further the theory of variable hybrids.

According to the letter dated 3 July 1867, Mendel was still experimenting with *Mirabilis jalappa* to try to throw light on the subject of whether the plant ovule was fertilized by one or more pollen grains—a question which was still disputed (Correns 1905, pp. 235–6). Mendel does not mention that a similar investigation of the mechanism of fertilization had already been made by Kölreuter, who concluded that only a single ovule in the ovary and a single pollen grain were sufficient in *Mirabilis* for fertilization, which was pointed out again by Mayr (1986). Mendel based his theory of the origin of variable hybrids on the axiom of fertilization of a single maternal cell by a single paternal cell. In his book *The variation of animals and plants under domestication*, Darwin (1868a) describes experiments carried out by C. Naudin, who considered a single pollen grain to be insufficient for fertilization of the ovule. In this eighth letter of 3 July 1870, Mendel mentions an eye ailment owing to which he was unable to start any other hybridization experiments (Correns 1905, p. 235). However, the experiment with *Mirabilis jalappa* seemed so important 'that I could not bring myself to postpone it to some later date'. The experiment yielded 18 well-developed seeds, and from these an equal number of plants, ten of which were already in bloom. 'The majority of the plants are just as vigorous as those derived from free self-fertilization. A few specimens are somewhat stunted thus far, but after the success of all the others, the cause must lie in the fact that not all pollen grains are equally capable of fertilization, and that, furthermore, in the experiment mentioned, the competition of other pollen grains was excluded. When several are competing, we can probably assume that only the strongest ones succeed in effecting fertilization.' Mendel decided to repeat the experiment, since he was unable to agree with Naudin's view that at least three pollen grains were required in order for fertilization to take place. In his ninth letter he mentions that on repeating the experiment he obtained the same results. The plants acquired from fertilization using one pollen grain could not be distinguished in any way from those produced by self-fertilization.

Mendel carried out a further experiment with *Mirabilis* to find out whether two pollen grains could participate simultaneously in fertilization (Correns 1905, pp. 239–40). He crossed varieties with various coloration, and the hybrids which first resulted from the crosses crimson + yellow and crimson + white showed no variations in their characteristic coloration. In the crimson variety a fairly large number of fertilizations were undertaken in such a way that two pollen grains were simultaneously put on each stigma, one of the yellow and one of the white variety. By this time Mendel knew the results of the crosses crimson + yellow and crimson + white. The next year he wanted to see whether a third colour would appear in addition to the hybrid colours. However, there is no further mention of this experiment in the next, the tenth letter, written more than two years later.

On 3 July 1870 Mendel wrote of 200 uniform specimens of the hybrid *Lychnis diurna* × *L. vespertina*, expecting the first generation to flower in August (Correns 1905, p. 241). After another two months, in his next letter, he reports 'a curiosity in the numerical ratios in which the male and female plants of this hybrid form'. He fertilized three flowers of *L. diurna* and planted the seeds of each capsule separately, and they produced 203 plants, of which 151 were female and 52 male. This result led him to ask: 'Is it mere chance that the male plants occur here in the ratio 52:203, or 1:4, or has this ratio the same significance as in the first generation of hybrids with varying progeny?' This led him to have doubts 'because of the strange conclusions which would have to be drawn in this case'. But the problem is easily dismissed 'if one considers that the *Anlage* for the functional development of the pistil alone or the anthers alone must have been expressed in the organization of the primordial cells from which the plants developed, and that this difference in the primordial cells could possibly be due to the ovules as well as the pollen cells being different as regards the sex *Anlage*. Therefore I do not want to dismiss the matter completely.' Here we see Mendel's reasoning in considering recently observed facts concerned with the determination of the origin of sex in plants. It was only after 1900, when Mendel's theory was further developed, that the sex ratio of 1:1 was proved. Mendel himself did not continue his investigation of the matter.

Enigmatic *Hieracium* hybrids

Mendel won renown in the scientific world mainly for his experiments with *Pisum*. His second publication, describing experiments with *Hieracium*, remained largely unnoticed even after the turn of the century. Fewer translations of it were published, compared with the *Pisum* paper (Matalová 1973, 1974). Though twentieth-century views of the *Pisum* research altered, it was never actually considered to have been erroneous, as was the case with Mendel's work with *Hieracium*. The major extent of these experiments, which are summarized in Table 8, was pointed out by Weiling (1969) and more recently Cetl (1973).

When, in 1863, Professor Makowsky published separately in book form his work *Moravian flora*, where he noted the occurrence of hybrids of *Hieracium*, an appendix to the book contained Mendel's first meteorological publication. Not only had the two authors taught together at the same school, but they shared natural scientific interests and saw each other often.

At the end of 1862 the herbarium of the Natural Science Society contained 20 000 species. The society members, including Mendel, used it for the classification of plants they picked. In the first letter to Nägeli, Mendel mentioned hybrids of *Hieracium*, saying they were very difficult to handle and unreliable as far as artificial fertilization was concerned, because of the small size and peculiar structure of the flowers. In the summer of 1866 he tried to combine *Hieracium pilosella* with *H. pratense, praealtum,* and *auricula* (Correns 1905, p. 196). Similarly, he considered crossing *Hieracium muro-*

rum with *H. umbellatum* and *H. pratense*. He obtained viable seeds, but was afraid that in spite of all precautions self-fertilization would occur. He states with satisfaction that *Hieracium* species can easily be grown in pots and set abundant seed, even if confined in a room or a greenhouse during the flowering period. In the first letter he makes the assumption that hybrids of *Geum* produce variable progeny, and writes that he believes that 'some species of *Hieracium*, if hybridized, would behave in a fashion similar to *Geum*'. His reason for supposing this was that in the previous summer he

Table 8 Mendel's experiments with *Hieracium*. (After Cetl 1973)

Combination/Year	1866	1867	1868	1869	1870	1871	1872
[1] *H. Pilosella* × *H. pratense*	C	F_1					
[2] *H. Pilosella* × *H. praealtum*	C	F_1					
[3] *H. Pilosella* × *H. Auricula*	C	F_1					
[4] *H. murorum* × *H. umbellatum*	C	F_1					
[5] *H. murorum* × *H. pratense*	C	F_1					
[6] *H. praealtum* var. *obscurum* Rchb. × *H. stoloniflorum* [*H. subcymigerum* × *H. flagellare*]	C	F_1	F_2	F_3	F_4		
[7] *H. praealtum (Bauhini)* × *H. aurantiacum* [*H. magyaricum* × *H. aurantiacum*]		C	F_1	F_2	F_3		
			BC	BC_1			
				C	F_1		
[7a] (*H. praealtum (Bauhini)* × *H. aurantiacum*) × *H. Pilosella* (B) [(*H. magyaricum* × *H. aurantiacum*) × *H. Pilosella* (B)]			C	F_1			
[8] *H. praealtum (?) echioides* × *H. aurantiacum*		C	F_1	F_2	F_3		
			BC	BC_1			
[9] *H. Auricula* × *H. Pilosella* (B)		C	F_1	F_2			
			BC	BC_1			
				C	F_1		
					C	F_1	
[10] *H. Pilosella* (B) × *H. Auricula*		C	F_1				
[11] *H. praealtum (Bauhini)* × *H. Pilosella* [*H. magyaricum* × *H. Pilosella* (B?)]		C	F_1				
[12] *H. Auricula* × *H. aurantiacum*		C	F_1				
				C	F_1	F_2	
					C	F_1	(F_2)
[13] *H. cymosum* × *H. Pilosella* (B?)			C	F_1			
[14] *H. Auricula* × *H. pratense* (var.)			C	F_1			
[15] *H. Auricula* × *H. cymosum*				C	F_1		

Table 8 *Continued*

Combination/Year	1866	1867	1868	1869	1870	1871	1872
[16] *H. XII* × *H. Pilosella* (B)] [*H. cymigerum* × *H. Pilosella* (B)]				C_1	F_1		
[17] *H. XII* × *H. Pilosella vulgare* (M) [*H. cymigerum* × *H. Pilosella vulgare*]					C		
[18] *H. Auricula* × *H. Pilosella vulgare* (M)					C	F_1	
[19] *H. Auricula* × *H. Pilosella niveum* (M)					C	F_1	
[20] *H. praealtum* (M) × *H. Pilosella incanum* (M) [*H. florentinum II obscurum* × *H. Pilosella relutinum*]					C	F_1	
[21] *H?* × *H. umbellatum* [*H. barbatum* × *H. umbellatum*]				C	F_1		
[22] *H. vulgatum* × *H. umbellatum*				C	F_1		
No. of individual operations	6	12	13	17	16	6	1

C, Crossing; F_1 to F_4, Hybrid generation; BC, Backcrossing; BC_1, Backcross generation; [ ], Nägeli's redetermination; (B), Brünn; (M), München.

had observed in seedlings of *H. stoloniflorum* W.K. bifurcation of the stem, which must be considered transitional among the piloselloids, and may appear as a perfectly constant trait.

At the start of his correspondence with Nägeli, Mendel was aware that in researching hybrids of *Hieracium* he was entering into a field in which the celebrated professor had 'the most extensive knowledge, knowledge that can be gained only through many years of zealous study, observation and comparison of the manifold forms of their genera in their natural habitat'. He therefore turned to him for advice. In the second letter he posed a concrete question. The summer before he had found a withered *Hieracium*, with the seed colour of *H. prenanthoides*, but which did not resemble any of the herbarium specimens of this type. He added that: 'our botanist declared it to be a hybrid (Correns 1905, p. 207).' Most probably the botanist in question was Professor Niessl. On this occasion Mendel also informed Nägeli that in the following year he was going to 'roam the sandy lignite country which extends eastwards from Brno for several miles'. As far as *Hieracia* were concerned, he considered the Moravian plateau to the west of Brno to be *terra incognita*. He promised Nägeli that if he found 'anything noteworthy' he would send it immediately.

In the introduction to his letter dated 6 November 1867, Mendel apologizes that his 'project of studying the *Hieracia* of this locality in their natural environment' had unfortunately 'been carried out only to a very limited extent' (Correns 1905, pp. 211–12). He was prevented from doing more by a lack of time and by no longer being quite fit enough for botanical field trips, since heaven had blessed him with an excess of weight which became very noticeable 'during long travels afoot'. Instead of wild forms of *Hieracia* he was sending 'some material from my garden'. In the first place he mentions the hybrid *Hieracium praealtum* × *H. stoloniflorum*, both parental forms of which he encloses, having found them in the surroundings of Brno. Mendel also gives a brief description of his method of artificial pollination. In the material *Hieracium praealtum* he tried 'to inhibit pollen development or at least to prevent it from reaching the stigma'. More than half the involucral bracts of the young, incompletely developed heads were cut off, and all the small flower buds, except for 10–12, removed; the latter were split open in several places with a fine needle so as to expose the style completely.

Mendel immediately performed fertilization with pollen from *H. stoloniflorum*, and later repeated the procedure. In spite of this drastic treatment he obtained four well-developed seeds which produced plants the next spring. Three of them totally resembled *H. praealtum*, whereas the fourth showed considerable divergence, and doubtless represented the hybrid form *Hieracium praealtum* × *H. stoloniflorum*. Mendel believed that 'at least once in four cases self-fertilization had been prevented by the procedure described; the latter seems to be useful, although it is most complicated, and strains and tires the eye'. Before this Mendel had verified the method of artificial fertilization for crossing *Cirsium* plants.

The *Hieracium* hybrid achieved was a healthy and luxuriant plant. Mendel compared it with the hybrids *Hieracium pilosella* × *H. auricula* and *Hieracium pilosella* × *H. praealtum* produced by F. Schultz using artificial fertilization; the latter were described as sterile. He obtained from it 'a number of good seeds' and therefore paid 'some attention' to it. In a total of 14 heads he counted 1044 flowers, and of these 624 furnished seeds which were to all appearances good. The majority were not viable, and only 156 plants developed from them. To this Mendel adds: 'Whether they will retain the characteristics of the hybrid or whether they will show variations will be determined by next year's observations.'

Mendel waited expectantly to see the results of experiments which were supposed to answer the question of whether *Hieracium* hybrids were also variable. In the same letter he gave further remarkable data. 'On average the number of flowers per head is (according to 14 different counts) 39 in *Hieracium praealtum*, 145 in *stoloniflorum*, and 75^2 in the hybrid.' He adds that 'the latter number thus does not represent the mean, which should be 92; it is, however, almost the exact geometric mean, since 75^2 is approximately 39×145' (Correns 1905, p. 214). In the following year Mendel was intending to observe this hybrid plant further, and expected to find hybrids in the combination *Hieracium pilosella, Hieracium stoloniflorum*, and *Hieracium aurantiacum*. He enclosed in his letter another species which he found in Moravia and which he does not identify. He makes particular mention of 'bifurcate types, which are probably of hybrid origin, in association with *Hieracium praealtum* and *Hieracium pilosella*' in his fourth letter to Nägeli, dated 9 February 1868. Again he turned to Nägeli for advice on where he might buy 14 species of *ArcHieracia* he required (Correns 1905, p. 219).

On 4 May 1868, Mendel thanks Nägeli for sending seeds, and looks forward to sending the flowers he has promised, adding: 'I shall do my utmost to produce all the possible hybrids among the species, and if they should be fertile, their progeny will be observed for several generations' (Correns 1905, p. 220). Immediately after this Mendel writes that on 30 March he was elected head of the monastery for life. Then he moves on to the subject of his experimental plants, saying that they have overwintered well. Most of the piloselloids and some of the *ArcHieracia* already showed flower buds. He denotes as successfully produced hybrids *Hieracium auricula* $\times$ *H. pilosella*, *H. praeatlum* (Bauhini) $\times$ *H. aurantiacum*, and probably also *H. pilosella* $\times$ *H. auricula*, adding that about 100 of the autumn seedlings of last year's hybrid *H. praealtum* $\times$ *H. stoloniflorum* have survived the winter, and remarking that these very small plants are uniform in the structure and hairy covering of the leaves, and resemble the hybrid seed plant. He awaits their further development 'with some suspense'.

Shortly afterwards, on 12 June 1868, Mendel informed Nägeli of his experiments with *Hieracium* (Correns 1905, pp. 221–4). He complains that though he had told the gardener to 'handle with great care' the plants Nägeli sent him, after returning from a long tour of inspection he had found that about half the potted plants had died, probably as a consequence of excessive watering. Only five of the species sent had survived. Nine species emerged from seed. Mendel immediately adds that he has obtained 112 offspring from the hybrid *Hieracium praealtum* $\times$ *H. flagellare*, and that 'all the plants are alike in the essential traits'. This was a major difference from the behaviour of variable bastards, whose progeny segregated. In the same year Mendel obtained a further five hybrid plants from various combinations of the parental species, but they were not yet developed far enough for him to be able to speak of success.

Mendel's seventh letter to Nägeli was written on 15 April 1869, two months before he reported his *Hieracium* experiments at the meeting of the Natural Science Society. He sent Nägeli the promised 'hybrids of *Hieracium, Cirsium,* and *Linaria,* obtained by artificial fertilization'. The *Hieracium* hybrids are mentioned first. Out of 14 samples, four were parental species. Number 14 is the hybrid form 112, seedlings of which had in the previous year been reminiscent of the hybrid numbered 11, i.e. *Hieracium inops* N., *H. flagellare* + *florentinum subcymigerum* C. In the same letter Mendel mentions having had success in obtaining hybrids of *ArcHieracia,* and supposes 'this year's sowing will yield results'. Out of the samples which Nägeli had sent he had been able to make use of only six, which overwintered well, so that he was able to extend his experiments 'exactly according to the plan which you, honoured friend, were kind enough to send me'. This was a programme of crossing of species which Nägeli had sent with his letter of 11 May 1868, and which was published in 1905 by Correns, along with Mendel's letters to Nägeli (Correns 1905, pp. 218–21). In this letter Mendel offers Nägeli proof of variable hybrids in *Cirsium, Geum,* and *Linaria,* and at the same time points out a new enigma. According to him 'hybrids of *Hieracium* show, strangely enough, very different behaviour in the production of their progeny'. He reported on this mystery in more detail at the meeting of the Brno Natural Science Society on 9 June 1869.

In 1865 it was stated in the *Proceedings* of the Natural Science Society that Mendel gave lectures on his research into *Pisum,* and that these lectures were published in an appendix to the *Proceedings.* In the same year the lectures were reported in the daily press. In 1869 the report on the meeting, presided over by Mendel himself, merely states: 'The worthy prelate Gregor Mendel showed several *Hieracium* hybrids obtained through artificial fertilization, together with the original forms of parental plants in the living state in pods. He gave a brief description of one compared with the other (i.e. the hybrids and the parents), and at the same time gave some information on their behaviour in successive generations.'

This was not, then, a lecture, but merely a report on his research, just as other members, such as Niessl or Makowsky, reported on plant hybrids. Their reports were published in the *Proceedings.* Niessl (1867c) published an eight-page contribution in small print in the first part of the *Proceedings,* where reports of the monthly meetings were to be found. Mendel's contribution on *Hieracium* was, however, again in the second part of the *Proceedings,* as had been the *Pisum* paper.

The introduction to the *Hieracium* paper gives a brief account of how Mendel succeeded in obtaining only the following six *Hieracium* hybrids, and only from one to three specimens of each:

Hieracium auricula + *H. aurantiacium*
Hieracium auricula + *H. pilosella*
Hieracium auricula + *H. pratense*
Hieracium echioides + *H. aurantiacum*
Hieracium praealtum + *H. flagellare* Rchb.
Hieracium praealtum + *H. aurantiacum.*

The first form in all cases was the maternal plant. Mendel paid most attention to a description of the method of artificial fertilization of the minute flowers with their peculiar structure. He gives a clear account of how difficult it was to achieve artificial fertilization by the method he had brought to perfection, and then goes on to describe the extraordinary profusion of distinct forms, such that 'no other genus of plants can be compared with it'. Some of these forms may be taken as type-forms, while all the rest represent intermediate or transitional forms by which the type-forms are connected together. The difficulty in separation and delineation of these forms, according to Mendel, provoked major controversies among experts, who had not as yet come to a definite conclusion.

A key problem was the elucidation of the intermediate or transitional forms. A matter of dispute was 'whether and to what extent hybridization plays a part in the production of this wealth of forms'. In the next section of his report Mendel gives a brief account of his views on the matter. Some researchers claimed that hybridization had a far-reaching influence, while others cited the Swedish botanist Fries, supposing that the richness of forms in *Hieracium* had nothing to do with hybridization. In the middle was the view that hybrids between species in a wild state are not rare, but cannot be of major importance since they are of short duration. The causes of this were, in Mendel's view, partly their restricted fertility and partly the experimental finding that hybrid self-fertilization is always prevented if the pollen of one of the parent forms reaches the stigma. It was therefore said that *Hieracium* hybrids could not maintain themselves as fully fertile and constant forms when grown near their progenitors. An attempt at the explanation of the origin of the middle form of the hybrid *Asplenium adulterinum* Milde was discussed in Brno by Niessl (1868*b*) at the meeting of the Natural Science Society, and a specimen of this hybrid was, according to the published report, sent 'for observation to Vienna, Prague, and to Dr J. Milde in Breslau'. Towards the end Niessl says that Brno 'took over the cultivation of this hybrid Herr Prälat Mendel'. Treating the origin of middle forms in *Hieracium* two years later Mendel did not mention the middle form of *Asplenium*.

The question of the origin of constant intermediate forms had, according to Mendel, attracted the attention of natural scientists *inter alia* because 'a famous *Hieracium* specialist' had, in the spirit of Darwinian teaching, defended the view that these forms are to be regarded as 'arising from the transmutation of lost, still-existing species'. Mendel had Nägeli in mind, though he does not mention him by name. He knew well enough that the problem could only be clarified through 'an exact knowledge of the structure and fertility of *Hieracium* hybrids and the conditions of their offspring through several generations'. Mendel added: 'If by the experimental method we can obtain a sufficient insight into the phenomenon of hybridization in *Hieracium*, then by the help of the information that has been collected in respect of the structural interrelations of the wild forms, a satisfactory judgement in regard to this question may become possible.' Only then did Mendel venture to relate the very slight results which he had obtained in his experiments, under the following four points:

1. The striking phenomenon was that the forms hitherto obtained by similar fertilization are not identical. 'Some forms of the hybrid *Hieracium praealtum* × *H. aurantiacum* and *Hieracium auricula* × *H. aurantiacum* and *Hieracium auricula* × *H. pratense* sometimes present an intermediate structure and sometimes are so near to one of the parent traits that the corresponding trait has receded considerably or almost evades observation.' Mendel's conclusion was that 'we have here only single terms in an as yet unknown series which may be formed by the direct action of the pollen of one species on the egg-cells of another'.

2. He divided hybrid forms into five categories according to their ability to be fertilized:

fully fertile—*Hieracium echioides* + *H. aurantiacum*
fertile—*Hieracium praealtum* + *H. flagellare*
partially fertile—*Hieracium praealtum* + *H. aurantiacum*
        —*Hieracium auricula* + *H. pratense*
slightly fertile—*Hieracium auricula* + *H. pilosella*
infertile—*Hieracium auricula* + *H. aurantiacum*

Mendel noted that among the seedlings of the partially fertile hybrids *Hieracium praealtum* + *H. aurantiacum* there was one plant which possessed full fertility.

3. The offspring produced by self-fertilization of hybrids did not vary, but agreed in their traits both with each other and with the hybrid plant from which they were derived. Mendel added the offspring of four combinations of hybrids, from which 14 to 112 plants flowered.

4. In fully fertile hybrids the pollen of the parental plant was not able to prevent self-fertilization, though it was applied in great quantity to the stigma protruding through the anther-tubes when the flowers opened. In one experiment with a partially fertile hybrid *Hieracium praealtum* × *H. aurantiacum* Mendel found that those flower-heads in which pollen of the parent type or of some other species had been applied to the stigmas developed a considerably larger number of good seeds compared with those which had been left to self-fertilization alone. A microscopic examination showed that a large proportion

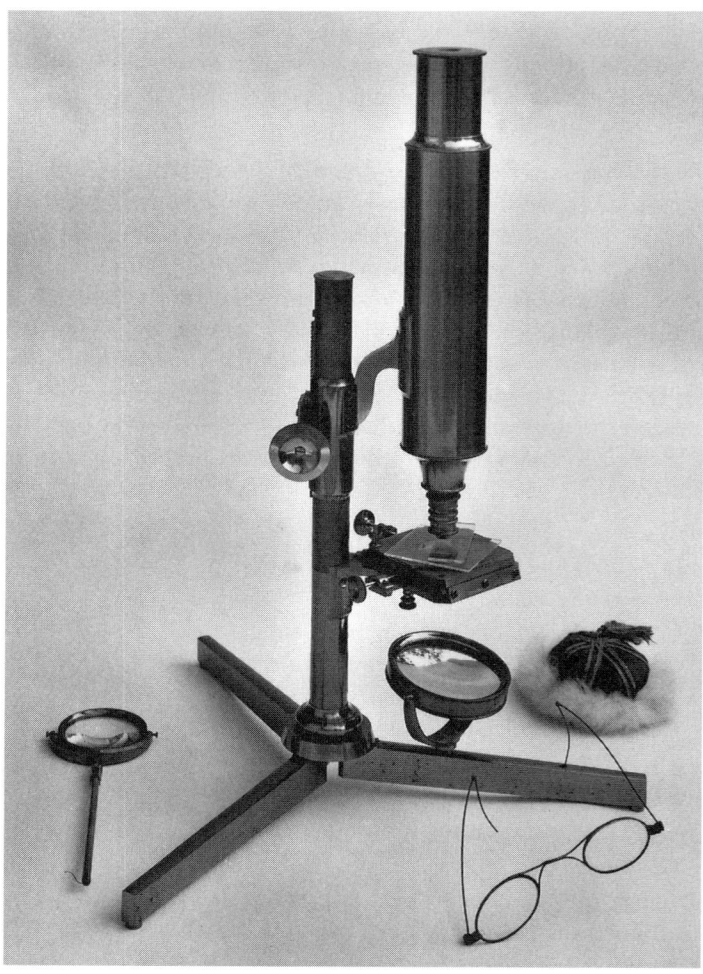

Fig. 5.11 One of Mendel's microscopes, and his glasses.

of the pollen grains of this hybrid had a defective structure, and thus their ability to fertilize hybrids with their own pollen was reduced. The frequent occurrence of seeds in plants of the fully fertile species in wild formation which did not develop a single good grain was explained by Mendel as being due to fertilization by foreign pollen, probably transferred by 'the industrious *Hymenoptera*'. In his conclusion Mendel expressed his satisfaction that Nägeli had sent him further species of *Hieracium*, so that he would be able to extend his experiments and in the following year would be able to report 'something more by way of extension and confirmation of the present account'.

In 1869 Mendel pointed out that in his 1865 lecture he had remarked on the fact that there were also hybrids whose progeny was not variable, and that, for example, according to Wichura the hybrids of *Salix* reproduce themselves like pure species. He was of the opinion that *Hieracium* was a similar case. After four years he was unable to demonstrate uniformity of hybrid traits and segregation in the hybrid progeny of parental traits His statement that 'in *Hieracium* according to the present experiments exactly the opposite phenomenon seems to be exhibited' is frequently repeated in the literature, often along with the view that the *Hieracium* experiments were a failure. But Mendel concluded his report very cautiously, as follows: 'Whether from this circumstance we may venture to draw the conclusion that the polymorphism of the genera *Salix* and *Hieracium* is connected with the special condition of their hybrids is still an open question, which may well be raised but as yet not answered.'

So far little attention has been paid to the fact that the *Hieracium* paper took a quite different form to that of the *Pisum* paper. In 1869 Mendel wrote to Nägeli that he had prepared preliminary experiments with this genus, and in the very next year he reported having obtained the first results, pointing out, however, that 'the work scarcely extends beyond its first inception'. He added perspicaciously: 'But the conviction that the prosecution of the proposed experiments will demand a whole series of years, and the uncertainty whether it will be granted to me to bring them to a conclusion have determined me to make the present communication.' Mendel's premonition proved well founded. His new post as abbot did not allow him to concentrate on his research.

Mendel stated wrongly in 1869 that Wichura had proved the existence of constant hybrids in *Salix*. In fact Wichura (1865) had only reached the conclusion that the progeny of self-fertilized *Salix* hybrids were 'the same or similar'. Wichura believed in the notion of blending inheritance and in the light of this belief and of Darwin's theory he supposed it possible for new species to be created through hybridization. Olby (1986) has shown that Mendel may have been under the influence of Kerner von Merilaun (1860), who published a study on willows, referring to the valuable experiments performed by Wichura. According to Kerner von Merilaun's ideas these were constant hybrids.

After his report on *Hieracium* experiments in 1869, Mendel continued his experiments with this genus, describing the results in his letters to Nägeli. To this day we do not know whether he sent the professor a copy of the *Hieracium* paper. On 3 July 1870, he mentioned neither the lecture nor any publication (Correns 1905, pp. 229–37). However, the opening paragraph of his letter complains of his having been very busy with economic matters. He states: 'Only in the past few days have I been master of my own time and in a position to resume my favourite occupation, which I had to discontinue about the end of June last year owing to an eye ailment.' Because of the difficulty of artificial fertilization of the tiny flowers of *Hieracium*, he made himself an apparatus for illuminating them, made up of a mirror with a convex lens, without suspecting what damage might be done with it.

During May and June Mendel made experimental crossings of *Hieracium auricula* × *H. praealtum*, and a peculiar fatigue and exhaustion of the eyes reached a serious degree, so that he was obliged to interrupt his work, and it was almost a year before he was able to continue. On 3 July 1870, he sent Nägeli 'some living *Hieracium* hybrids', again with the parental forms 'where necessary'. He recalled some of the data contained in the published *Hieracium* paper. A great enthusiasm for continuing the experiments in that year is apparent. But Mendel regretted, in spite of numerous attempts, not yet having succeeded in obtaining a single hybrid of certain species of piloselloids using fertilization with foreign pollen. He explained this as being due to the latter's inability to overcome the influence of the autogenous pollen.

Mendel was more easily able to achieve fertilization with foreign pollen in the varieties of *H. praealtum*. He repeatedly showed that *H. auricula* is a completely reliable experimental plant for fertilization. In the previous year he had fertilized more than a hundred heads of this species with the pollen of *H. pilosella, cymosum*, and *aurantiacum*. About half of the flowers dried up because of injuries suffered, and only two to six seeds were obtained from each of the surviving plants. In the last combination 98 plants survived, and Mendel expected them to flower the next month. He enclosed in his letter a form designated No. XII, which he could neither name nor classify. He found it in large numbers in forest clearings. His very first experiment with fertilization using the pollen of *H. pilosella* was completely successful. Among the 29 plant hybrids he noted very striking variations representing types ranging from a close resemblance to one parental species to a close resemblance to the other, so that no one would take the extreme forms for siblings if he found them growing in the wild. Mendel asked for Nägeli's opinion on form XII, which he considered the best experimental plant for fertilization with foreign pollen, since a fairly large number of hybrids could be obtained from it quite easily. At the same time he emphasized the methodological element of the research, which he had already demonstrated in the *Pisum* experiments: 'The variations occurring among hybrid individuals can only be interpreted in cases where a fairly large number of hybrids is obtained from the same fertilization.' In his eighth letter Mendel also mentioned the second and third generations of three hybrid forms, and did not observe variation. His conclusion was that 'these hybrids of *Hieracium* exhibit a behaviour exactly opposite to that of *Pisum*'.

In 1869 Mendel anticipated in *Hieracium* a constant hybrid form, and was disappointed to observe variability in the hybrid generation and uniformity of traits in the progeny of hybrids, instead of uniformity of the hybrid generation and variability of traits in the hybrid progeny. This behaviour differed radically from that of the variable hybrids he had found not only in *Pisum*, but also in a whole series of other plant species. Nor did the observed variation in the hybrid generation conform with the idea of constant hybrids. He expressed his point of view as follows: 'Evidently we are here dealing only with individual phenomena that are the manifestation of a *higher, more fundamental law*' (Correns 1905, pp. 231–2).

In his research into hybrids of *Hieracium*, Mendel became more and more interested in the problem of the differences in the fertility of hybrids. He explained to Nägeli that, 'If, in flowers of partially fertile hybrids, the stigmas are covered with the pollen of other, not too distantly related species, they always produce more seed than when kept isolated and dependent upon self-fertilization; that this is due exclusively to the action of the foreign pollen can easily be demonstrated by cultivation of seeds. Careful isolation is, however, not necessary for completely fertile hybrids. Experiments with *H. praealtum*(?) + *H. aurantiacum* have shown that foreign pollen, even that of the two parental species, may be put upon stigmas in quantity without interfering with self-fertilization. All seeds produce the original hybrid form' (Correns 1905, p. 234).

Three months later, on 27 September 1870, with his ninth letter Mendel sent Nägeli 29 enigmatic hybrids denoted Number XII—*H. cymigerum* + *H. pilosella* (var. Brünn), asking whether this might not be *H. poliotrichum* Wim. At the same time he gave details of the hybrid *H. auricula* + *H. aurantiacum*, as he had promised in his previous letter. Some of the 84 specimens of this hybrid had died, while others had not yet flowered. Among the rest he observed considerable variation. 'Each hybrid trait appears in a certain number of variants which represent different transitional stages between one ancestral trait and the other. It seems that the variants of the different traits may occur in all possible combinations. This seems probable because in the available hybrid plants the assortment of variants of the traits is exceedingly diverse, so as hardly ever to be the same in any two instances. If this assumption is correct, many hundreds of possible hybrid types should result because of the large number of differences between *H. auricula* and *H. aurantiacum*. The observed number of hybrid types is too small in the case of parental species as distant as these to determine the true facts' (Correns 1905, p. 238). Mendel was of the opinion that success would be attained more easily with the hybrid *H. auricula* + *H. pilosella vulgare*, from which he hoped to obtain about 200 specimens the next year. In this letter, Mendel also mentions a remarkable proportion in the varying fertility of hybrids of *H. auricula* + *H. aurantiacum*. About a quarter of them were completely fertile, one half partially fertile, and one quarter sterile. He never mentioned this phenomenon again. Nägeli replied on 30 April 1871, apologizing for the delay, which had been due to a visual disorder. He advised Mendel to simplify his method of castrating the flowers, and enclosed his own photograph. This was a friendly gesture on the part of the professor.

Mendel next wrote about his experiments, in the tenth letter, dated 18 November 1873 (Correns 1905, pp. 242–7). He regrets in the opening paragraph 'having had to neglect my plants and bees so completely', adding 'since I have a little spare time at present, and since I do not know whether I shall have any next spring, I am sending you today some material from my last experiments in 1870 and 1871'. He enclosed five determined hybrid specimens with their various fertilities designated; they had been grown in isolated beds, and thus did not disturb each other by the growth of their stolons. It is remarkable that 84 hybrids from sample No. 2, 25 from sample No. 3, and 35 from sample No. 4 were all uniform. But around 90 hybrids from sample No. 5 were very variable. For the first time Mendel described in one hybrid generation uniformity and in another one variability.

Along with the tenth letter, Mendel sent Nägeli three completely fertile *Hieracium* hybrids which were transplanted to separate beds for further experiments. Two of the variants sent to Nägeli were closer to *H. aurantiacum* and could not be distinguished with certainty. The third resembled the parental species *H. auricula* more closely. Since these variants were completely fertile, they were, according to Mendel, to serve for studies of later generations. He adds: 'But these experiments were not executed.' But in his view it could probably be assumed with a good deal of certainty that the progeny originating from the self-fertilization of these variants would not be subject to the same variation shown in the original hybrids. It can thus be seen that he assumed some segregation.

In conclusion Mendel mentions Gärtner's view of the prepotency of the parental pollen over the pollen of hybrids in several species of plants. He performed a simple back-crossing experiment. The partially fertile hybrid *H. praealtum* + *H. aurantiacum* was fertilized with the pollen of *H. aurantiacum*. In the following year two types of plants were obtained from the seeds of the artificially fertilized heads: some completely resembling the hybrid seed plant, and others much closer to *H. aurantiacum*. He did not wish to deny here Gärtner's assertion that the pollen of the hybrid is ineffectual in competition with the parental pollen, since he had no evidence of this. He was seeking an explanation in the peculiar structure of the florets and the reaction of the organs of fertilization.

In another part of his letter Mendel points out the effect of the environment, which could cause 'sexual weakening or complete sterility, wherein the male organ always suffers first, as in animals in captivity'. He mentioned from his own experience the example of the fertility of *H. pilosella incanum* (Fig. 5.12), which could not adapt itself very well to the local climate. In 1870 the May and June flowers were completely sterile, but partially fertile in the following year, and towards autumn individual heads

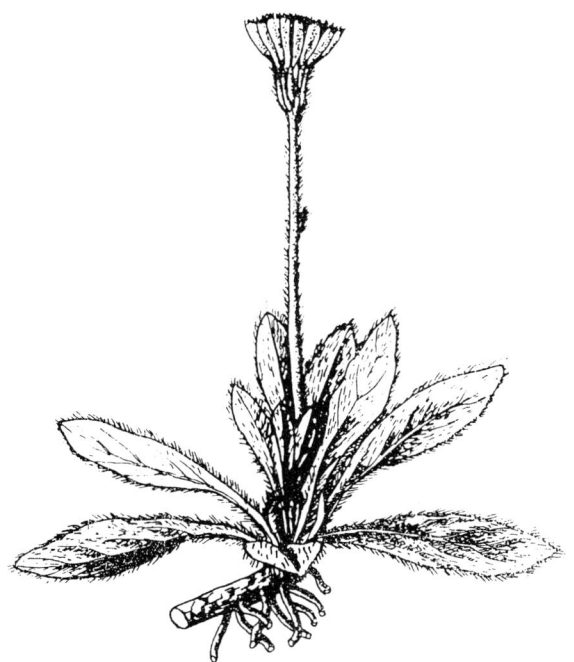

Fig. 5.12 Mendel's last experimental plant, *Hieracium pilosella* L.

appeared to be completely fertile. Mendel explains this as being due to the effect of a drought and high temperature acting to the detriment of pollen quality. He has a theoretical explanation to hand: 'If this were actually the case, the naturally occurring hybridization in *Hieracium* should be ascribed to temporary disturbances which, if they were repeated often or became permanent, would finally result in the disappearance of the species involved, while one or another of the more happily organized progeny, better adapted to the prevailing telluric and cosmic conditions, might take up the struggle for existence successfully, and continue it for a long stretch of time, until finally the same fate overtook it' (Correns 1905, p. 247).

Mendel associates the observed differences in fertility of hybrids under the influence of the environment with the origin and decay of these hybrid species in nature, adding another remarkable explanation: 'Species of which numerous hybrids have been shown to exist I would consider decrepit or would at least assume to be well past their prime.' He gives as an example *H. auricula* and *H. praealtum*. In this way he hints at the existence of decrepit species, ascribing a decisive role to the action of natural selection. This was Mendel's last standpoint regarding the enigmatic *Hieracium* hybrids. It was shared by the greatest expert on the subject, Professor Nägeli. The last information the academic in Munich received from the researcher in Brno was of exceptional interest to him. According to Correns' account (p. 247) he sent a letter to Mendel on 23 June 1874, and shortly afterwards another one. He was particularly interested in the question of whether the polymorphic *Hieracium* hybrids which Mendel had mentioned in his ninth letter had the same mothers and the same paternal plant.

Mendel never replied to Nägeli's last two letters. To this day we do not know why not. Thus ended a remarkable correspondence between a solitary pioneer of a new line of research and a leading European scholar, who published several hundred pages on the subject at around the same time. Both of them considered a question which was coming to the forefront of natural scientists' interest, the problem of the development of organic forms in connection with hybridization, generation, and heredity. But each of them had an entirely different point of departure, and used in his experiments different methods.

THE GENESIS OF A THEORY

The first people to study Mendel's work after its rediscovery in 1900 were biologists, and later those who were called geneticists. The latter soon generalized the theory contained in the *Pisum* paper as Mendel's laws of heredity. Later, historians of science began to look for his forerunners in the field of plant-hybridization research. Even later, philosophers and socio-logists of science concentrated on specific aspects of the interpretation of Mendel's theory in connection with a reconstruction of his experiments and of the thought processes involved in making his discovery.

One can gain a new insight into Mendel's scientific achievement by a study of scientific revolutions. T.S. Kuhn (1962) explained the birth of a scientific discovery using the example of how Copernicus' theory originated. According to him the process of scientific discovery passes through the sequence: normal science, crisis, revolution, normal science. He called nor-mal science a paradigm, which in the course of the development of know-ledge is replaced by another paradigm which responds better to the ques-tions the previous one could not find an answer to. Other methodologists have in their study of the development of science laid more emphasis on the internalist's creative process, which operates to provide a rational explana-tion for the growth of objective knowledge, or an externalist approach, dealing with the influence of economic, social, and cultural factors on research activity. Later, philosophers of science offered alternative theories to explain the origin of scientific discovery, mostly relating to an interaction of the two approaches. Research into the origin of scientific discoveries was first applied to the fields of astronomy and physics, and later to chemistry. As regards the biological sciences, the first to be subjected to this kind of scrutiny was Darwin's Theory of Evolution.

Mendel's own scientific achievement attracted more attention from the point of view of the study of the origin of his discovery, chiefly after the publication of new information about him after 1965. New stimuli were created by an analysis of two fragments of the records of Mendel's research into plant hybridization, for which an explanation was offered by J. Heimans (1969, 1970). Olby (1966, 1971, 1979) suggested examining Mendel's discovery in the broader context of the mid-nineteenth-century knowledge of plant hybridization and plant physiology. He also drew attention to the need for a more detailed investigation into Mendel's symbolic notation of traits in the *Pisum* experiments. His remarks encouraged others to make a reassess-ment of the significance of Mendel's research, especially of the *Pisum* experi-ments. One of these was Kalmus (1983), who wanted to investigate the origin of Mendel's theory from the viewpoint of stochastic philosophy, with which Mendel was well acquainted.

Mendelian mythmaking

From the 1980s sociologists also took up Olby's challenge and began to consider Mendel's research. Brannigan (1979, 1981) gave Mendel as an example of a purely empirical researcher into plant hybridization, where hybridization was considered an explanation for the process of the evolution of organic forms. Callender (1988) even reached the conclusion that Mendel, contrary to the accepted opinion, 'did not enunciate a law of segregation which he thought might prove applicable to all the plant hybrids with which he experimented'. Shortly after this Corcos and Monaghan (1990) made a radical departure from the traditional view in supposing that 'Mendel was not a theorist, but an empiricist or an empirical worker, as he defined himself in a letter to de Vries'. There are two errors implicit in this statement. First of all Mendel, in his second letter to Nageli—*not* to de Vries—was explaining the manner in which simple and combination series were derived, admitting that in this case he could be considered an empirical worker. But in the next line he adds: 'If I then extend this combination of simple series to any number of differences between two parental plants, I have entered the rationale domain.' Mendel points to the theoretical generalization, concluding that 'the behaviour of each pair of differing traits in a hybrid association is independent of all other differences in the two parental plants' (Mendel 1866*b*, p. 23). Corcos and Monaghan (1990), in their attempt to present Mendel as a purely empirical researcher, also admitted that he was guided 'by certain basic philosophical ideas derived from his unique interests and background'. Recently Corcos and Monaghan (1993) in their book *Gregor Mendel's experiments on plant hybrids — a guided study* came to the conclusion that Mendel's objective in his research was to find the empirical laws that describe the formation of hybrids and the development of their offspring over several generations, and not the laws of inheritance that are generally credited to him. This view was rejected by Falk and Sarkar (1991), who explain that 'Mendel was studying and reporting on inheritance simply because hybridization is conceptually inseparable from inheritance'. Defending their view Monaghan and Corcos (1993) wrote that they had scrutinized Mendel's paper carefully and analysed it 'sentence by sentence again and again for the past 9 years'. Acknowledging the historical context of Mendel's experiments with intensive breeding activity in Moravia in the first half of the nineteenth century, they nevertheless concluded that the breeders in Moravia were not looking for the laws of heredity and 'like most of the breeders of plants or animals before and after them, were not concerned with the theoretical underpinning of their practice'. The supposed lack of interest in the theory of heredity seems at odds with the great endeavours of the Moravian breeders to disclose the theory of breeding practice, culminating in the formulation of the 'genetical laws' in 1819 and the research question of heredity by Napp in 1837, to which attention was recently paid by Orel and Hartl

(1994). On the one hand Corcos and Monaghan say that Mendel was not a theorist while on the other they attribute to him basic philosophical ideas and remark that he was a brilliant experimenter, which is a contradiction.

An extreme interpretation of Mendel's research was offered by Di Trocchio (1989, 1991) in a monograph published in Italian and later in his paper 'Mendel's experiments: a reinterpretation'. According to Di Trocchio (1991), 'the original aim of Mendel's experiments was to check whether or not new species could be produced by hybridization'. Ignoring Mendel's description of the manner in which individual experiments were carried out, Di Trocchio states that he 'began to cross 22 varieties in all possible ways, in order to have the largest possible quantity of data and the highest probability of observing the birth of a new species'. Elsewhere Di Trocchio states that 'the best and easiest way of analysing the collected data [of Mendel] would have been the forked-line system, which allows the progressive disaggregation of complete data (e.g. the numerical results of poly-hybrid crosses) in order to find the scores of single characters'. The paper argues on this basis that Mendel never carried out his monohybrid experiments in the garden, but 'only on the pages of his notebooks'.

This view is inconsistent not only with Mendel's emphasis on describing in the *Pisum* paper the methods he used in his scientific work, but also with what he states in his letters to Nägeli. The conclusion in Di Trocchio's paper is, however, that 'we must still consider him [Mendel] the father and founder of genetics'.

The interpretation of Mendel's research as merely empirical, in the tradition of plant hybridizers, was rejected by Iris and Laurence Sandler (1985), who stressed that 'Mendel tackled and solved the problem in purely genetic terms, and produced a correct and amazingly complete answer, but to an as yet unformulated question.' MacRoberts (1985), analysing the neglect of Mendel's paper in the nineteenth century, mentioned the critical opinions of Olby, Brannigan, Corcos, and Monaghan, according to which Mendel was applying normal science to typical mid-nineteenth-century problems. But he considered it paradoxical that nearly all those commentors agreed that Mendel's approach was 'brilliant, rigorous, and systematic'. According to MacRoberts 'normal science of this calibre is simply not ignored'. Soon afterwards Bowler (1989, p. 103), in his monograph *The Mendelian Revolution*, reached the conclusion that 'a number of historians have begun to argue that the orthodox interpretation of his [i.e. Mendel's] research programme has been influenced by the geneticists who desire to see Mendel as a pioneer student of their discipline.' According to his view, 'the traditional image of Mendel is a myth created by the early geneticists

to reinforce the belief that the laws of inheritance are obvious to anyone
who looks closely enough at the problem'.

Theory in the *Pisum* experiments

M. Grmek (1981) drew attention to the possibility of a quite different
approach to the genesis of the great discoveries in the natural sciences.
He set out from the definition of a discovery as 'a statement, characterized
by a novelty, where novelty is understood as what cannot be deduced but
must be constructed within the sphere of logic, or observed in the course
of experimental sciences, whether physical or biological'. Grmek was led
to reject the often repeated idea that Mendel's discovery was a typical
example of inductive empiricism. On the contrary, in his view this discovery was acquired while the experiments were being carried out. The
theoretical nature of Mendel's research was also stressed by F. Ayala
(1988), in considering the nature of scientific discovery in general. He
emphasized that Mendel's discovery was 'a far cry from inductive general-
izations derived by role extension from observation'. According to Ayala,
'not only was Mendel extremely imaginative in the careful design of his
experiments, and in developing a theory to account for their results, but
he had a remarkable insight into the nature of scientific enquiry.' A new
methodological view of Mendel was also taken by M. Campbell (1985),
who examined separately the questions of theory generation and theory
acceptance.

In Chapter 2 Mendel's predecessors in this research have been considered. Gavin de Beer (1965) recalled Newton's phrase 'If I have seen further it is by standing on the shoulders of giants'. Accepting Fisher's (1936)
idea that Mendel must have thought out his system of particulate inheritance before he subjected it to experimental test, de Beer came to the conclusion that 'Mendel had no precursors at all to help him in his discovery of
a principle on which is founded the whole science of genetics'. In this context he believed that Mendel knew nothing about the experiments by
Sageret (1826), Naudin (1863), and Colladon (1822). We can now say
that Mendel was aware of the experiments of the first two authors. If 'A
predecessor is a man who, at an earlier date, makes a discovery which his
successor is able to enlarge to a general principle of universal validity',
then, according to Catcheside (1966), Mendel had no predecessors.
Considering Newton's phrase, Merton (1973) added that Newton had to be
a genius 'to find the way up to those shoulders'. The same goes for Mendel.
In his search for the resolution of the problem he was faced with, he set up

a demanding research programme by means of which he considered it to be soluble. He concentrated entirely on research, and the result was a new theory, which since 1900 has been described in different ways. Scholars who have looked into Mendel's theory have had different levels of knowledge at their disposal. But one must take into account the different periods in which these evaluations were made, each with its own level of scientific knowledge.

No attempt to explain the internal creative act of a discoverer can take in his substantial achievements to their full extent. Thus in the literature on the subject one finds throughout the development of knowledge quite different interpretations of many different discoveries. The evaluation of the achievements of Lamarck may serve as an example here. Burkhardt (1977) showed that Lamarck's chief merit was not the concept of the heredity of acquired characters, as had been stated repeatedly in previous works, but a new system of inorganic and organic sciences, opening up a new approach to the study of the origin and development of life. Similarly, one may point to quite different explanations of Mendel's discovery since the 'rediscovery' in 1900.

An analysis of Mendel's reasoning as he carried out his experiments into plant hybridization and built up his theory gives a new insight into the generation of a theory, arising not only out of the *Pisum* experiments, but also out of his experiments with other plant species.

When, in 1964, J. Kříženecký (1965b, p. 72) was preparing the plan of activities of the Mendelianum in Brno, he denoted Mendel's idea a 'spiritual mutation'. By this he meant a sudden moment of enlightenment in Mendel's reasoning, which according to Grmek (1981, p. 25) C. Bernard described as 'a subconscious maturation process marked by assimilation of new data', based on the observations of the researcher and leading through hypothesis to experiment. In the case of such observation a certain theoretical framework is involved. One can set out from this picture when trying to reconstruct the *Pisum* experiments and seeking to create a new theory of the course of individual experiments. In agreement with Grmek, one can see a sequence of theoretical views in Mendel's research; these are in a state of flux, producing experiments and being modified according to their results.

After returning from university, Mendel arrived at a recognition of a set of phenomena surrounding hybridization for which no satisfactory explanation existed. This was the basis of his long-term research programme, in whose design he used knowledge gained during his studies in Brno, and knowledge gained particularly later at the university in Vienna. He realized

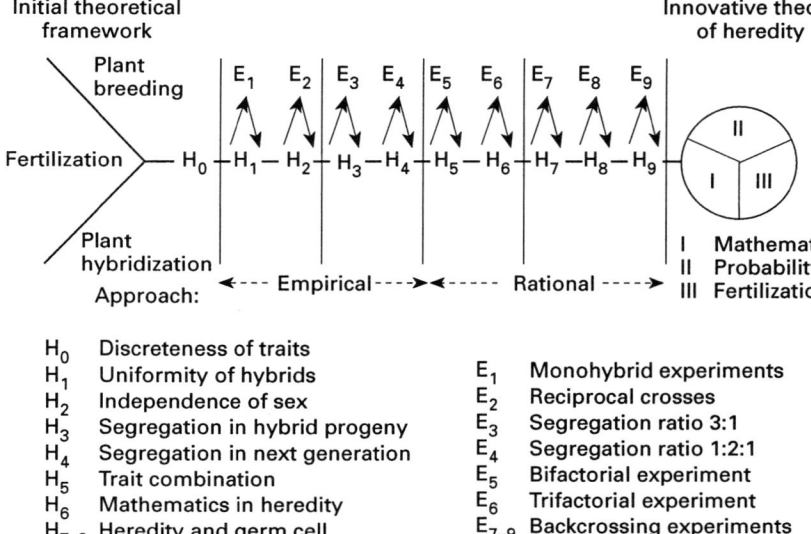

Fig. 5.13 Scheme of Mendel's experiments (E_i) and successive hypotheses (H_i) combining the traditions of plant breeding and plant hybridization with mathematics and probability to account for the numerical proportions of traits observed among offspring in successive generations.

that it was the kind of problem which lent itself to experimental solution, and created a whole new concept of research from the viewpoint of methodology, unparalleled among both his predecessors and his contemporaries, who were also by then trying to clear up the mystery of hybridization and heredity.

Mendel may have found inspiration in the physics textbooks of his teachers at Vienna University. Baumgartner and Ettingshausen (1842, p. 5) stressed that the aim of research was to seek the principles and laws of the appearance of forms in general. An appropriately arranged experiment was, according to these authors, 'the surest means for learning the operation of nature from all sides and for getting a clue to the laws and the interaction of phenomena'. In the field of research into plant hybridization, the only person in the nineteenth century to plan for the long term and to perform gradually experiments with such a suitable model, which represented the atomization of the plant organism into discrete plant traits, was Mendel. The model of discrete pairs of traits was his *initial theoretical framework*, which he could modify into the idea of numerical proportions of the parental traits in the hybrid progeny. The starting-point was a rejection of the concept of blending inheritance; Serre (1981) emphasizes this as being the most important factor.

Mendel's reasoning in creating hypotheses and testing them in experiments is depicted in Fig. 5.13. The hypothesis of the discreteness of traits is shown as H_o, Mendel having arrived at this before he began the experiments described in the *Pisum* paper. The hypothesis of the uniformity of hybrids is designated H_1, and the experiments to verify it E_1. Mendel crossed plants differing in one alternative trait pair in the seed, and in the very first year proved the uniformity of dominant traits in maternal plants. At the time natural scientists were still unclear as to the respective roles played by the paternal and maternal plants in the passing on of traits, even though Kölreuter had shown the equal contribution of both parents to the origin of the offspring and the identity of reciprocal crosses (Mayr 1986). But in fact, ninety years later this was still not generally accepted. Thus Mendel examined the outcome of crossing the maternal plant with the pollen of the paternal plant, and at the same time used the pollen of the maternal plant to fertilize what was originally the paternal plant. This move already implied a further hypothesis (H_2), and these experiments can be designated E_2. The conclusion was that 'it is entirely immaterial whether the dominating trait belongs to the seed or the pollen plant; the form of the hybrid is identical'.

The proportion of parental traits in the progeny of hybrids reproduced by self-fertilization, and the proportion of these traits in subsequent generations were defined by further hypotheses, designated H_3 and H_4. Mendel proved their validity in experiments denoted E_3 and E_4. In the progeny of hybrids in the first two experiments the segregation of the seed traits was demonstrated using an analysis of 15 537 seeds. Mendel reached the same conclusion when he evaluated a further five pairs of plant traits. He generalized his conclusion in *the numerical segregation ratio 3:1*. This was a quantitative evaluation of the results of an investigation into the transmission of qualitative traits from the generation of parental plants to the progeny. As regards the first two seed trait pairs, Mendel proved in the first year the uniformity of dominant traits, while in the second year he demonstrated segregation of the dominant and recessive traits. He later revealed the same pattern in a further five plant traits. This was *the first significant step in his research*.

The originality of this approach can be seen in contrast with the published views of Nägeli (1866a) in the paper 'Theory of hybrid formation', written before he received a reprint of the *Pisum* paper from Mendel. Nägeli states that the hybrid combinations designated AB and BA could not be identical. At the same time, the author states that no one has caused self-fertilization of individual plants during successive generations. Nägeli describes three types of plant in the hybrid progeny, the first of which corresponds to the type of the hybrid plants, while a further two are similar to the parental forms. He adds that these forms have low stability, and readily change among themselves. Under the influence of *Naturphilosophie*, Nägeli experimented with species hybrids and saw a smooth process of gradation in nature; he did not believe that the traits of the parents might appear in the offspring in an unchanged form. The obstacle to this conclusion was his idea of blending inheritance. Mendel's universe, by contrast, was discrete, and he attributed the changes he observed in nature to a process of the repetition and replication of discrete units that corresponded to the physicist's idea of nature.

When he received a reprint of the *Pisum* paper, Nägeli did not realize that it contained the detailed analysis of hybrid progeny he had called for in his own paper. A closer study of Mendel's publication and careful reading of the researcher's second letter could have revealed an entirely new methodological approach, leading to the explanation of a problem to which Nägeli himself had drawn attention in his speculative works on plant hybrids.

The segregation of the traits of the parental plants in the progeny of hybrids had also been described by the Frenchman Charles Naudin (1863). Crossing different plant species, he believed that the hybrid plant was like a mosaic, in which different essences of both parental species were united. According to him, a hybrid formed the germinal cells of both parents, with the result that there were three combinations in the next, the F_2 generation: one of the hybrid forms, the other two being a reversion of the hybrid forms to the parental species. In this conception, however, Naudin failed to arrive at the idea of numerical segregation of traits. In his book on animal and plant domestication Darwin (1868*b*, vol. II, p. 70) reported prepotency and latency of traits when describing the results of crossing the peloric snapdragon with the common form. He states that the crossed plants perfectly resembled the common snapdragon. In the progeny of self-fertilized hybrids he observed 88 to be of the common variety, two in an intermediate condition between the parental forms, and 37 perfectly peloric forms—hence a segregation ratio of parental traits close to 3:1—without comment. Darwin sought the explanation in the provisional hypothesis of pangenesis, where, anticipating blending inheritance, he was unable to find an explanation for the transfer of traits from parents to offspring. Later, under the influence of Darwin's theory and the development of cytology, Hugo de Vries (1848–1935) observed the proportion of parental traits in the hybrid progeny, as is explained in detail in Chapter 8. But it was not until 1900 that he actually mentioned numerical segregation ratios of 3:1.

It can now be concluded that Mendel was the only person in the nineteenth century to analyse the occurrence of traits in the hybrid progeny and the following generations, and to demonstrate that *the segregation ratio of 3:1 resolves itself into the ratio of 2:1:1*. This was *the final refutation of the concept of blending inheritance*, and at the same time *Mendel's second important step in his research programme*.

In his first lecture Mendel used the terms 'trait' (*Merkmal*) and, generally, 'form'. In the conclusion to his most simple experiments he introduced the practice of designating the dominant trait using a capital letter, such as A, and the recessive trait using a small letter, a, the hybrid thus being Aa. The hybrid progeny was expressed as a *simple series* (*Entwicklungsreihe*) A + 2Aa + a, whose terms refer to the forms of the hybrid progeny. Mendel was thus introducing mathematical symbolism into his research, and used it to design his subsequent experiments. The experiments with the crossing of peas differing in a single pair of traits were concluded with an explanation of the reversion of hybrids to the parental forms.

The results of the experiments denoted E_1 to E_4 led Mendel to an explanation of the free combination of two and three pairs of traits. His new hypotheses, denoted H_5 and H_6, were tested in experiments with the same numerical designations. He generalized hybrid progeny combining several trait pairs as *combination series* (*Kombinationsreihe*). He made use of his knowledge of combinatorics and in particular of the series described in a mathematics textbook by his teacher, A. Ettingshausen (1826), to which attention was drawn by Jindra (1971b). But it was not simply a matter of adopting a technique described in a book. Mendel introduced into his research a new kind of notation developed in symbolic logic, which was later to be the start of the new scientific language of research into heredity.

At the end of his first paper, Mendel demonstrated the use of mathematics in research into plant hybridization and heredity. In doing so he went beyond the empirical realm, providing in his explanation a rational framework from which new combinations of plants traits could be deduced.

Mendel's contemporary in England, Francis Galton (1822–1911), also tried to apply mathematics in research in heredity. He was an abstract thinker, believing that gemmule-like determinants do exist. In 1875, Darwin asked him how, in his view, one might explain the fact that a hybrid may have every part of the stem, leaf, and flower intermediate, and will produce millions of buds, all of which reproduce exactly this intermediate form. Galton replied: 'If there were two gemmules only, each of which might be either white or black, then in a large number of cases one-quarter would always be quite white, one-quarter quite black, and one half would be grey.' In this way, according to Olby (1965) Galton derived the 1:2:1 ratio of structural units which constitute the hybrid individual, and thus Galton made the closest approach to Mendelian theory that was achieved during the last century. Lints and Delcour (1968) rejected this position, pointing out the fact that Galton's ratio refers to the relative frequencies of the white, grey, and black structural units which constitute the F_1 hybrid generation, and not the relative frequency of the classes in the F_2 progeny of the hybrid generation. Olby (1968), even so, sees every reason for recognizing Galton's view of 1875 together with his papers during the period 1872–5 as the closest approach to Mendelian theory any nineteenth-century biologist made.

According to Darlington (1966, p. 98), for the successful use of mathematics in this area of research it was crucial 'in which order the researcher makes quantitative and qualitative studies'. A purely quantitative approach led Galton—in Darlington's words—'a little astray', 'while Mendel's quantitative examination of qualitative traits was successful'. Later, Kalmus (1983) added that Galton had also failed because he did not proceed from combinatorial speculation to experimentation because of his 'lack of practical horticultural skill'. It can therefore be concluded that only Mendel was successful in the application of mathematics in research into heredity, and this was *the third important step in his research programme.*

After the turn of this century little attention was paid to Mendel's second lecture for a considerable period of time. This can be illustrated, *inter alia,* from the collection of classic papers in genetics edited by J.A. Peters (1959). The two final sections of the *Pisum* paper are omitted from this, the stated reason being that 'these paragraphs have little bearing on the principles Mendel proposed', since experience with students indicated that 'these pages serve primarily to confuse rather than to clarify'. In fact, in these fourteen pages, comprising almost a third of the *Pisum* paper, Mendel tried to explain his deeper understanding of the mechanism of transmission of parental traits to offspring, and of fertilization, in concordance with the cell theory. This procedure was in full accord with Schleiden's axioms for research into scientific botany. In the second lecture Mendel wished to throw light on these more complex problems of heredity first introduced in the concluding sections of the *Pisum* paper.

The explanation of the theory in the concluding remarks of the *Pisum* paper was also incomprehensible to Mendel's contemporaries, partly because he used the terminology of the natural scientists of the first half of the nineteenth century, and was tackling a problem which came to prominence in the second half. In the title of his paper he was using the term 'hybridization' after the publication of Darwin's Theory of Evolution. By then hybridization was understood differently than it had been in the period up to 1850. For Gärtner, hybridization was already a subject of investigation relating to the origin of new plant forms and new plant species. But

Mendel additionally probed the question of the origin and development of hybrids, and in effect also the origin and development of plant forms, in the context of the latest knowledge on plant reproduction. And thus in his model of discrete trait pairs he was also able to examine the transmission of traits from parents to offspring.

Darwin (1859) stated that the laws of inheritance were unknown, thus providing a stimulus for their investigation. Nägeli (1865a) stated that 'the topic of the origin of hybrids is also very important from the reproductive point of view', thus further underlining the need to investigate the transfer of traits from parents to offspring, i.e. heredity. He did not, however, use the term 'heredity'. Mendel's first letter to Nägeli stresses the importance of 'new experiments, in which the degree of kinship between the hybrid forms and their parental species is precisely determined' rather than the mere estimation of traits from general impressions (Correns 1905, p. 195). The degree of kinship between forms was in fact heredity. In his second letter Mendel emphasizes that segregation traits were unchanged, and there is 'no indication that one of them has either inherited or taken over anything from the other' (Correns 1905, p. 201). It was the only occasion on which Mendel explicitly used the term 'inherited'.

In the introduction to his paper, Mendel uses the term *Entwicklungsgeschichte*, translated into English as 'evolutionary history'. A more appropriate translation would be 'developmental history'. This was the *Naturphilosoph* conception of ontogenic and phylogenic development, and it included the enigma of fertilization and the associated idea of heredity. The Sandlers (1985) realized that Mendel had 'succeeded in completely divorcing the transmission from generation to generation of the hereditary units from their behavior during ontogeny'. At the annual meetings of The Sheep Breeders' Association in Brno Professor Nestler had separated the problem of heredity from the enigma of generation back in the year 1836. But no one was capable of finding an explanation based on the level of knowledge current at the time. Only Abbot Napp in 1837 made a fairly detailed definition of the question 'what is inherited and how?' (Orel 1975a). Later, Wagner (1853), in the course of a philosophical speculation on the enigma of generation and the cell theory, proposed a method of investigating heredity to give an indirect explanation of the enigma of generation. Six pages are devoted to questions for research into heredity. He assumed crossing of animals differing in various traits, and the investigation of parental traits in hybrids and their progeny. He considered the possibility of a statistical evaluation of traits and the establishment of accurate numerical data. Wagner himself tried to undertake such experiments, involving a large number of observations, with fishes and frogs. He soon found, however, that he did not have the necessary conditions in the laboratory for carrying them out. But he was aware that the same problem could be investigated in plant generation experiments, which also meant plant-hybridization experiments.

In the second half of the the last century the isolated problem of heredity appeared from time to time in the deliberations of naturalists. In a retrospective analysis of the nineteenth-century search for a way of investigating heredity, Churchill (1987) showed that between 1883 and 1885 five German biologists, relying on notions of structural continuity, felt compelled to offer an explicit theory of heredity, and also did not succeed. Having been trained at Vienna University in plant physiology, Mendel set out from the assumption that one pollen grain fertilized one egg cell. He attributed the traits he was investigating to determinants inside the germ cells. This was a new hypothesis, depicted in Fig. 10 in three parts, denoted H_7–H_9, which he proved in simple reciprocal back-crossing experiments denoted E_7–E_9.

In his 'concluding remarks' Mendel first mentions hybrids in the offspring, the segregation of whose traits was mentioned by Kölreuter and Gärtner. In the following lines he also draws attention to offspring which 'remain exactly like the hybrid and propagate unchanged'. He wanted to evaluate the features of hybrids in general according to his observations in the *Pisum* experiments. According to Mendel, Gärtner himself had admitted that precise determination of whether a form bears a greater resemblance to one or the other of the two parental types often presented great difficulty, since much depended on the subjective viewpoint of the observer. He also points to other circumstances which could contribute to making the results variable and uncertain, such as the number of traits observed in the experiments of the conditions of the plant's cultivation. In the following lines Mendel tries to explain why Gärtner did not observe the two parental types among the offspring of the hybrids in experiments involving a small number of experimental plants. Assuming that 'development of hybrids follows the law valid for *Pisum*, then the series obtained in each separate experiment must comprise very many forms, because the number of terms is known to increase as the third power of the number of differing traits' (Mendel 1866*a*, p. 40).

In the next paragraph Mendel outlines his theoretical idea of constant and variable hybrids in connection with the cell theory. This was a new concept, based on his model of hybrids that recombine defined pairs of plant traits. He supposes that the constant progeny of a hybrid could be proved only when germinal cells and fertilizing pollen were alike both provided with the *Anlage* for creating identical individuals, as in the normal fertilization of pure strains (Mendel 1866a, p. 24). (The term *Anlage* is rendered in the first English translation by Druery, in 1901, as 'material', in the second translation, by Stern and Sherwood in 1966, as 'potential'.) Mendel adds, 'We must therefore consider it inevitable that in a hybrid plant identical factors act together in the production of constant forms.'

Unlike his predecessors Kölreuter and Gärtner, Mendel planned his experiments in the light of up-to-date knowledge of the cell theory, and it was on this plan that he developed his theory. He referred to the opinion of 'famous physiologists', without mentioning their names, according to whom the propagation of higher plants was 'initiated by the union of one germinal and one pollen cell into one single cell, which is able to develop into an independent organism through incorporation of matter and the formation of new cells' (Mendel 1866a, p. 40). Here he explains very briefly and aptly the determinants of traits. He adds: This development proceeds in accord with a constant law based on the *material composition and arrangement of the elements* that attained a viable union in the cell. This introduced the new term *element* to denote the material character of the determinant of heredity.' He could say no more at the time.

In the subsequent part of his exposition, Mendel first formulates the hypothesis of the occurrence of *constant hybrids*, then of *variable hybrids*, as follows: 'When a germinal cell is successfully combined with a dissimilar pollen cell, we have to assume that some compromise takes place between those elements of both cells that cause their differences. The resulting mediating cell (*Vermittlungszelle*) becomes the basis of the hybrid organism, whose development must necessarily proceed in accord with a law different from that for each of the two parental types. If the compromise be considered complete, in the sense that the hybrid embryo is made up of cells of like kind in which the differences are *entirely and permanently mediated*, then a further consequence would be that the hybrid would remain as constant in its progeny as any other stable plant variety. The reproduction cells formed in its ovary and anthers are all the same, and like the mediating cell from which they derive.'

Mendel goes on to discuss hybrids whose offspring are variable. He assumes that a compromise takes place between the different elements of the germinal and the pollen cells 'great enough to permit the formation of a cell that becomes the basis for the hybrid; but this balance between the antagonistic elements is only temporary, and does not extend beyond the lifetime of the hybrid plant'. Mendel's conclusion is that the different elements of this hybrid form 'succeed in escaping from the enforced association as late as the stage at which the reproductive cells develop'. He adds to this an explanation regarding 'the production of as many kinds of germinal and pollen cells as there are combinations of potentially formative elements'. Mendel goes on to say that 'This attempt to relate the important difference in the development of hybrids to *a permanent or temporary association of differing cell elements can, of course, be of value only as hypothesis, which, for lack of well-substantiated data, still leave some latitude'.*

A decisive factor in the explanation of the determinants of the transmission of traits through germ cells was Mendel's knowledge of plant physiology, acquired at the Vienna Institute of Professor Unger. The emphasis on this physiological explanation was also apparent in the report published after Mendel's second lecture in Brno, on 8 March 1865. According to this Mendel 'spoke about cell formation, fertilization, and seed production-in general, in the case of hybrids in particular' (Olby 1966: 1985 edn, p. 221).

In his lecture Mendel encouraged his audience, and in the published version his readers, to verify his findings by repeating the most important of his experiments regarding the germ cells of hybrids. Plant physiologists of the day were engaged in a debate over new discoveries in the field of generation and fertilization in lower plants. Plant physiology, according to Olby (1971), provided Mendel with a framework for a mechanistic type of thinking in relation to cell processes. At the start of his deliberations stood the research of Pringsheim on plant fertilization. But in 1971 Olby doubted that 'Mendel ever knew of Pringsheim's work'. Now we can show that he may have been assisted by his friend J. Nave, who studied law in Vienna while Mendel was there, and attended Professor Unger's lectures in plant physiology at the same time. Nave (1858), speaking to natural scientists in Brno, explained the latest findings relating to algae, and in this connection mentioned the research of Pringsheim, who in the mid-fifties of the last century explained how the spermatozoid—the male organ—touched the female organ and was dissolved in its substance. This was the first proof of fertilization in plants as the process of the joining of one maternal cell with one paternal cell. Krausse (1983) illustrated how the research of Pringsheim attracted the attention of Unger. He evoked in Mendel an interest in algae that can be proved in the microscopic preparations described by Milovidov (1968) and by Orel and Čunderlík (1985). According to Nave (1864), 'there are no better verified principles of the theory of beings, of the development and reproduction of plant cells than those offered by a thorough study of algae, where one has before one's eyes so clearly the significance of the existence of individual cells'. With his microscopes, described by Procházka (1985), Mendel was able to examine the structure of algae from the standpoint not only of the ideas of Nave, but also of those of Unger.

Niessl (1866) also pointed to Pringsheim's research in connection with the occurrence of hybrids and the explanation of plant fertilization. Later, another of Mendel's friends, A. Tomaschek (1871), Professor of Natural Science at the Brno Institute of Technology, gave an even more detailed account of procreation in plants, also referring to Pringsheim. (He too had studied plant physiology at Vienna University under Professor Unger.) In his paper on experimental growth of plants from one pollen grain *in vitro*, he wrote the footnote: 'Pringsheim, a professor at Jena, discovered in 1855, *as everybody knows*, the first two different organs in *Alga vaucheria sessilis* Lyrgb.—in one of them the protoplasm content of the unfinished cell is formed (*Eizelle, Prospor*) while in the other this content is shaped to one (in other plants several) minute seminal bodies (*Samenkörperchen*), both organs open themselves, and one seed body penetrates into the unfinished cell, and dissolves therein, and now the cell is completed through the formation of the actual cell wall; inside there begins a developmental process from which a new individual of the same kind arises' (Orel 1981). This was a very instructive interpretation of Pringsheim's research.

Tomaschek used to visit Abbot Mendel, and one may suppose that they spoke together of the research, and, in this connection, of a problem in which both of them were interested, the fertilization of plants. Like Tomaschek, Mendel gave an explanation in a footnote. In addition, he refuted Schleiden's idea of the pollen tube penetrating into the embryo sac being the origin of the embryo, writing:

'If the influence of the pollen cell were only external, if it merely played the role of a foster-mother, then the outcome of each artificial fertilization would have to be that the resulting hybrid resembled the pollen plant exclusively or very closely. Experiments have in no way confirmed this up to now.' R. Wunderlich (1982) concludes from these words that at the second sitting of his teachers' examination Mendel could have explained the origin of embryos in this manner, which may have led to a conflict with Professor Fenzl, and might also have formed the motivation for Mendel's research after he returned to Brno.

Mendel's theorizing in his second lecture aroused a series of learned discussions in the 1950s. C.D. Darlington (1951) pointed out that geneticists had not realized until 1950 the importance of Mendel's concept of 'the elements inside the cell which segregate, recombine and above all determine the traits'. In his view the previous confusion regarding Mendel's idea of units of heredity derived from his denoting homozygotes A or a, while he reserved double letters for the hybrid form Aa—which Darlington considered 'a super example of the attempt to separate facts from hypothesis'. Two years later Darlington (1966, p. 95) presented Mendel's concept of the determinant of heredity, the gene, as 'rather the primary law of biology'. Olby (1979) was later to regard this as a glorification of Mendel, whom he considered lacking in thoroughness in not denoting the determinants of heredity in the *Pisum* paper using two letters representing the two alleles—a practice introduced by early twentieth-century geneticists.

The explanation can be sought in Mendel's notation of segregation of traits in connection with the germ cells transmitting determinants of the traits investigated. Mendel (1866a, p. 24) saw 'an important clue' in the fact that 'in *Pisum* constant forms appear among the progeny of hybrids and that they do so in all combinations of the associated traits'. In his experiments Mendel became convinced that 'constant progeny can be formed only when germinal cells and fertilizing pollen are alike both endowed with the *Anlage* for creating identical individuals, as in normal fertilization of a pure strain. Therefore we must consider it inevitable that in a hybrid plant also identical factors are acting together in the production of constant forms.' In this context Mendel introduced the idea of the mediating cell, and on this basis sought a theoretical explanation of the existence of variable and constant hybrids. In his *Pisum* experiments he demonstrated variable hybrids, leaving the second kind, constant hybrids, for further research. This theoretical idea of Mendel's led to a discussion on the character of Mendel's theory and his idea of the determinants of the transfer of the parental traits to the offspring.

Heimans (1962), in evaluating Mendel's theory and that of his teacher de Vries, stated that Mendel 'occupied himself only with the phenotype difference, that is with directly recognizable characters, for which he almost invariably uses the term *Merkmal*'. Heimans ascribes the idea of the unit of heredity to de Vries. Later, analysing a fragment of the record from Mendel's experiments, Heimans (1969) comes to the conclusion that Mendel also did not know the physical character of the elements determining the transmission of traits. Olby (1971) assumed that 'what Mendel described as the two contrasting characters, the character pair, was later projected back into the cell's elements or factors'. In addition he pointed out that Mendel did not propound the idea of duality of elements for traits. Later, while taking Heiman's views into account, Olby (1979) supposed that Mendel 'was committed to a materialist and determinist explanatory framework, i.e. that the characteristics of living organisms are determined by material entities in the cell'. He confirmed at the same time Heiman's view that

Mendel did not arrive at a notion of duality of determinants of traits, pointing out by way of evidence the following quotation from the *Pisum* paper: 'In the formation of these cells all the elements present participate in a perfect free and equal arrangement, whereby only the differing elements are mutually exclusive' (Mendel 1866a, p. 42). This text is said by Olby (1979) to have been the reason for the conflict of the concept of units of heredity 'with classical genetics', in the sense that if dual units existed only in hybrid traits, then 'the number of like elements determining a character would increase every time the germ cells fused in fertilization'. Olby's conclusion was that Mendel cannot 'have had the conception of a finite number of hereditary elements, which in the simplest case is two per character', and therefore 'Mendel was clearly no Mendelian.' Soon after, Guédés (1981) stresses that 'Mendel was indeed the first Mendelian.' According to him, Mendel 'set out from the idea of the transmission of traits through the determinants in the germ cells and the fertilization process when two determinants for constant traits are united'. Guédés shows how Mendel imagined this process in terms of a finite number of elements.

The interpretation of Mendel's text should be sought in the context of his hypothesis of the variable hybrid denoted symbolically *Aa*. Mendel set out from the observation that no changes in the characteristics of hybrids can be noted throughout the vegetative period, which was consistent with his primary interest in hybrids and his idea that hereditary determinants emerge unchanged after their association together in hybrid union. In his paper Mendel wrote in German *dass es den differirenden Elementen erst bei der Entwicklung der Befruchtungszellen gelinge aus der erzwungenen Verbindung herauszutreten*. The first English translation by Druery changed the meaning of Mendel's explanation as follows: 'It is only possible for the differentiating elements to liberate themselves from the enforced union when the fertilizing cells are developed.' The text is correctly translated by Stern and Sherwood (1966, p. 43), as follows: 'The differing elements succeed in escaping from the enforced association only at the stage at which the reproductive cells develop.' The difference in meaning is evident. After this explanation Mendel wrote that 'only the differing elements are mutually exclusive' (in German *nur die differirenden Elementen sich gegenseitig ausschliessen*). In the whole paragraph Mendel was dealing only with the hybrid in the progeny whose parental traits segregate, and in this context he did not use consistently enough the formulation of the behaviour of different elements of hybrids. This is not, however, definitive proof that he had no idea of the segregation of like elements.

The concept that one character pair is determined by one pair of *Anlagen* was not, in Olby's view, introduced before Correns (1900c, p. 163), who, interpreting Mendel's theory, wrote that the latent *Anlage* in the hybrid form 'is preserved, and prior to the definitive formation of one sexual nucleus complete (*glatte*) separation of the two *Anlagen* occurs'. On page 166, however, Correns states that 'the hybrid produces sexual nuclei that unite the *Anlagen* for individual traits of parents in all possible combinations, but not those of the same trait pair itself' (original German: *nur die desselben Merkmalspaares nicht*). Correns' text is imprecisely translated in the book by Stern and Sherwood (1966, p. 130) as 'but both *Anlagen* for the same pair of traits are never combined'. This interpretation of Correns' text, quoted by Olby (1979, p. 58), cannot be put forward as a first proof of the duality of elements as determinants of traits by Correns.

Recently Monaghan and Corcos (1990) even concluded that Mendel did not arrive at the notion of particulate determinants of heredity, even though he stated that the germinal and pollen cells are endowed with *Anlagen*. They take him to mean only germinal and pollen cells. The authors suggest a revision of Mendel's text to leave out the phrase 'both endowed with *Anlagen*' and the German word *Faktoren*. According to this the revised sentence from page 24 would read: 'In our experiments we find everywhere confirmation that constant progeny can be formed only when germinal and pollen cells are alike [omitting the phrase "both endowed with the *Anlage* for creating identical individuals"], as in the normal fertilization of pure strains. Therefore we must consider it inevitable that in a hybrid plant also identical [germinal and pollen cells] [instead of using Mendel's term "factors" the authors give in brackets the term "germinal and pollen cells"] are acting together in the production of constant forms.' This is the first time that Mendel's text itself has been revised in such a way as to make his experiments more comprehensible if taken as the work of an experimentalist. Reasonable people may differ in their views about Mendel's motivation or understanding of his work, but most would agree that judgement must be based on the words in which Mendel chose to express himself (Orel and Hartl 1994).

H. Kalmus (1983) tried to throw new light on the concept of the units of heredity in Mendel's paper. He had studied in Prague, and knew the milieu of ideas in educational establishments in the erstwhile Austro-Hungarian Empire. His primary assumption was that at the outset of his experiments Mendel was familiar with Aristotelian logic, which may have affected his theorizing when he began his research. Kalmus had already drawn the attention of geneticists J.B.S. Haldane and T. Dobzhansky to this fact, but his conclusion was that no one had followed up this train of thought. Darlington (1966, p. 97), only mentioned in passing Haldane's view that Mendel was a Thomist. He supposed, however, that a misunderstanding had arisen on the basis of shortcomings in the English translation of

Mendel's work in respect of the transfer of traits from the parental genera-tion to the progeny. In describing three classes of individuals in the progeny of hybrids as *A*, *Aa*, and *a*, Mendel, in Darlington's view, evaded the unproven duality which the use of two letters for constant traits would imply. According to him, the unclear explanation of the determinants of heredity in Mendel's paper is at odds with the ideals inherent in Thomism, which is uncompromisingly deterministic. Dobzhansky (1964) expressed his opinion of Haldane's views. He considered them to pay 'a compliment to Thomist philosophy', but was unconvinced that it was warranted, and saw no parallel between the gene and any of St Thomas's ideas known to him.

J. Mausbach (1930) pointed to the influence the ideas of St Augustine (354–430) may have exerted on Mendel's experimental approach. In his work *De Trinitate*, St Augustine mentioned the hypothesis of evolution, describing the body elements of the world (*seminum semina*) which 'cannot be seen but are grasped through the reason'. B. Volodin (1968) also referred to the possibility of Mendel's research having been affected by the teaching of St Augustine, stating that it directly inspires 'rational investigation'.

Kalmus assumed that in the highly intellectual milieu of the Augustinian monastery in Brno the monks discussed Aristotelian ideas in connection with their interest in the natural sciences, and in this connection new stimuli for Mendel's research may have arisen. The monastery library contains a copy of a textbook of logic written by W. Esser (1823), which was used by the philosophy teachers at the monastery, Bratránek and Klácel. Mendel studied theoretical and practical philosophy at the Philosophy Institute in Olomouc before coming to Brno, and he may well have met with the interpretation of Aristotle's categories and his metaphysics in this textbook. Kalmus therefore came to the conclusion that Mendel was 'thoroughly saturated in the then flourishing Aristotelian scholastic tradition'. He supposes that each of Mendel's seven trait characters can be classed as one or several Aristotelian categories. The first category, the essence or substance, to which the other attributes (predicates) are accidental, had previously been stated by Plato. This type of distinction appeared in the ideas on plant traits expressed by A.G. Agricola (1716), when he tried to explain the universal propagation of plants. Setting out from the religious–philosophical literature ascribed to Hermes Trismegistos, a mythical figure from ancient Egypt, Agricola ascribed to plants a material soul which could be divided into innumerable particles (Orel and Gabriel 1981). The being of the plant represented a *rationis essentia*. Agricola explained the differences between plants as *accidentiae*, consisting not in the being of the soul, but merely in its particular structure, which was responsible for the differences between them. This was the basis of Agricola's picture of the wonderful possibility of the transmutation of plants. Mendel's distinction between *Anlage* and *Merkmal* could, according to Kalmus, be considered a certain analogy with the categories of essence and attribute.

The second of Aristotle's categories was quantity, which could be either continuous or discontinuous. In the *Pisum* experiments Mendel's point of departure was discontinuous quantity in discrete trait pairs. Aristotle's distinction between the actual and the potential could be implicated in two of Mendel's notions, that of *Anlage* and that of dominance. *Anlage* as a potential relates to the manifest actual—a trait. The potential is not always realized, and this could explain why a recessive trait does not appear in a hybrid, but reappears in the hybrid progeny.

Mendel's fundamental recognition that the opposites of his paired characters do not blend in a variable hybrid could also be anticipated in the Aristotelian notion of the different kinds of combination, distinguishing between mixture and synthesis. In the former case two predicates came together to form a new predicate. From this point of view Mendel's concept of constant hybrids corresponded to the category of mixture, and the less intimate association of elements inside the mediating cells in variable hybrids would then correspond to the category of synthesis.

Kalmus, in appreciating the fundamental role of combinatorics in connection with original symbolic logic, was led to the conclusion that Mendel may also have known G. Boole's (1874) system of algebra of classes and their properties. Boole's symbols denote discrete classes, the relations between which are stated in the form of propositions. Dominance is an example of the propositional principle of equivalence or the algebraic notion of *idempotency*. This principle can be stated as follows: inasmuch as D is concerned, it does not matter whether one or two dominant elements are presented in the fertilized germ cell. One exerts the same influence as two. Another logical twist of dominance and recessivity could be stated in terms of De Morgan's Law (1847), to which attention has already been drawn by R. Hončariv (1971) and Jindra (1971*a*). It can be explained by the verbal statement that the recessive character is manifested if and only if neither germ cells carries a dominant element. The idea of idempotency could thus be implied in the Aristotelian concept of potential. Whether the potentiality of a dominant character is present in a developing organism once or twice makes no difference—in neither case would it become actual.

The average course of fertilization of hybrids, when two differing traits are associated in them, was visualized by Mendel (1866*a*, p. 30). It implies that A equals $\frac{A}{A}$, a equals $\frac{a}{a}$, and 2Aa equals $\frac{A}{a} + \frac{a}{A}$, which does not conform to the rules of arithmetic. Mendel could have arrived at the

identification of A with A + A and a with a + a in the axiom of idempotency in symbolic logic. The axiom refers to classes, and has no numerical con-
notations. The class of plants producing angular peas, the recessive trait,
contains all such plants. Thus, in symbolic logic expressions like A + A are
superfluous, and this could explain why Mendel described constant traits
using only one letter, e.g A or a. He was, however, not consistent in applying idempotency. In the visualization of the process of fertilization Knievel
(1994) saw the weakest point in Mendel's paper, because Mendel did not
consider mathematical exactness as well as appropriate biological explanation. Knievel even added: 'From mathematician Mendel we should not
expect it.' Anticipating that Mendel's research goal was the investigation of
the origin of species Knievel used this weakness for his theory that Mendel
was no Mendelian.

Kalmus pointed to two distinct statements:

A = 1		A
2Aa = 2	are the numbers in classes	Aa
a = 1		a

The terms on the right-hand side symbolize classes, which are not
definable by numbers, whereas the same symbols on the left-hand side
denote frequencies. The terms used by Mendel could have been based on
the Aristotelian idea of potential, and it makes sense if we define A as the
class of plants which when self-pollinated can produce only offspring of the
dominant character, a as the class of plants which can produce offspring of
the corresponding recessive character, and Aa as the class which can produce either. In this conception Kalmus was in favour of the view that
Mendel did not conceive of pairs of hereditary particles, and that 'his differing, mutually exclusive cell elements' responsible for the differences in
the development of the hybrids 'can aspire only to the status of hypothesis'.

According to Kalmus, Mendel's contemporaries did not consider
Aristotelian categories. To prove his point he cites the research of C.
Naudin (1863), who tried to explain the segregation of traits in the hybrid
progeny through the action of an *essence specifique*. In his conception, however, this was some sort of chemist's essence, a liquid suspension, representing the mosaic of all traits of the plant, which are intimately mixed, and in
the development of germ cells segregate in a random process. C. Meijer
(1983), however, is convinced that Mendel saw the elements as chemical
fluids, like Kölreuter and Gärtner. In that case, according to Kalmus,
Mendel would have to write the simple series as $A_2 + 2Aa + a_2$. His notation A + 2Aa + a, however, corresponds to the concepts of physics and the
physical character of Mendel's elements.

An open question is what Mendel *thought* he had discovered. He was certainly under the impression that his research was important. The key point can be found in the fact that his paper was written for oral presentation, where the principal technique for emphasis is repetition (Hartl and Orel 1992). One idea repeatedly expressed is as follows (p. 29).

... pea hybrids form germinal and pollen cells that in their composition correspond in equal numbers to all the constant forms resulting from the combination of traits united through fertilization.

Mendel's discovery in his *Pisum* experiments is *a synthesis of scientific ideas stemming from three basic fields of knowledge:*

(1) the application of mathematics in research into heredity;
(2) the elucidation of the basic mechanism of fertilization in connection with heredity;
(3) the application of the theory of probability in the production of germ cells in the fertilization process, and in the transmission of parental traits to offspring.

Each of these three components was the result of imaginative conjecture proved in experiments. Such experiments can be described using Kuhn's (1981) term 'thought experiments', considered as 'one of the essential analytic tools which are deployed during crisis, and which then help to promote basic conceptual reform'. Putting together previously unrelated scientific and empirical ideas was the essence of Mendel's creativity, which Kalmus denoted 'multiassociation', as being typical of innovation in scientific research. It includes not only what were called after 1900 the Law of Segregation and the Law of Independent Assortment of Traits, but also the explanation that traits of plants are inherited through determinants in germ cells. Different germ cells united in the hybrid are produced in equal numbers and participate in the fertilization process with equal frequency. Chance determines which of each of the male germ cells fertilizes each of the female germ cells, and also governs the heredity of traits. Mendel's approach can also be explained as gradual scientific growth and as the product of a systematic selective mechanism, according to the naturalist—cognitive approach of Giere (1988) and Efron and Fisch (1991). Mendel's scientific reasoning was explained by E.L. Tatum (1977) as being 'the product of the hybrid vigour resulting from cross-fertilization between disciplines. The new theory was created in a special social environment and cannot be considered the product of Mendel's mind only. The analysis of the preconditions of Mendel's achievements illustrates that he was also able to make full use of scientific ideas from his background and was prepared for the research' (Orel and Kuptsov 1983).

Considering a different interpretation of Mendel's discovery, Fisher (1936) came to the conclusion that 'Each generation, perhaps, found in Mendel's paper only what it expected to find; in the first period a repetition of the hybridization results commonly reported, in the second a discovery in inheritance supposedly difficult to reconcile with continuous evolution.' To these two early periods two others can now be added: the period of Mendel's glorification, which coincided with genetics becoming pre-eminent among the biological sciences, and the period of Mendel's diminution, reflecting the general iconoclasm and hero-bashing of more recent times (Orel and Hartl 1994). In principle it is in agreement with the critical view of Sapp (1990), according to whom 'To understand geneticists' reconstructions of Mendel's intentions is to understand the divergent and sometimes conflicting definition of what Mendelian genetics signifies or connotes'.

Mendel's theory was in fact the answer to the question formulated by Abbot Napp in Brno before Mendel was accepted in the monastery. It was also explanation of what I. and L. Sandler (1985) called 'Mendel's eccentric definition of the problem of heredity'. Nägeli and his contemporaries, according to the Sandlers, in considering Mendel as an empiricist, were 'unable to perceive Mendel the theorist', not recognizing that 'his approach to experimentation was hypothetico-inductive'. This new approach of Mendel's was perceived even by E.M. East (1923), who emphasized that Mendel attacked the problem as a physicist at a time when 'those who were endeavouring to investigate inheritance by means of hybridization were not prepared for their task'.

In his synthesis of ideas Mendel was exceptionally successful, and this is often referred to in the literature as his 'good luck'. If one is to consider this the work of chance, then one might recall Pasteur's words: 'In science, fortune favours only those who are prepared.' With his basic and university education and his work at the Brno monastery, Mendel was well prepared for his research. In addition he was able to make use of his gift of keen observation and his imaginative capacity, which was clearly in evidence when he described an unusual natural phenomenon. An intuitive ability is to be found in his brilliant explanation of the origin and course of the whirlwind which hit Brno in 1870 (Mendel 1871). As an eye-witness to this phenomenon, exceptional in Central Europe, Mendel immediately elaborated a remarkable theoretical explanation with mathematical and geometrical reasoning, though he considered it to be no more than an 'airy hypothesis'.

In the *Pisum* paper Mendel offered an innovative theory for explaining the transmission of parental traits to offspring for variable hybrids. His method of researching the transfer of traits of pea seeds and plants from generation to generation was also innovative. In experiments with other plant genera he further elaborated his theory according to the same method.

At present there are various views on how Mendel imagined the elements determining the transmission of traits through germ cells. In the past a similar lively discussion was entered into on the subject of his 'too good to be true' results in the *Pisum* experiments. George (1983), at a workshop on the role of Mendel in the foundation of genetics, stated aptly that 'We are lucky to have limited information, and we are completely free. We can speculate to our hearts' content because nobody can say we are wrong. They can only say "I do not agree with you".' It is an exaggerated view, but expresses neatly the difficulties which arise when the details of different aspects of Mendel's research are discussed. Similar disagreements are likely to arise in the critical evaluation of his experiments with other plant genera.

In 1865 Mendel spoke in his lectures about experiments with peas, and only mentioned other experimental plants in passing. However, he did not consider the experiments with peas to be closed. The fragment record called *Notitzblatt*, dealt with below, see pp. 204–5, shows that he could return even after fifteen years to a re-examination of data dealing with the transmission of seed-coat colour of peas.

Theory in experiments with other plant genera

In his second lecture, Mendel first of all briefly described his experiments with the hybrids of different species of *Phaseolus*, and merely mentioned in passing *Dianthus caryophyllus*, derived from a white-flowered variety. At the time Mendel (1866*a*, p. 38) expressed the conviction that 'Anyone surveying the shades of colour that appear in ornamental plants as a result of like fertilization cannot easily escape the conviction that there, too, development proceeds according to a certain law which possibly finds its expression through the combination of several independent colour traits'. Information on Mendel's experiments on species other than *Pisum* escaped the attention of scholars because no one published Mendel's letters to Nägeli until Correns (1905), and then they were published only in German. It was not until 1950 that translations were made into English, and subsequently into Russian, Spanish, and French. According to Sturtevant (1965, p. 12), only the letters to Nägeli show how actively Mendel was engaged in genetic studies on several other kinds of plants. 'The picture that emerges is of a man very actively experimenting, aware of the importance of his discovery, and testing and extending it on a wide variety of forms. None of these results were published; it is difficult to suppose that his work would have been so completely ignored if he had presented this confirmatory evidence, even though it was not enough to convince Nägeli.' (Fig. 5.14.) Sturtevant shows not only the great extent of Mendel's experiments with other plant species, but also his approach to the investigation of ever more complex problems.

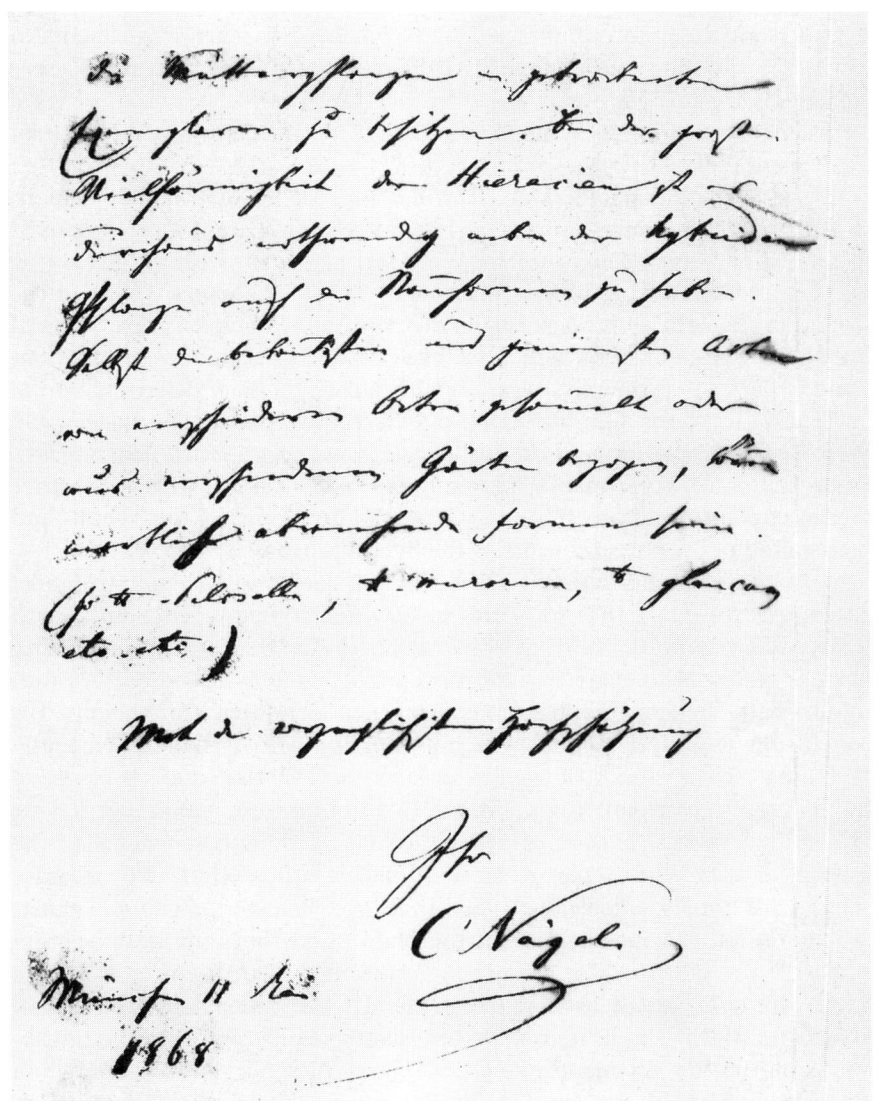

Fig. 5.14 Remainder of Carl Nägeli's letter to Mendel dated May 11th 1868 dealing with the exchange of experience in the hybridization research and *Hieracium* plants.

In the *Hieracium* paper Mendel only reports the results of his first experiments, a fact which contrasts with the detailed manner in which he evaluated the results he obtained with *Pisum*. His early experiments, performed in the 1850s, had been undertaken in quite a different milieu. At the time Brno naturalists were not particularly interested in the occurrence of hybrid plants in nature. But when he began his *Hieracium* experiments in 1861, the situation had changed. At the meetings of the National Science Society discussions got under way from the start of 1862 on various hybrid forms in the Brno area, and Mendel began investigating some of them experimentally.

In the first experiments with *Phaseolus* Mendel (1866a, p. 32) demonstrated that 'the numerical relationships in which different forms occurred in individual generations were the same as in *Pisum*', without giving the numerical data relating to the experiment. This was a *simple series*. In the second experiment with *Phaseolus* he met with more complex segregation of traits of flowers and seeds, which he was not able to explain even as a *combination series*. He promptly came up with a hypothesis of determination of one pair of traits through the action of two creative potentials (*Anlage*), and developed it in the form of *composite series* (*zusammengesetzte Hybridreihe*). Later on in his experiments he demonstrated variable hybrids in other species of plants only according to the proven uniformity of the hybrids and segregation of the parental traits in the hybrid progeny.

In the experiments crossing *Phaseolus multiflorus* × *Ph. nanus*, Mendel attempted to explain that the crimson flower is compounded from several independent colours. Olby (1979) raised the critical question of why Mendel made no apology for putting both A_1 and A_2, denoting crimson colour, with the same contrasted character, a, denoting white colour. His conclusion was that Mendel was not thinking 'of the pair of mutually excluding factors'. From Olby's discussion it is clear that there is plenty of room for disagreement about Mendel's intentions in summarizing his explanation in this manner. Although Mendel undoubtedly associated the symbol a with white colour, he also clearly stated that two or more colours are 'totally independent', and that they 'behave individually exactly like any other constant trait in the plant', and that hybrid association $A_1a + A_2a + \ldots$ would be formed by fertilization with germ cells from a white plant. It can be assumed that Mendel had independent factors in mind, and that the a written as partner for A_1 is different from the a written as partner for A_2. Mendel's matrix (Mendel 1866a) and the symbol aa make sense if he were using positional notation; otherwise consistency would require the white plants to be symbolized simply as a (Hartl and Orel 1992).

In experiments with further plant genera Mendel was especially engrossed by the great difference in the flower colours of hybrids and their offspring in the genus *Matthiola*. On the basis of the results he had obtained, which he considered preliminary, he outlined the possibility of the explanation lying in *more complex composite series* than those he had demonstrated in the *Phaseolus* experiments. In further experiments he adopted the same approach, considering the proportions of traits in the hybrid progeny, even physiological traits such as the capacity for fertilization in *Hieracium* or the occurrence of sex in *Lychnis*.

Mendel's research on *Hieracium* has a more remarkable history than that on *Pisum*. In 1869 he published a short report on his initial experiments, where he hinted that the results were different from those obtained for *Pisum*, but on the theoretical level left the problem open for further research. The early-twentieth-century scholars noted the revolutionary character of Mendel's *Pisum* research, but for many years the *Hieracium* experiments were regarded as unsuccessful owing to an unfortunate choice of model plant. However, the *Hieracium* experiments reached their climax after the publication of Mendel's initial report. He had already mentioned constant hybrids in the *Pisum* paper, without giving *Hieracium* as an example. Gärtner (1849), in the chapter on natural plant hybrids, paid great attention to *Hieracium* and *Cirsium* hybrids. On page 598 *Hieracium* hybrids are mentioned as being intermediate and constant, arising, in the author's view, 'through cosmic and climatic effects'. Gärtner went on to say that the genetic source of bastards in the wild could only be the subject of hypothesis according to their resemblance, since it is not the kind of one of the stock parents which is produced, but there is a typical predominance of species types. In his report Mendel (1870a) places *Hieracium* hybrids among the intermediate types. In Brno naturalists had been reporting the appearance of *Hieracium* hybrids in nature since 1863. They showed a knowledge of the latest literature on the problem, and were well acquainted with the papers by Nägeli. In his third letter to Nägeli Mendel gave the name of F. Schultz (1854–5), who first attempted to cross *Hieracia*. The *Hieracium* paper also cites E. Fries, who was an honorary member of the Brno Natural Science Society. His monograph *Epicrisis generis Hieraciorum*, published in 1862, was in the Society's library (Fries 1862). Mendel was thus acquainted with the literature on hybrids of *Hieracium*. This genus was at the time at the forefront of his research interest (Orel 1973b). He paid no less attention to these than to *Pisum* hybrids, and this investigation was much more taxing. However, the trouble he had was associated with the researcher's suspense over the outcome of his experiments, as had indeed been the case with *Pisum*.

It is clear from Mendel's last letters to Nägeli that he knew that he was crossing very distant species, and that the variants of traits he was observing in hybrids were myriad. Gaissinovitch (1988, p. 147), rightly expresses the view that Mendel could not therefore demonstrate uniformity in hybrid traits, as had been the case in the previous experiments. Indeed, in these letters Mendel expresses himself in very general terms, without going into any detail regarding his theoretical speculation. Matalová (1990) even reached the conclusion that at the end of his *Hieracium* experiments Mendel supposed the problem of hybrids could be further clarified in experiments with crossing races of bee. Mendel experimented with bees right up to the end of his life, as is described in Chapter 6, pp. 231–7.

The impetus for experimenting with *Geum* came from Gartner's monograph. Cetl (1973) stresses that he included the *Hieracium* experiments in his research programme on the basis of his own deliberations during his research. He may have reflected that if the forked stems were a transitional constant character in piloselloids, then this trait might arise by means of crossing, leading to hybrid constancy. In this way Mendel could have arrived at the idea of investigating the constancy of *Hieracium* hybrids. It was also a taxonomical problem, as transitional forms were very difficult to classify within the highly polymorphic genus *Hieracium*. Later Mendel abandoned the notion that *Hieracium* hybrids were constant, assuming the possibility of proving the existence of a more complex type of variable hybrid. But he was, in Cetl's (1973) words, 'gradually obliged to move into the sphere of broader questions of fertilization from the viewpoint of physiology and ecology, and, one might say, evolution'. In publishing Mendel's letters to Nägeli, Correns (1905, p. 191) stresses: 'What Mendel published is far from being all he did'. In addition it can be pointed out that the *Hieracium* experiments were conducted under conditions quite different from those under which he performed the *Pisum* experiments. As a teacher at the *Realschule* he was able in his free time to concentrate almost entirely on his *Pisum* research. As abbot he had to perform daily the duties of head of the monastery, both internally and in public. He was frequently away from Brno, leaving his experimental plants in the gardener's care.

Mendel's choice of *Hieracium* as a new experimental plant was, however, in no way 'unfortunate', as Dunn (1965*b*, p. 14) supposes. It was a logical step forward in the research that had begun with the *Pisum* experiments. In Brno Professor Niessl was considered an expert on *Hieracium*, and he had set up a collection of Moravian hawkweeds. Even scientists abroad were aware of his intention. Kerner von Marilaun thanked him in a letter dated 6 December 1872 for sending *Hieracium* plants. He was especially fascinated by the form designated No. 15, which seemed to him to be a hybrid 'in which *Hieracium echioides* plays a part' (Iltis 1966, pp. 204–6), and wanted to get hold of more abundant material in order to carry out a more detailed determination. It follows from Niessl's letter that the form was common on walls in the Brno area. Kerner adds: 'perhaps next summer Herr Mendel will be able to obtain a greater number of specimens of this plant, and if so I hope he will not forget me'. We can see from this that Mendel was in contact with this scientist, who was also interested in research into *Hieracium* hybrids. But no more detailed information regarding their cooperation is available.

In the *Hieracium* paper Mendel mentions six hybrids. Correns (1905) indicated 21 hybrid forms. He based this figure on the letters to Nägeli. In spite of the trouble he had with fertilization, Mendel obtained remarkable results. In one combination he managed to grow as many as 200 specimens, which he was then able to study in detail. Soon, however, he became engrossed in the question of hybrid fertility. He met up with the mysterious apogamy found in the genus, subsequently explained by C. Ostenfeld (1904*a,b*). Correns (1905, p. 248), stated that *Hieracium* hybrids 'soon got into discredit' and that scientists supposed that those occurring in the wild were mutant forms. But Correns points out that Mendel was already creating his own *Hieracium* hybrids, adding: 'It must not be forgotten that this was the result of quite exceptional work, which it will not be easy to repeat one day.' Much later, L. Powers and R.C. Rollins (1945) explained that apogamy in these hybrids is genetically determined and, as a result, in Mendel's experiments the hybrid progeny sometimes exhibited uniformity, sometimes variability.

The fertility of *Hieracium* hybrids was a precondition for the examination of the variability of traits in the hybrid progeny. In accordance with his *Geum* experiments, Mendel was able to assume that variability can be proved even in one recessive trait in the hybrid progeny. He also wished to examine the progeny in further generations. In the literature there is still a prevailing view that Mendel only observed the non-uniformity of *Hieracium* hybrids and the constancy of the progeny after self-fertilization. Matalová (1990) has expressed the view that Mendel could have proved the constancy of hybrids in agreement with the theory of the purity of elements.

In fact, Abbot Mendel did not lose interest in research into hybridization even after his experiments had come to an end. As late as the mid-1870s he

Fig. 5.15 *Notitzblatt 2*, proving Mendel's re-examination of the data from *Hieracium* experiments in the context of the results obtained in *Pisum* experiments.

returned to experiments with both *Pisum* and *Hieracium*. This is shown by fragments of records, known in the literature as *Notitzblatt 1* and *Notitzblatt 2* (Fig. 5.15). In the second fragment Mendel compared the results obtained with *Pisum* and *Hieracium* to those obtained with hybrids of the genus *Salix* reported in the scientific literature. Heimans, (1970) concludes from these fragments that later Mendel had become convinced that *Hieracium* hybrids were not constant, and saw this to be in accordance with his theory of variable hybrids. In the fragmentary record Mendel, instead of the notation 'constant hybrids' wrote of 'no changeable reproduced combinations' (*unveränderlich sich fortpflanzende Combinationen*). He explains this by supposing that cross-fertilization of polymorphic species of *Hieracium* yields segregating multifactorial traits which, given the extent of their variability, Mendel was unable to take into account because of the limited number of specimens he observed. Heimans' conclusion was: 'Indeed, Mendel on this notepaper, as nowhere else in his published studies, is handling the "Mendelian laws" himself, as if he were setting an example for all future geneticists.' Heimans saw in this a striking parallel with Mendel's desperate fight against the taxation of the monastery, to which our attention will be given in the next chapter. In his view Mendel's battle in the scientific field 'was fought with equal obstinacy, against the elusive phantom of the allegedly constant, intermediate hybrid. Equally lonely and steadfast, always relying on the fixed fundamental rules he laid out before, temporarily checked, it is true, but not disappointed by the conflicting results of his *Hieracium* research, he consequently and unfailingly pursued his ultimate objective.' Heimans added: 'It is tragic indeed, especially considering that Mendel did not live to witness the vindication of his views.'

On investigating the exchange of *Hieracium* plants between Mendel and Nägeli, Weiling (1969) found that the hybrids sent to Nägeli were included in a monograph on *Hieracium* published by Nägeli and Peter (1885). The authors acknowledged Mendel's contribution to the interpretation of the *Zwischenformen* in this genus, but did not consider his original conception in the *Hieracium* paper. In *Hieracia Nägeliana* Peter (1885–6) published 28 specimens originating from Mendel. These data were adopted by other authors prior to 1900, but none of them gave any details of Mendel's experiments described in the *Hieracium* paper.

Mendel's experiments with *Hieracium* were left unfinished. Only part of the experiments is described in the *Hieracium* paper. 'Each new method brings new results, and every new result is a source of new method' wrote Nägeli (1853) in a paper whose reprint he sent Mendel. In this spirit Mendel improved the method in the course of his *Hieracium* experiments, and might have obtained new results. He did not, however, leave any information relating to his understanding of the question: what part has hybridization to play in the origin of constant intermediate forms?

Hybrids and the idea of evolution

After 1900 there was much speculation on Mendel's attitude towards evolutionary theory, especially towards Darwin's theory. Before long, contradictory conclusions had been drawn. Some said Mendel rejected Darwin's theory, while others claimed he had supported it. The former based their claims on a biographical notice on Mendel published by Bateson (1902), which states: 'With the views which were at that time coming into prominence, Mendel did not find himself in full agreement.' At the time Bateson opposed the Darwinian theory of continuous variation, and in his criticism he made use of only half of a letter from Mendel's nephew, Ferdinand Schindler, written in 1902, where he states: 'He [Mendel] read with great interest Darwin's work in German translation, and admired his genius, though he did not agree with all the principles of this immortal natural philosopher. He often told us nephews that we would find in his bequest papers for publication which he could not publish during his lifetime. But we did not get anything from the monastery' (Coleman 1967). Writing many years afterwards, Mendel's nephew probably exaggerated his uncle's attitude to evolution. In the same year as Ferdinand wrote this letter, his brother Alois published a speech given in Mendel's native village, in which he stated that Darwin's theory had aroused interest among both scientists and theologians. But while many scientists merely launched into further daring theories Mendel took the only correct way, which was to examine the theory through experimentation (Kříženecký 1965*b*, p. 8). The two Schindler brothers both studied medicine in Vienna, and they were equipped to understand Darwin's theory even while Mendel was still alive. But it was not until many years afterwards that Ferdinand recalled a manuscript of Mendel's relating to Darwin. His brother Alois did not mention it. According to Mendel's biographer, H. Iltis (1966, p. 103), when Darwin's name came up, Mendel 'said that the theory was inadequate, that something was still lacking'.

Later, Mendel was referred to as an opponent of the Darwininan theory for ideological reasons. As an example one might cite a piece of information which came to light in 1933. In his book, A. Orel (1933, p. 175) came to the following conclusion: 'Into the important, but dark, realm of causes and laws according to which changes follow, about which he knew nothing at all, his contemporary, the Brno hermit Gregor Mendel (1822–1884), introduced the first light through his brilliant discovery of the laws of heredity, which has practically the same import as Copernicus' laws, and brought to an end the scientific catastrophe of Darwinism.' This is a quotation from a book which was clearly designed to propagate Nazi ideology. O. Richter (1941) challenged Iltis' presentation of Mendel as a free-thinking natural scientist, and tried to show him as an opponent of Darwin's theory. Later, Richter (1943), in his biography of Mendel, stated in the concluding part that Mendel offered the most convincing evidence as 'the most important witness for the prosecution' against the principles of Darwin's theory. B. Matoušková and O. Matoušek (1959) even stated that, immediately after Makowsky's lecture on Darwin's theory, Mendel defended the constancy of species. In reality, not only did no one produce any evidence to that effect, but it is not even proved that Mendel attended the meeting in January 1865 where Makowsky gave his lecture. It might, however, be assumed that he would have attended his friend's lecture if he could. In the period after 1948, Lysenko and his disciples also claimed Mendel to have been opposed to Darwin's Theory of Evolution, without putting forward any evidence (Herčík and Novák 1952).

Sinoto (1971) drew attention to Mendel's motivation for his hybridization research into obtaining new colour variants in ornamental pea plants, which was in fact 'nothing else than a problem of evolution'. The *Hieracium* research was also according to Sinoto an attempt by Mendel to solve the problem of the diversity of species by means of the doctrine of hybridization, thus again dealing with the problem of the mechanism of evolution here too. And according to Olby (1979) and other historians of science, such as Bowler (1989) an Di Trocchio (1989), and Corcos and Monaghan (1990), in his research into plant hybridization Mendel was mainly interested in explaining the role of hybridization in the appearance of a new species which was of evolutionary interest. Examining the emergence of the concept of heredity in nineteenth-century science Olby (1993) even later attributed heredity, which geneticists conventionally attribute to Mendel, to F. Galton. Callender (1988) wished to present Mendel as an opponent of Darwin's concept of descent with modification, because he defended Linnaeus' idea of production of new plant species by hybridization against the views of Kolreuter and Gärtner, which Meijer (1983) considered 'an exaggeration, as much as Richter's in 1943'. According to Bowler (1989, p. 105), Mendel presented the production of constant *Hieracium* hybrids as potentially new species. For this reason, the author supposes, he regarded his *Hieracium* experiments as being more important than the *Pisum* experiments. But a detailed study of Mendel's letters to Nägeli refutes this supposition.

Mendel's attitude to evolution should be understood in connection with early-nineteenth-century ideas on the subject. On page 87 of Gärtner's monograph Mendel made a note referring to page 72 of the journal *Flora* dating from 1848. In an extensive paper on the transformation of species, Hornschuh (1848) stated that the problem of species transformation is the weakest link in plant science, and that no one had as yet explained the transformation of species or refuted the idea. He was particularly interested in the question of what the difference was between the seeds of vetch (*Vicia sativa*), pea (*Pisum sativum*), and lentil (*Errum lens*). In this context Hornschuh drew attention to the fact that in the progeny of crossed plants rigorous segregation of forms of flowers and plants came to the fore, without the occurrence of any intermediate forms. For Mendel this was a serious problem both from the viewpoint of breeding and from that of taxonomy, physiology, and evolution. He sought an explanation in a law of the origin and development of hybrids.

Mendel wrote about the origin and extinction of species as early as 1850, in his geological essay, at his first attempt at the examination for teachers. This essay was published by Orel *et al.* (1983). At the time he based his views on literature that had been influenced by C. Lyell and by German *Naturphilosophie*. His fellow monks sought in the study of natural sciences along with philosophy an explanation of the phenomena they met with in nature. Klácel (1843) saw the whole of nature 'in action'. He was acquainted with the German translation of a four-volume work by Erasmus Darwin, *Zoonomia*, published in 1805, which contained an explanation of the origin of life from the original viable filaments, and the development of simple forms to more complex ones. The origin of life was explained in similar terms by M.J. Schleiden (1848) in his book *The plant and its life*. In the copy which has survived in the monastery library there are pencilled marginalia, most probably Mendel's. Schleiden stated that life arose 'at least once from the struggle of inorganic elements'.

Mendel may also have acquired a new methodogical view of the evolution of living nature from a study of Schleiden's masterpiece *Principles of scientific botany*. In the extensive introductory section on methodology the author emphasizes that 'only developmental history can open up the way to a knowledge of the plant; yes, the whole arrangement of the plant can be recognized with certainty only when not only its parts are compared, but also the whole of its developmental history' (Schleiden 1849–50, pp. 141–6). In the introduction to the *Pisum* paper Mendel expressed his conviction that he was contributing by his research to the answering of an important question of developmental history. The content of this concept was explained by Schleiden in *Principles of scientific botany*, which Mendel studied. According to him, *Entwicklungsgeschichte* (developmental history) was 'a leading maxim for the investigation of the laws of organisms'. It originated from the beginnings of organic creation, and was to develop into a driving mechanism, 'a natural process', based on the fact that developmental history set out from a given embryo (Orel 1979).

The distinguished traits of two plants, according to Mendel (1866a, p. 42), 'can after all be caused only by differences in the composition and grouping of elements in dynamic interaction in their primordial cells'. This explanation is reminiscent of the recommendation of Unger (1852, pp. 4–6) to investigate the 'physics of the plant organism' using the methods of physics and chemistry. Unger was of the opinion that Nägeli had gone furthest with such research. Nägeli (1866a) tried to explain the differences in plant traits by the physical and chemical composition of substances within the cell. He assumed that the 'sleeping *Anlage* and dispositions' determining the traits were subject to both internal and the external environmental influences.

In Nägeli's understanding of the term, the *Anlage* lost their identity in the process of reproduction. Mendel rejected the idea of the influence of environment on changes in traits, introducing a concept of *Anlage* that retained their identity, and were not endowed with a force determining some sort of graduation, as had been envisaged by Nägeli. Mendel returned to Unger's original notion that daughter cells were replicas of parental cells, and his 'mediating' cells complied with Unger's insistence on the need to explain the concentration of the whole being of the plant in the cell. M. Campbell (1985) was therefore right in her conclusion that Mendel's *mediating cells* were the equivalent of Unger's source of variation. Both Nägeli and Mendel sought an explanation for evolution in connection with hybridization and cell theory, applying the latest knowledge in various different branches of science. But each set out from his own scientific conception.

Mendel may have first heard about Darwin's Theory of Evolution from the report of K. Schwippel, who in September 1861 lectured in Brno on the chapter 'On the geological succession of organic beings' (Orel 1971d). Darwin wrote 'all the chief laws of palaeontology plainly proclaim, as it seems to me, that species have been produced by ordinary generation: old forms having been supplemented by new and improved forms of life, the products of variation and survival of the fittest'. A copy of the German translation of Darwin's *The origin of species* is very likely to have reached the hands of the Brno naturalists in 1862. The minutes of the monthly meeting of the Natural Science Society contain a list of forty-three new books donated by Franz Czermak on

14 January 1863. Among them is the German translation of Darwin's *The origin of species*, published in Stuttgart in 1862. The monastery library contains a second German edition of this book, published in 1863, which was bought by Mendel. It is bound, and contains Mendel's marginalia.

At a meeting of the Natural Science Society in January, Makowsky (1866) spoke about Darwin's theory, and the lecture was subsequently published under the title 'On the theory of organic creation'. He interpreted the nucleus of Darwin's theory as follows: 'Natural selection and the differentiation of traits which follow from it form the centre of gravity of Darwin's theory, by which it stands or falls. It is based on enormous experience in horticulture and animal breeding, where the organic form in the hands of rational man is as plastic as wax'. Makowsky closed his lecture with the words: 'Gentlemen, however much Darwin's theories may be opposed to current views, they are at least as justified as those of the fixity of species. They open up a wide, almost virgin field of investigation into the evolutionary processes, and their interaction with environmental influences. They should put us on our guard against mental indifference, which is bound to lead to ignorance.' The lecture was reported in the newspaper *Neuigkeiten* on 13 and 14 January, and according to the report it met with 'general approval'. The news item does not mention any opposition to the theory. Gavin de Beer (1966), publishing an English translation of the reports, drew attention to a very appropriate quotation of the following words by Goethe at the end of the report: 'The task of scientific research in the future should, for instance, not be to investigate for what purpose an ox has horns, but how it has come to have them.'

In his *Pisum* paper, Mendel expressed a belief that the understanding of the course of variation was an important contribution to the understanding of the evolution of organic forms, a process which he takes for granted. Mendel was acquainted with *The origin of species* before he wrote the *Pisum* paper. But he explained his theory of the origin of hybrids without mentioning Darwin's name. Later, however, in the *Hieracium* paper, Mendel (1870*a*, p. 28), wrote: 'The question of the origin of the numerous and constant intermediate forms has recently acquired no small interest since a famous *Hieracium* specialist has, in the spirit of the Darwinian teaching, defended the view that those forms are to be regarded as from the transmutation of lost or still-existing species.' Mendel drew attention to Nägeli's view, though he did not mention him by name.

In his letter to Nägeli dated 3 July 1870, however, Mendel mentioned having eye trouble, but stressed that even in such a situation he could not bring himself to postpone an important experiment relating to the views of Naudin and Darwin that a single pollen grain does not suffice for the fertilization of one ovule (Correns 1905, p. 235). In the same letter Mendel mentions Darwin's name a second time, describing hybrids of *Matthiola annua* and *glabra*, and *Zea* and *Mirabilis*.

A month later Mendel returned to the question of evaluating hybrids of *Mirabilis*, which in his experiments behaved like hybrids of *Pisum*. He added that Darwin and Virchow had pointed to the high degree of independence which was typical of individual characters and whole groups of characters in animals and plants. In this case the data were in agreement with Mendel's research results. In his tenth letter to Nägeli he even used Darwin's phrase 'struggle for life' when explaining the origin and extinction of *Hieracium* species (Correns 1905, p. 247).

Indirect information regarding Mendel's view of Darwin's theory can also be found in his notes on the German editions of Darwin's books. In his copy of *The origin of species* Mendel noted twelve pages where Darwin's text seems to have caught his attention. Some of these may be singled out. On the first page which Mendel marked, Darwin writes that 'species are not immutable', and on the next line that 'natural selection has been the most important, but not the exclusive means of modification'.

On page 17 he marked the text: 'it seems clear that organic beings must be exposed during several generations to new conditions to cause any great amount of variation'. Gavin de Beer (1964) expressed the view that in the *Pisum* paper Mendel, in criticizing the idea of changes occurring in plant traits when living conditions were changed, had Darwin in mind. In fact Darwin accepted Gärtner's view, and Mendel may have polemized with the latter author, whose book he had studied in detail at the start of his experiments with *Pisum*. On pages 62–3 Darwin states that 'certainly no clear line of demonstration has yet been drawn between species and subspecies', and that the differences 'blend into each other by an invisible series'. The same attitude to the differences between species and subspecies interested Mendel, but he was not able to accept the blending of traits as an explanation. He also marked with double lines the text on page 302: 'The slight degree of variability in hybrids from the first cross or in the first generation, in contrast with their extreme variability in the succeeding generations, is a curious fact and deserves attention'. Here Mendel must have felt some gratification in the thought that his theory was soon to explain this curious fact.

There are also remarkable marginalia by Mendel in Darwin's *Variation in animals and plants under domestication*, published in German translation in Stuttgart in 1868. In the first volume only five entries by Mendel can be found. The book is unbound and mostly uncut. On the last page, numbered 530, Mendel pencilled a note at the end of the concluding text:

412 Muster
312 dtto
489
505

On page 412 there is a description of Knight's experiments with *Pisum*.
There follows Master's opinion that the nature of the soil has an influence
on the loss of the character of plant varieties. Mendel added in pencil
Wiederlegung, meaning 'contradictory'. On page 312 Darwin described
traits of different poultry breeds. Mendel may have been interested in
Darwin's description of alternative traits of poultry, and on pages 412 and
312 he pencilled the note *Muster* (model). This may have been connected
with his first ideas of the experimental crossing of animals. This would be
confirmed by a surviving letter addressed to Abbot Mendel and dated 9 May
1869 (Mendelianum, sign. no. 203). The headmaster of the *Gymnasium* in
Brno, A. Král, explains how he is able to check the fertilization of eggs at
the start of incubation. No other evidence of Mendel's interest in this
subject has ever come to light.

Mendel paid more attention to the second volume of Darwin's *Variations*.
There are fifty-seven marginalia, mostly in Chapter XXVII 'On the provisional hypothesis of pangenesis'. Mendel expresses strong disagreement
with Darwin's views in a note on page 497, doubly underlined and with an
exclamation mark: '*As each unit, or group of similar units* [underlined by
Mendel] throughout the body, casts off its gemmules and as all are
contained within the smallest egg and seed, and within each spermatozoon
or pollen-grain, their number and minuteness must be something in-
conceivable.' In the footnote Mendel pencilled 'to succumb to an impression
without giving the matter proper thought' (*sich einem Eindrucke ohne
Reflexion hingeben*). A similar polemic remark can be found on page 525,
where Darwin writes: 'a certain number of gemmules being requisite for the
development of each character'. On that same page he writes of 'several
spermatozoa or pollen-grains being necessary for fertilization'. This may
have been instrumental in persuading Mendel to perform his experiments
with *Mirabilis*. On page 126 there is a passage where Darwin notes the
appearance of different characters in the progeny of the same parents,
which he could not understand, but attributed to superfetation. The fertilization of one ovum by one male cell was, however, an axiom in Mendel's
research.

Mendel also carefully studied Darwin's book *The effect of cross-fertilization in
the vegetable kingdom*, published in German translation (Darwin 1877). He
had the opportunity to read this book twelve years after publishing his *Pisum*

paper. At the time he was fully occupied with his duties as abbot. On page 20 he marked the text: 'As the flowers which were crossed were never castrated, it is probable or even almost certain that I sometimes failed to cross-fertilize them effectually, and that they were afterwards spontaneously self-fertilized.' Mendel, with his experience of techniques of artificial fertilization of plants, could not leave this passage unnoticed. He reacted similarly to Darwin's text on page 24, where the author speaks of 'the whole subject of hybridism' as 'one of the greatest obstacles to the general acceptance and progress of the great principle of evolution'. Darwin was unable to understand the appearance of parental traits in hybrid progeny. It contradicted his concept of the occurrence of hybrids due to blending inheritance.

According to Niessl's information, published by E. Proskowetz (1902), Mendel performed transplantation experiments with *Ranunculus caethaefolius* Bluff, and on 14 April 1869 Professor Makowsky showed experimental plants at a meeting of the Natural Science Society, commenting that Mendel had not proved any changes due to different cultivation conditions. In the published report it is stated that Mendel intended to continue these experiments into further generations. But Mendel never mentioned them again. In this connection Niessl recalled Mendel's statement: 'This much already seems clear to me, that Nature does not modify species in any such way; some other force must be at work.' This was also the basis of Mendel's conclusion that there was 'something still lacking' in Darwin's theory. He hoped his research would fill this gap. In this connection Niessl noted that Mendel began his research at a time when Darwin's Theory of Evolution was at the forefront of natural scientists' interest. Niessl, who was sixty-three at the time, did not realize that Mendel had begun his experiments four years before Darwin's theory was published.

A remarkable document relating to Mendel's opinion of Darwin's theory is a surviving copy of a letter from the claustral prior of the monastery, A. Tkadlec, written on 28 November 1917 to an unknown addressee in Darmstadt. A copy of this letter can be found in the Mendelianum (sign. no. 6943). The letter reads as follows: 'Allow me to add the following *private* information. I am the last member of the monastery to have been received by Mendel in 1883; in the following year he died. After his death no letters were found, or even notes, relating to his crossing of plants. This is understandable. His discoveries did not meet with acclaim; indeed, he was even attacked and his theory suspected of being contrary to the revealed truths of the Christian religion, the untruth of which was proved by Weismann's scientific works. In bitterness he burned everything which reminded him of his previous activity, and devoted his attention to the economic affairs of the monastery.'

When, on 15 October 1883, Mendel received Tkadlec into the monastery, he was already seriously ill, and the monastery was pervaded with the atmosphere of gloom described by A. Rambousek in a letter to P. Křížkovský dated 8 May 1883 and published by Eichler (1904, and Orel 1971c). Mendel was criticized for his dispute over taxation of the monastery, and at that time criticism of his scientific views may also have appeared. This could have been the origin of Tkadlec's views, and the latter possibly associated Mendel's work with Darwin's theory. It is not out of the question that the tired and sick abbot burnt the notes on his research, realizing that no one would understand them anyway. On 24 July 1924, looking back over several decades, Alois Schindler wrote to A. Matoušek that half a year after Mendel's death Abbot Rambousek had told him he had been in a dilemma as to what to do with the papers relating to Mendel's scientific work. In the end he decided to burn them. At the time Schindler regretted not having shown an interest in his uncle's work earlier, so that he might have preserved very valuable scientific documents (Kříženecký 1965b, pp. 102–4).

In connection with Mendel's relation to Darwin the idea has even been put forward that Mendel met Darwin during his trip to the Great Exhibition in

Fig. 5.16 Mendel in the middle of a group of Moravian visitors to the Great Exhibition in London, taken in front of the Grand Hotel in Paris in 1862.

London in 1862 (Fig. 5.16). Mendel wrote to his brother-in-law L. Schindler on 14 July 1862, that on 24 July he was travelling via Vienna and Paris to London, and that he would be returning around the middle of the following month. A large party of enthusiasts travelled from Brno to see the Exhibition, and they had their photograph taken in front of the Grand Hotel in Paris. No list of participants has yet come to light (Kříženecký 1965b, p. 120).

O. Richter (1943, pp. 172–3) asked Darwin's son Leonard to search the documents for any indication that Mendel might have visited Darwin's home. Gavin de Beer (1964) published Leonard Darwin's letter to his niece, from which it follows that at the time Mendel was in London, he, Leonard, was seriously ill, and his parents were living outside London and receiving no guests. He could find no mention in his father's correspondence relating to any contact with Mendel. It has not previously been taken into account that at the time Mendel was in London he had not published any results of his research into plant hybridization. In addition, he was known to be a very modest man, and one cannot imagine he would try to visit a man who was already internationally known as a prominent naturalist. Last, but not least, Mendel could not speak English.

It is worth investigating the question of whether or not Mendel's friend J. Nave also took part in the trip to London. It has not so far been possible to confirm this from the group photograph mentioned above. At the time, Nave (1864, 1867) was putting the finishing touches to the manuscript of a book on algae published in German and in English. The translator and editor was the Revd. W. W. Spencer, MA, Fellow of the Royal Microscopic Society. Nave's book was received very well in England. A second edition appeared the same year, a third in 1881, a fourth in 1896, and even a fifth in 1904. This clearly-written book from the provincial town of Brno, dealing with questions which were of more interest to experts in England was a great success. Mendel's paper, containing quite a new theory, remained unnoticed at home, though the *Proceedings* of the Natural Science Society were delivered to the libraries of the Royal Society and the Linnaean Society in London, the Natural History Society, the Royal Geological Society in Dublin, and later also to Cambridge University. Nave died in 1864, and no correspondence relating to his contact with natural scientists in England has been found. Previously, Nave (1858) had lectured in Brno on the latest findings relating to the problem of generation and fertilization. This question was also related to evolution. If Nave was in London with Mendel in 1862, he may also have been interested in Darwin's theory. In that case he would have talked to Mendel about it. But for the time being we have no information to go on.

Mendel may also have spoken to Klácel about Darwin's Theory of Evolution. His fellow monk left for the USA in 1869. In his lecture on

Darwinism in America, partly published in English by Matalová (1979), Klácel recalled his own experiments with crossing plants to investigate the variability of plants traits. In this connection he pointed to the importance of seeking the smallest particles of matter determining heredity with the aid of a microscope. In the USA, Klácel obtained the English editions of Darwin's most important books, and these copies can still be seen in Prague. It is remarkable that he marked almost exactly the same passages in the text as Mendel did in his German translation. One might assume that they had previously discussed the theory together in Brno. Darwin's theory was also of great interest to Bratránek. In the first volume of the correspondence of J. W. Goethe with leading natural scientists, Bratránek (1874) compared Goethe's theory of metamorphosis with Darwin's Theory of Evolution. To the question put forth by zoologist O. Schmidt 'was Goethe a Darwinian?' he replied that Goethe's concept of type was in basic contradiction with Darwin's concept of evolution.

Mendel came across Darwin's theory as his *Pisum* experiments were drawing to a close. From his notes and from indirect evidence one can suppose that he did not see any conflict between this theory and his own. But we also have no detailed knowledge of how Mendel understood the individual parts of Darwin's theory, particularly the concept of descent with modification. Whatever the case, he was firmly against the theory of pangenesis, according to which new species arose through the action of the environment.

Questions of whether Mendel was a proponent or an opponent of evolution did not start to appear in the literature until after 1900. Biologists were seeking the mutual relations between Darwin's and Mendel's theories, and in this confused situation the two ideas met late in the day and through the interpretation of scholars. Mendel himself made a thorough study of Darwin's books during his abbacy. He was surely interested in their impact on his own scientific work. But he had formulated his own research problem before he came across Darwin's work. In his experiments into plant hybridization he investigated his own problem, which according to him seemed to be 'the one correct way of finally reaching the solution to a question whose significance for the developmental history of organic forms must not be underestimated' (Mendel 1866a, p. 4). He tackled experimentally the transmission of traits of organisms from generation to generation. It was the problem of heredity successfully divorced from the term generation in the sense suggested by Wagner (1853). In this connection Mendel had, before coming across Darwin's Theory of Evolution, also investigated the role of hybridization in the origin of new species. That was Mendel's own attitude to the evolution of organic forms in nature.

Honesty in presenting experimental data

A hundred years after the publication of Mendel's work on *Pisum*, a spate of papers appeared which drew attention to a previously unknown problem: Mendel had carried out experiments with peas according to a theory conceived in advance, and either he or an assistant 'adjusted' the numerical results from individual experiments to fit the expected segregation ratios. The first to suggest something of the sort had been R.A. Fisher in 1911, to whose comments attention is drawn by Norton and Pearson (1976). According to them, Fisher, while still an undergraduate, had pointed out that Mendel's data fell within the limits of probable error, adding: 'It may have been just luck; or it may have been that the worthy German abbot, in his ignorance of probable error, unconsciously placed doubtful plants on the side which favoured his hypothesis.' (Edwards 1986). When Fisher later (1936) published his critical paper on Mendel's research, the editor of the journal said that he expected to find strong evidence that the data 'had been cooked'. Twenty years later, in the marginal comments to a new English edition of the *Pisum* paper, Fisher, (1955*b*) stated that the data in the *Pisum* paper had been 'systematically sophisticated'. But he no longer doubted that Mendel had been deceived by a gardening assistant, who knew only too well what his master expected from each trial he made. In 1936 Fisher had written to E.B. Ford that he could not conceive 'that Mendel himself had any hand in it'. Thus Fisher took up different stances at different times in explaining the falsification of the *Pisum* paper data.

Fisher was most fascinated by the experiments in which Mendel tried to find the proportions of heterozygous to homozygous forms resulting from crosses between parents differing in one trait. He grew ten seeds from each plant to be tested and counted the number of seedlings which did not exhibit segregation of plant traits into dominant and recessive forms. When no recessive form appeared, Mendel considered the parent to have been homozygous dominant, and when a recessive form appeared, to have been a heterozygous dominant. Fisher (1936) drew attention to the fact that if each plant had a probability of 0.75 of exhibiting the dominant trait, the probability of all ten seeds from one plant doing so is not 0.75, but 0.75^{10}, which is 0.0563. This means that with samples of ten seeds a number between 5 per cent and 6 per cent of heterozygous plants will be mistaken for homozygous plants because they are likely to have no recessives among the offspring, and the production of recessives is taken as the test of their heterozygous nature.

Fisher used the chi-square test on Mendel's data from crossing experiments using plants differing in one, two, and three pairs of traits, and also in experiments with the germ cells of hybrids. His conclusion was that the results of the individual and pooled tests showed that the published data were too good in favour of the expectation, so that if one were to reproduce Mendel's experiments exactly, the probability that one would arrive at the same results is about one in thirty thousand.

No one has been able to explain why the criticism made by Fisher (1936) remained unnoticed for so long, until Zirkle (1964) noted that modern statisticians, without mentioning Fisher's name, supposed Mendel to have manipulated his results to fit the theory by chance, without having convincing data to go on. Later Zirkle (1966) lectured in Brno at the Mendel Memorial Symposium without mentioning bias in Mendel's data. A year previously Sir Gavin de Beer (1964) had also referred to Fisher's criticism. His view was also that some assistant who knew what Mendel expected had altered the figures. Olby (1966, pp. 182–3) compared Mendel's data with those of the researchers who after 1900 repeated Mendel's experiments, and found that the data of Tschermak were even more precise than those of Mendel. Together with Kříženecký, he considered the explanation to lie in the fact that Mendel could stop evaluating his experimental material as soon as he was convinced that he had achieved the required reliability of data. Beadle (1967) also admitted the possibility that Mendel stopped counting peas when he obtained results close to those expected. But he very carefully explored, as Sapp (1990) noted, the phenomenon of unconscious bias to account for Mendel's results, using experience from his genetic experiment with maize to illustrate his point. Later Campbell (1985) pointed out that Mendel denies this possibility in the *Pisum* paper itself, where he states: 'it is necessary to observe without exception all members of the series of offspring in each generation'. Dunn (1965*b*, p. 12) took into account the too-good-to-be-true results of Mendel's experiments, but at the same time reminded readers that there was no evidence to support Fisher's conclusion that the results had deliberately been falsified by Mendel. Sturtevant (1965, pp. 13–16), writing at the same time, considered three possible ways in which the data might have turned out too well. Mendel may have unconsciously evaluated poorly expressed traits in favour of the expectation; he may have left such traits out of his evaluation; or, finally, the results may have been altered by a helper who assessed unclear traits in favour of the theory. Sturtevant's conclusion was that the best answer to this new problem was that Mendel was right.

Fisher's 1936 paper was republished by Curt Stern and Eva R. Sherwood (1966) in *A Mendel source book*. In the foreword Stern writes: 'Why Mendel's specific data are too good from a statistical point of view remains unknown, but comments which throw some light on this question have kindly been provided by Professor Sewall Wright. These follow the reprint of Fisher's remarkable paper.' In a three-page contribution, Wright (1966), an acknowledged geneticist and statistician, concludes: 'Taking everything into account, I am confident, however, that there was no deliberate effort at falsification.'

In 1966 publications began to appear which asserted that Fisher's method of analysing Mendel's data had not been entirely appropriate. Thoday (1966) pointed out that pollen cells occur in tetrads, and this may have affected the results of segregation of traits in Mendel's experiments, for which Fisher had assumed a mere binomial distribution. Most attention was paid to the question in the period after 1965 by F. Weiling (1985, 1989, 1991), a biometrician, who summarized his investigation of this enigma. He not only made a new study of Mendel's publication, but also sought new evidence regarding the scientist's personality. He computerized Fisher's analysis, and pointed out that the latter based his calculations on Mendel's having obtained ten plants from ten seeds. If one supposed a realistic germination rate of 80–90 per cent, the re-computed chi-square test came out lower than that of Fisher. Weiling (1966a,b) also computed the same test for experimental data in repeated experiments with the crossing of peas by Correns, Tschermak, Bateson, Kilby, and Derbishire after 1900, and found no major difference between their data and Mendel's. Lamprecht (1968b) reached the same conclusion. Van der Waerden (1968) undertook a statistical re-evaluation along the lines suggested by Weiling and came to the conclusion that 'some low values of the chi-square test can be explained without assuming the data to be biased'. He admitted the possibility of the data from the experiments with germ cells of hybrids having been influenced by the expected result. His conclusion was: 'However, the evidence is not so strong as Fisher thought it was.' In agreement with Weiling, Orel (1969), in defending Mendel's precision in presenting experimental data, opened the question of how long 'the story of too-good-data' in Mendel's research would continue. Later, Cock (1980) also rejected the idea of any intention on Mendel's part of adjusting the data.

Alongside the authors who pondered over the question of how Mendel's results came to be 'too good' in Fisher's estimation, there were those who questioned the suitability of the statistical methods Fisher used to prove his point. These latter views were sparked off by Weiling's papers, repeatedly drawing attention to the rigorousness of Mendel's data. Later Root-Bernstein

(1983) made a detailed study of Mendel's paper, along with the whole story of 'too-good-to-be-true' results, and came down on the side of those asserting that 'there is something wrong with Fisher's analysis'. According to him Mendel's results are prefectly valid. Root-Bernstein concentrated his attention on the subjective character of Mendel's trait classification in seeds and plants. In his view what Mendel published was not a 'real' description of his peas, but his perception of how peas could be categorized into 'ideal' discrete groups. He was led to this idea by Campbell (1985), who cited some of the recent psychological literature concerning observer bias in science. According to this new approach, Fisher did not analyse Mendel's experiments *per se*, but rather Mendel's perception of his results. Root-Bernstein also took into account difficulties in Mendel's classification of traits which Fisher did not consider, quoting Mendel himself, who wrote in the *Pisum* paper that 'some of the traits listed do not permit a definite and sharp separation, since the differences rest on a "more or less" which is often difficult to define' (Mendel 1866*a*, p. 7). Root-Bernstein also considered the analysis of the fragment of Mendel's notes from the *Pisum* experiments, from which Heimans supposed that 'all calculation can be explained as a laborious attempt to find the true position of the boundary line'. The conclusion of Root-Bernstein's analysis was that 'Mendel's results are statistically unlikely and must therefore be used as a clue to his method of classification, not as a basis for a normative judgement about the validity or non-validity of his results'. Pilgrim (1984), inspired by Weiling's analysis of Mendel's data, considered anew the too-good-to-be-true paradox, coming to the conclusion that 'there is no evidence that Mendel did anything but report his data with impeccable fidelity'.

From 1965 onwards, Weiling published fifteen different papers pointing to the rigorousness of Mendel's results, and in 1985 he summed up his point of view, reaching the conclusion that Fisher's model of binomial distribution did not fit in with the biological character of Mendel's experiments, and pointing to the more suitable model of hypergeometric distribution. Later Pilgrim (1986) agreed with this assessment, and stated that what Fisher should have been asking was: What is the probability that the data in Mendel's paper represent a truly random sample? The chi-square method does not answer this question. Lebel (1987), confirmed the view of Weiling, and drew attention to the comments of Sewall Wright on this subject (1966) in favour of Mendel's honesty.

Fisher's pupil A.W.F. Edwards (1986), after a detailed study not only of Fisher's publications on the subject, but also of his notes from a study of Mendel's research, made a reassessment of Mendel's data from the statistical point of view in the light of Weiling's assertions. He concluded that his overall impression from reading all the commentaries since Fisher's in

1936 was that 'a good deal of special pleading, not to mention downright advocacy, has failed to make a substantial impact on Fisher's conclusion'. Edwards supposes that 'The segregations in Mendel's paper are in general closer to Mendel's expectations than chance would dictate'. Edwards's examination of Mendel's data was recently analysed by D.L. Hartl. He draws attention to two series of experiments consisting of a progeny test in which plants with the dominant phenotype were self-fertilized and their progeny examined for segregation to ascertain whether each parent was heterozygous or homozygous. Fisher (1936) and Edwards (1986) over-looked the fact that only in the first series of experiments does Mendel explicitly state that he cultivated 10 seeds from each plant. The observed result of these experiments is not significantly more deviant from the true expectations and this series of progeny tests yields no evidence that the data had been adjusted. More problematical seems to be the second series of progeny tests; the reported data differ very significantly from the true expectation. According to Mendel 'this experiment was conducted in a **manner quite similar** to that used in the preceding one'. Mendel is usually very precise, and if he had done the experiment in exactly the same manner as before it would have been easy for him to say so. Hartl shows that if Mendel had cultivated more than 10 seeds per plant, then the statistical test would be quite different and the insinuation of data tampering evaporates. Thus the uncertainties in the experiments and ambiguities in this analysis discredit any inference of deliberate manipulation or falsification of data (Orel and Hartl 1995).

Weiling (1991) returned to the controversial evaluation of the data from Mendel's experiments, and offered convincing new arguments for Mendel's honesty in publishing his papers. According to him Mendel's ratios are not random samples of adequately large and thus binomially distributed total-ities, as presupposed by the chi-square test, but rather the sum of several or numerous ratios based on single plants (in the case of seed characteristics) or ratios of offspring (in the case of plant characteristics). This fact 'makes it impossible to add the chi-square values for the different experiments on the results of which Fisher based his judgement'.

Weiling's viewpoint on the one hand and Edwards' on the other are based on different scientific approaches. One can suppose that in future there will be further differences of opinion. Whatever these may be, they will not alter the basic assessment of Mendel's achievements in the history of science, which were aptly expressed by Fisher (1936) as follows: 'The facts available in 1900 were at least sufficient to establish Mendel's con-tribution as one of the greatest experimental advances in the history of biology.'

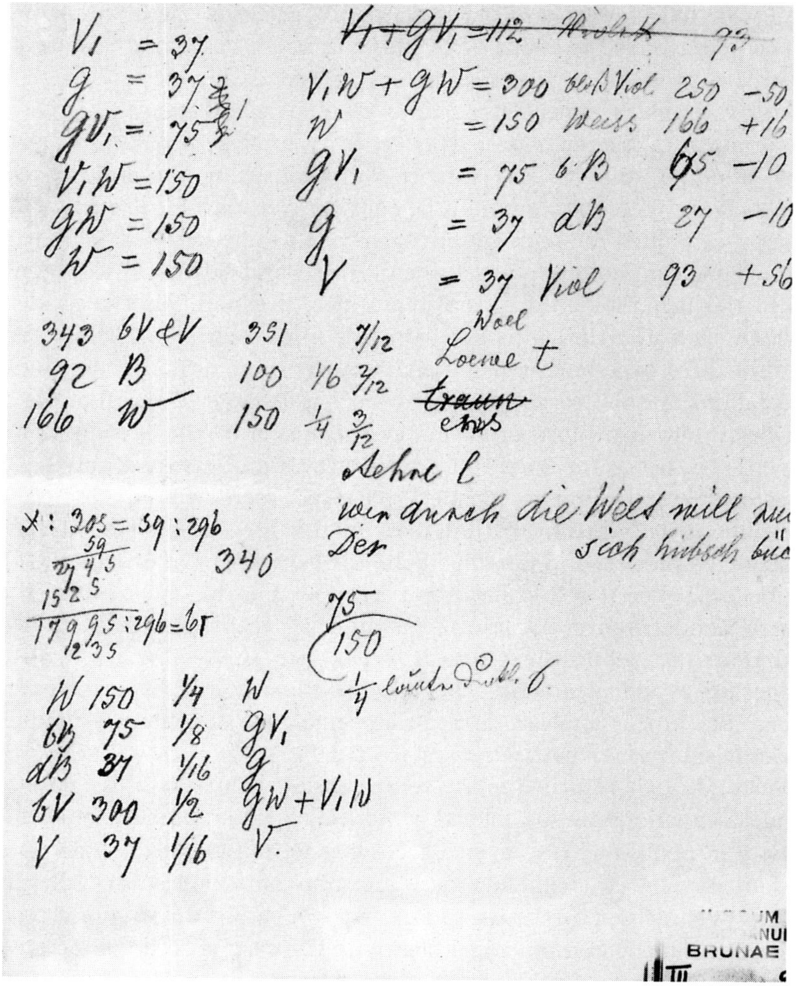

Fig. 5.17 *Notitzblatt 1*, illustrating Mendel's notes relating to his re-examination of numerical data from the experiments in plant hybridization.

Aside from any scientific evaluation of Mendel's experimental results, there have also been sensation-seeking articles accusing Mendel of deliberately falsifying the record. In Mendel's home town of Brno, biologists F. Herčík and L. Novák (1952), anxious to support Lysenko's pseudo-scientific ideas, asserted that the fragment of the researcher's notes, known in the literature as *Notitzblatt 1*, found by Richter (1924) (see Fig. 5.17) show how the abbot manipulated the data in order to prove his 'deceitful' theory. The sheet, measuring 33.3 cm × 21.0 cm, is covered on both sides by notes on the tax problem, written in pencil by Mendel. The text on the reverse side and partly on the front side evidently represents a draft of a letter written by Mendel. The experimental data cover three-quarters of the front side. They are a series of letters representing classes of the offspring of hybrids. The *Notitzblatt* suggests that Mendel was playing with the data and apparently trying various ways of grouping the phenotypes into classes. When publishing the fragment Richter (1924) assumed that Mendel had been trying to analyse segregation data of a dihybrid and trihybrid cross involving a blue- or violet-flowering plant, probably the cross *Linaria vulgaris* × *L. strata*. Mendel mentioned this cross in his third letter to Nägeli in November 1867.

Richter also published a letter by Professor F. Frimmel, a noted plant geneticist in Brno, according to whom Mendel was trying to explain data from flowers or seeds derived from crossing *Lathyrus odoratus*. An additional note by Frimmel states that the *Notitzblatt* should be considered 'a hasty sketch of an idea, comparable with the study sketch of a painter'. A photocopy of the *Notitzblatt* was later published by Darlington and Mather (1950), with an explanation by Fisher, who assumed that the numbers 343:92:66 represented the ratio 9:3:4 characteristic of segregation for two genes related in action by recessive epistasis involving seed colour, probably of beans. Lamprecht (1966) took another view, assuming that these data referred to segregation of the seed colour of *Phaseolus vulgaris*. Olby (1966) inclined to Fisher's explanation of factor interaction, but thought that the data were probably related to the cross *Linaria vulgaris* × *L. striata*.

Heimans (1969) analysed the *Notitzblatt* in detail, and concluded that Mendel's calculation related the trifactorial experiment with the crossing of peas differing in round or wrinkled seed form, yellow or green seed colour, and grey-brown or white colour of the seed-coat. In his paper Mendel describes the colour of the seed-coat as grey, grey-brown, or leather-brown, with or without violet spotting. This differential pigmentation of the seed-coat is described in 687 seeds of his 24 hybrids. According to Heimans, the offspring of the hybrids included 166 peas with a white seed-coat, which Mendel denoted by C. The number 166 was the key to Hiemans' original explanation of the *Notitzblatt* data. After subtracting 166 white forms from the total number of 639, the number of other forms should be 473. However, in the *Notitzblatt* Mendel gave the number as 435, a difference of 38. Heimans assumed that in investigating the non-uniform appearance of the non-white seed-coat, Mendel had to base his classification on the labels on

his seed bags, some of which may have been lost by then. There are five
classes listed in the *Notitzblatt*.

blas Viol	250
Weiss	166
bB	65
dB	27
Viol	93

According to Heimans, Mendel could distinguish four subdivisions: dark
brown and light brown, both with and without spots. He united the 250
blass Viol with the 93 Viol, and the 65 light brown with the 27 dark
brown, thus giving a theoretical ratio of 343:92:166. After correction he
simplified it to the ratio 7:4:3. He thus came to the interaction of factors
without giving any explanation. According to Heimans, the *Notitzblatt* illustrates a 'great master at his work'.

   Olby accepted Heimans' opinion and added that 'there is, therefore, no
escape from the conclusion that Mendel's conception of the character pair
did not lead him to the conception of mutually exclusive pairs of factors
also'. This argument was later used by those who rejected the duality of
trait determinants in Mendel's conception. Heimans' interpretation of the
Notitzblatt (Fig. 5.17) as an examination of data dealing with pea experiments seems to be correct. Nevertheless, it should not be claimed—to use
Frimmel's words—that this sketch gives the final theoretical explanation of
traits in the process of germ production and plant fertilization, which
Mendel treated in the conclusion of his paper in 1865.

   A significant contribution was made to this interpretation of the
Notitzbatt by Weiling (1991), who reconsidered Richter's and Heimans'
important question: when was it that Mendel wrote the notes? It has been
accepted that Mendel wrote them in the 1870s, because on the reverse side
of the sheet of paper is pencilled text written by Mendel dealing with the tax
law, issued on 7 May 1874. Heimans assumed that this text was written by
Mendel in connection with the memorandum Mendel submitted to the
Governor's office in 1876 which dealt with the tax law, issued on 7 May
1874. Weiling showed, however, that between lines 12 and 13 of the text
on the reverse side, written in pencil, the following line has been inserted in
ink:

   Zu der am 26. Juni abgehaltenen Baum
   (On the tree established on 26 June).

Mendel had made a note about a meeting dealing with fruit trees without
giving the year. On examining the archives of the Agricultural Society,

Weiling found that on 7 July 1880 Mendel reported the results of the examination of experts on fruit-tree growing at the meeting of the committee of the Agriculture Society. Weiling published Mendel's report, and we can read that the examination took place on 26 July 1880. Thus the *Notitzblatt*, which analyses numerical data from research into plant hybridization, could have been written after this date, three years before Mendel's death. It seems that at that time Mendel returned to a re-examination of the data from his most important experiments into plant hybridization. The preserved documents are still open to further analysis, and the fragmentary data should not be relied upon to make definitive inferences about Mendel's theory in the *Pisum* experiments. They are also certainly not open to interpretation as deliberate distortions of the experimental data, as suggested by Herčík and Novák (1952).

In connection with Fisher's criticism from 1936, G. G. Doyle (1968) published an article entitled 'Too many small Xs or hanky-panky in the monastery?' Later M. Gardner (1977), in the introduction to his paper 'Great Fakes of Science', states: 'Yes, even Brother Mendel lied.' Similar views were published by Broad and Wade (1984) and an anonymous author (1989). These selected examples show how great scientific achievements can be discredited by dilettantes who claim a combination of two incompatibles: the rigorousness of a meticulous scientist, and falsification of the results.

Another subject of controversy was Mendel's explanation of the colours of flowers and seeds in the progeny of hybrids of *Phaseolus multiflorus*. Olby (1979) examined the question from the viewpoint of Bateson and Saunders (1902b), who found Mendel's reasoning in writing that A_1 and A_2 are allelomorphic to a, 'obscure and not altogether valid'. Olby wondered why Mendel made no apology in explaining the colours of flowers and seeds in the progeny of hybrids for putting both A_1 and A_2 with the same contrasted character a. The main reason for obscurity according to Olby was that Mendel was thinking in terms of the white colour when he wrote a, using the same letter for a class of plants contrasting with A_1 and those which contrasted with A_2. It follows that at the time Mendel was also not thinking of the pair of mutually exclusive factors. Even though Mendel associated the symbol a with the colour white, he also clearly stated that the two or more colours are 'totally independent', and they 'behave exactly like any other constant trait in the plant', and that the 'hybrid association $A_1a + A_2a$ ... would be formed by fertilization with germ cells from a white plant'. It can therefore be assumed that he had independent factors in mind and that the a written as partner for A_1 is different from that written as partner for A_2.

Mendel may also have been lucky in his research. In his *Pisum* experiments he was working with seven pairs of traits and came to the conclusion that 'the behaviour of each pair of different traits in a hybrid association is independent of all other differences in the two parental plants' (Mendel 1866*a*, p. 22). After 1900, geneticists thought for a long time that the genes determining the traits Mendel investigated must have been distributed evenly, so that there was one gene located in every chromosome. Dunn (1965*b*, p. 12) estimated that the probability of selecting such traits was 1:163. But gene maps for *Pisum sativum* revealed the localization of the genes of the traits Mendel experimented with in five out of the seven chromosomes. This work was published in Swedish, and escaped the attention of geneticists (Nilsson 1951). Later Lamprecht (1968*a,b*) published in German a gene map with the localization of the genes for the traits in question on four chromosomes. More attention was attracted by a short note in English by Blixt (1975), pointing out that on chromosomes 2 and 3 there are no genes, and that chromosomes 1 and 4 carry more than one of the genes of Mendel's seven trait pairs. He also published a table explaining the localization and linkage. Two loci that Mendel tested extensively together are on chromosome 1. These, according to his symbolism, are B-b and C-c. Afterwards Novitski and Blixt (1979) showed that the great genetic distance separating them (204 units) precluded Mendel's detection of linkage.

Table 9   Mendel's trait pairs and their location on the chromosomes

Trait pairs	Symbols used by Mendel	Alleles in modern terminology	Chromosomes and locus
Mature seeds, smooth-wrinkled	A–a	R–r	7,60
Seed color, yellow-green	B–b	I–i	1,204
Seed coat and flowers, colored-white	C–c	A–a	1,0
Mature pods, smooth expanded–wrinkled indented	D–d	P–p or V–v	6,10 or 4,211
Unripe pods, green-yellow	E–e	Gp–gp	5,21
Inflorescences, from leaf axils–umbellate in top of plant	F–f	Fa–fa	4,78
Plant height, 1 m to around 0.5 m	G–g	Le–le	4,199

According to Novitski and Blixt (1979)

What is problematic is the determination of the pod shape locus, controlled by either of two genes, as illustrated by Novitski and Blixt (1979)—see Table 9. The difference between two simple dominant forms of the shape of mature pods, ppV- or P-vv, is only discernible to the trained observer through close inspection of the thick inner tissue of the pod, and the differences are not easily detected. It is assumed that Mendel must have been observing the action of one of these loci. Nilsson (1951) supposed that it was the D-locus on chromosome 6, while Lamprecht (1968*a,b*) believed it to be the B-locus on chromosome 4.

If Mendel examined the p-locus, linkage could have manifested itself in traits B-b and C-c on chromosome 1 and in traits D-d and F-f on chromosome 4. In these cases Piegorsch (1986), in considering the great chromosomal separation of these loci, concluded that frequent recombination would have rendered any evidence of linkage unnoticeable. If Mendel examined the pod shape determined by the v-locus, linkage would also have appeared in traits F-f and D-d. But Mendel analysed in detail the combination of only two seed traits, and one plant trait (A-a, B-b, and C-c), and only mentioned all possible combinations of all seven traits. In the case of other traits he merely found the occurrence of all expected combinations of the traits, without making a more detailed analysis of the segregation ratios.

W. Piegorsch, in the course of a critical evaluation of various statistical analyses of Mendel's numerical data, and the possible influence of linkage, inclined to the view that Mendel experimented with a trait controlled by the p-locus. He added that he was using pea varieties which were mentioned in German seed catalogues in the mid-nineteenth century, many of which no longer exist. It is thus no longer possible to answer the question of the localization of the genes of the traits he experimented with. Piegorsch's conclusion was that Mendel cannot be castigated for having falsified his data with respect to the linkage question.

Most of Mendel's experiments were performed during the period when he was teaching at the *Realschule*. The most demanding experiments with *Hieracium* were undertaken when he was already abbot of the monastery. He published the results of his research into plant hybridization, by which he wished to contribute to an explanation of a problem which was of immeasurable importance for the evolution of organic forms. During his abbacy he moved not only among the members of the monastery community, but also in church circles, among public officials, agricultural experts, and natural scientists, who were trying to develop pure science. We may now ask how Mendel the researcher appeared in his new role.

6

The abbot

From the very modest position of teacher of experimental physics I thus find myself moved into a sphere in which much appears strange to me, and it will take some time and effort before I feel at home in it.
—MENDEL, to NÄGELI, 4 May 1868 (CORRENS 1905, p. 220).

In four letters written to Nägeli between 31 December 1866 and 9 February 1868, Mendel writes of his experiments with further plant species. In a letter dated 6 November 1867, he apologizes for the fact that his project of studying *Hieracia* in their natural habitat has been carried out only to a very limited extent (Correns 1905, p. 211). The main reason, according to Mendel, is lack of time. At the same time he laments being 'no longer very fit for botanical field trips', since heaven has blessed him with 'an excess of avoirdupois' which becomes very noticeable when he travels long distances on foot, and 'as a consequence of the law of general gravitation, especially when climbing mountains'. On 4 May 1868, he thanks Nägeli for sending *Hieracium* seeds and living plants, saying that he is most grateful for the consignment. At the same time, Mendel promises to do his utmost to produce all the possible hybrids among the species, and, should these prove fertile, to observe their progeny for several generations. Only after this introduction does Mendel confide in his friend: 'Recently, there has been a completely unexpected turn in my affairs. On 30 March my unimportant self was elected lifelong head by the chapter of the monastery to which I belong.' Mendel the researcher adds at once: 'This shall not prevent me from continuing the hybridization experiments of which I have become so fond; I even hope to be able to devote more time and attention to them, once I have become familiar with my new position' (Correns 1905, p. 220).

This letter reveals certain misgivings on Mendel's part relating to his new role as abbot. At the same time he clearly nurtured the hope that, since he would no longer be working as a schoolteacher, he might have more time to devote to the experiments, with whose progress he was satisfied at that juncture. On 9 February he writes that after obtaining two years' experience with *Hieracium* he was about to undertake systematic experiments with the genus (Correns 1905, p. 218). In the next letter he only describes the *Hieracium* experiments. He had undertaken experiments with *Phaseolus* and other genera, and though he came across the problem of low fertility and a more complex determination of the traits he was investigating, he was further developing the theory he had proved in his *Pisum* experiments. In his experiments with other plant genera, Mendel managed in a very short time to explain the more complex determination of the heredity of the traits he was investigating. But in the *Hieracium* experiments he came across a new mystery. The success he had achieved in his previous experiments gave him the confidence to seek an explanation for the appearance of polymorphic traits in *Hieracium*, a problem which was becoming a matter of prime interest to the foremost botanists of the day, of whom Nägeli, his correspondent, was one.

Mendel was flattered that Nägeli replied to his letters and sent him reprints of publications on plant hybridization. He refused to be discouraged by the other's critical attitude to the *Pisum* experiments, and was sure that his continuing research would enable him to convince Nägeli of the justification of his approach. Today we find it difficult to imagine a secondary-school teacher with almost a hundred pupils in his class, teaching six days a week, and still somehow finding time to undertake research as demanding as Mendel's. Undoubtedly the lack of any private life outside the monastery walls made it possible for him to concentrate on his work in a peaceful atmosphere. He studied the literature, performed experiments, and in evaluating their results synthesized his individual conclusions into new hypotheses, at the same time planning ever more complex new experiments dealing with the same problem. The stimulus for this may have been provided by natural scientists' new outlook in the light of Darwin's Theory of Evolution, which left none of them indifferent, whichever side he was on. Mendel's appetite for research must have been further increased by the speculations on hybridization engaged in by the celebrated Professor Nägeli under the influence of Darwin's theory.

His election as abbot not only brought a complete break with the daily rhythm of the natural history teacher and enthusiastic researcher, but also reshaped the personal life of this humblest of members of the monastery (Fig. 6.1). A letter to his younger sister's husband, Leopold Schindler, dated 26 March 1868, shows that he had foreseen the possibility, though he considered it a remote one, of being elected to replace the deceased Abbot Napp

Fig. 6.1 Members of the Augustinian monastery in Brno in about 1862. Standing, from left to right: Benedikt Fogler, Anselm Rambousek, Antonín Alt, Tomáš Bratránek, Joseph Lindenthal, Gregor Mendel, and Václav Šembera. Seated, from left to right: Pavel Křížkovský, Baptist Vorthey, Cyrill Napp, and Matouš Klácel.

(Kříženecký 1965*b*, p. 123). At the time no one could predict who might be chosen. The letter states that 'should the choice fall upon me, which I scarcely dare hope, then you will receive a telegram next Monday. If you do not receive one, someone else has been elected.' He also mentions the industry of his sister's eldest son, studying at the Technical University in Brno. At the time Mendel was paying the cost of the boy's accommodation and studies in Brno out of his salary as supply teacher. His sister had two more sons, both of whom later studied at the Brno *Gymnasium*, followed by medical school in Vienna. His concern to support their education could have been one of the motives for his wish to be elected abbot.

The surviving records of the election show that after two rounds of voting there was still no clear winner (Marvanová 1968; A.K. Meijer 1984). It was not until the third round that Mendel won the support of a majority. His major rival for the abbacy was Bratránek (Fig. 6.2), who obtained two votes in the first round and four in the second. In the first round Mendel obtained six votes, in the second five, and in the third twelve. Bratránek was more interested in continuing as a teacher at Cracow University, and he stepped down to make way for Mendel. At the time the monastery community was split evenly between those who were of German ethnic origin and those who were Czech. Since 1848, tension between the

Fig. 6.2 Tomáš Villanova Bratránek (1815–84), a prominent Augustinian as rector of the university in Cracow. He was known as a university teacher and as the author of publications in natural philosophy.

Czechs and the Germans in Brno had been on the increase, and the dispute started to come to a head in 1867. The German minority occupied the leading positions in economic and political life, and the German language was given preferential treatment. The representatives of the Czechs were striving to achieve equal representation in the provincial institutions. Thus the abbatial elections were observed with a good deal of interest by the police authorities. A police memo from the time states that it was undesirable for Šembera, Rambousek, or Křížkovský to be elected, since they were 'Hussites'—in other words, their views on the question of Czech national rights were considered extreme. The most outstanding proponent of Czech rights was Klácel, whose name does not appear in the police report. The fact that he had been downgraded from his professorship of philosophy in the 1840s, and later persecuted, seems to have been considered sufficient guarantee that his election was not on the cards (Eichler 1904). In this situation the ethnic Czech friars probably voted for Bratránek to begin with. Mendel consistently voted for Klácel, thus expressing his neutral attitude towards the more electable candidates and awaiting his own chance. In the end Klácel played a role in Mendel's election; it was his view that Mendel with his liberal ideas would make a better abbot than any of the other candidates of either Czech or German origin. This choice may in the end have been influenced by Mendel's personal qualities. Klácel's friend, Professor Helcelet, mentions in a letter to his friend Professor Hanuš that the newly elected abbot of the Old Brno monastery was 'of mild to indulgent temperament, an enthusiast for the natural sciences' (Kabelík 1910, p. 542). This reflects attitudes to Mendel's election outside the monastery and church circles in general.

On 30 March 1868, Mendel's ambition to take over the post left vacant by the deceased Abbot Napp was fulfilled (Fig. 6.3). This assured him an income which allowed him to support his relatives, especially as regards the studies of his sister Theresa's three sons. But if he thought this event would

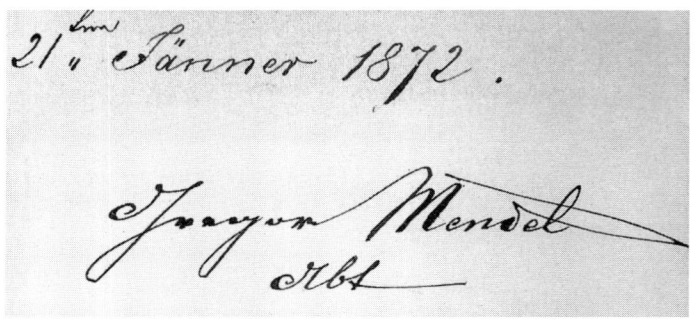

Fig. 6.3 The signature of Abbot Mendel, in German and Czech.

mean that he had more time for his experiments, he was soon to find the opposite was the case. There were many new duties attached to the office of abbot, including those outside the monastery, and their fulfilment led Mendel further and further away from the scientific work which twenty-five years earlier had motivated his entry into the monastery.

Right at the start of his term of office, Mendel realized that he would have to tackle complex economic problems facing the monastery. By 24 December 1869, the ministry of culture and education had rejected his application for the waiving of contributions to the religious fund due in respect of the period 1830–64 and totalling 34 000 guilders. Abbot Mendel was requested to inform the ministry as to how the debt would be paid. On 31 January 1870, Mendel wrote to explain that his monastery was not in a position to pay off such a sum; his reasons take up twelve pages, and include a detailed account of the monastery's development from the mid-eighteenth century (Richter 1943, pp. 109–13). During the period in question, he asserts, the community had often been under such pressure from outside that it was on the point of collapse. First of all it was occupied by the Prussian army in 1745. The main building and some parts of the church were demolished. Everything was rebuilt at great expense in 1777, mainly thanks to loans. But six years later the state confiscated the reconstructed building as offices for the provincial authorities, and the monastery was obliged, without any form of compensation, to move to the dilapidated premises of the defunct Cistercian monastery on the outskirts of Brno. Large-scale rebuilding had been essential before the community was able to function. The monastery got even deeper into debt. In 1805 Brno was occupied by Napoleon's army, and the effect on the monastery was again far from beneficial. The size of the community had by then fallen to a mere quarter of what it was before it had been forced to move out of its original building. Under the Emperor Josef II, contributions to the religious fund had been increased, and for a long time the monastery found itself unable to pay them. It suffered further major losses in 1866, when Brno was occupied by the Prussian army. In his memorandum Mendel points out that major reconstruction of the abbey church and the outbuildings is required, at an estimated cost of 30 000 guilders. He once again requests the ministry to waive back-payments to the religious fund.

A further section of Mendel's letter points to the monastery's services to the State, the chief one of which he considers to be 'the cultivation of science in all its aspects', which had always been considered 'one of its primary tasks'. Mendel adds that up to 1849 the monastery had provided professors of philosophy and mathematics for the Philosophy Institute, and that two members of the community were at the time teaching at Cracow

University and the *Oberrealschule* in Brno. He also states that half its members work as curates in the Old Brno parish or teach religion in the two
main schools of the city and one village school. In conclusion, the new
abbot mentions the dangerous work undertaken by his monks in the nearby hospital, two of them having died over the last two years as a result of
infections contracted there.

A NEW POLITICAL SITUATION

Mendel stepped into the shoes of the deceased Abbot Napp at the age of
forty-six. His predecessor had been only thirty-two years old when he took
office, and he held the post for a full forty-three years. The period had been
one of political calm, generally speaking, marred only by the revolutionary
events of 1848. Before he was elected abbot, Napp had worked for seven
years as a professor of Bible studies and oriental languages, and he was
preparing to sit his *rigorosum* for a doctorate. He was also interested in the
broad study of science even at that early stage, in connection with increasing the profitability of the monastery's landholdings. Before his election as
abbot, the monastery had been called upon to fulfil more energetically its
duty under imperial decree to provide teachers for the theological and
philosophical institute in Brno. After his election, Napp exploited this decree
to develop science in the monastery, and later extended this sphere of his
activities beyond the monastery walls. Napp's work has been described in
detail by Zlámal (1937, 1991) and Czihak and Sládek (1991).

In 1829 Napp had been appointed to the land committee of the aristocratic Diet, which dealt with questions of title in the province. Gradually his
industriousness and organizing talent brought him further offices. From
1832 to 1848 he also held the post of Director of *Gymnasia* in Moravia and
Silesia, and after 1848 he became a leading member of the education committee of the provincial Diet. He also held an important post in the
Agricultural Society, which fulfilled the dual roles of academy of sciences
and economic society for the province. From 1827 he was a member of the
central committee of this society, and soon began dealing with the Society
agenda as deputy to the chairman. In 1849 he became officiating director of
the Society, and in 1864 formal chairman. From 1827 to 1849 he was
chairman of the Pomological Association, which was subject to the
Committee of the Agricultural Society. He was also a functionary or member
of many other cultural and social societies and institutions. In the revolutionary year of 1848 he was an important political figure in the province, and he
oriented himself well in the stormy political atmosphere as a result of his

progressive views. In all the posts he held, his organizational talent and exceptional capacity for hard work marked him out. The newly elected Abbot Mendel was expected to follow in his predecessor's footsteps in this respect.

In 1867 the absolutist Habsburg Empire became the Austro-Hungarian constitutional monarchy, guaranteeing equal rights above all to the German and Hungarian ethnic groups. The rights of other nations within the monarchy, which as far as the province of Moravia was concerned meant the Czech majority, were for the most part left unrecognized. The vehicle of this new policy was the German Constitutional Party, also called the Liberal Party. The opposition was the Conservative National Party. In the latter a leading role was played by major landowners and members of the Church hierarchy, both Czechs and Germans. The Liberals stood chiefly for the rapidly developing industrial interests. Their chief spokesman in Brno was soon to be J. Auspitz, who in 1867 left his post as headmaster of the *Realschule* and became Inspector of Schools for Moravia. He also published the influential daily paper *Tagesbote*, which was the mouthpiece of the Liberal Party.

In 1870 the Liberal Party formed a new government, which began to implement a policy of reforms. Many of the responsibilities for schools and the family were taken out of Church hands and transferred to the State authorities, which gave rise to opposition in Church circles. It came as a great shock when the newly elected Abbot Mendel openly supported the Liberal Party in the provincial Diet (Vybral 1968). Abbot Napp had always toed the line of the Church hierarchy. Mendel's political stand earned him the disfavour of the bishop's consistory, and also that of the Czech members of his own monastery, who regarded the new party as representative of German nationalism. In September 1871, Mendel went even further by putting his signature to a protest led by the Liberal Party in the provincial Diet against the illegal election of a member of the Conservative Party. This move led to open hostility towards him. With the same resolve he had shown in tackling a scientific problem, he now expressed political views which were in contradiction to those of the Church hierarchy. He appears soon to have realized he might have gone too far in his political gesture. When, on 2 September 1871, Count Wladimir Mittrowsky offered him a candidature for the newly founded Moravian Centre Party, Mendel declined politely, referring to circumstances he himself was 'unable to change' (Richter 1943, p. 56). He carefully pointed out conditions in the monastery, where his political stand on behalf of the Liberals had apparently given rise to criticism. But he continued to take the Liberal Party newspaper, thus showing his sympathy with that political grouping.

In 1902 Mendel's nephew Ferdinand Schindler wrote to W. Bateson that his uncle did not like 'ultramontane propaganda', which one might suppose to be a sign of disagreement with the attitude of the higher echelons of the Church hierarchy to questions of education, to which Klácel had fallen victim when he was downgraded from his post of professor of philosophy (Coleman 1967). One might also place in this context Mendel's signature on the petition of six members of the monastery community in 1848, mentioned in Chapter 4. Mendel's political gesture in support of the Liberals marked the start of a period of tension between the new abbot and his superiors in the Church.

Some of the members of the monastery community criticized new measures introduced by the Liberal government which made inroads into the influence of the Catholic Church. Křížkovský, for instance, in a letter to his sister dated 23 February 1872, wrote of a bitter campaign waged against the Church by the State, and put the following interpretation on it: 'atheistic tendencies among all classes, stirred up by Liberal newspaper, imperiousness with the aim of enslaving the Church, envy and covetousness for the Church landholdings, and finally also the negligence and errors of individuals among the clergy' were, in his view, the causes of this state of affairs (Eichler 1904). Křížkovský's last remark might well be taken to include his own abbot.

When Mendel was elected head of the monastery, he was considered to be tolerant from the point of view of ethnic tensions. No one might have supposed he would later speak out in support of the German nationalist party. We cannot estimate the extent to which he himself was aware of the impact the Liberal Party would have in relation to the ethnic question. Apart from Křížkovský, there were both ethnic Czechs and ethnic Germans in the community who criticized his stance. Today we are inclined to explain it in the light of his social origins. He may have been influenced to a large extent by Dr Auspitz. Mendel was grateful to him for making it possible for him to teach at the *Realschule* without having passed his teacher's examinations, and also for his support in setting up the Natural Science Society. Circumstantial evidence of this does exist.

Auspitz's reaction to the unfavourable criticism evoked by Mendel's political gesture was to nominate the abbot for a state honour. In spring 1872 he received the Order of Franz Josef for 'outstanding political work and meritorious teaching at the *Realschule*'. The nomination emphasizes those of Mendel's activities of which the ex-headmaster of the *Realschule* and subsequently spokesman of the Liberal Party in Brno, Auspitz, was best informed (Vybral 1968). From then on Mendel was much more cautious in his public proclamations. He devoted himself more to scientific matters in the commissions of the provincial Diet, and put his natural scientific knowledge to use on the Committee of the Agricultural Society.

THE AGRICULTURAL EXPERT

In 1869 the Moravian and Silesian Agricultural Society approved a new constitution, and in January of the following year Abbot Mendel was elected a member of the Central Committee of the society. In August 1868 he and his friend Olexík had been made commissioners for examinations in pomiculture. This was an expression of recognition of his expertise in the field. Mendel held this office right up to the end of his life. At subsequent elections, held every three years, he was repeatedly re-elected to the committee, but later received fewer votes, a fact which may have been due to his support for the Liberal Party. From 1872 he took over the agenda of the society as deputy to the chairman, Count Dubský, who had to travel to Brno from Lysice, a good distance away. In the following year Mendel was officially made vice-chairman, and often signed the agenda on behalf of the chairman. In 1882 he was even offered the post of chairman, but declined on grounds of health. According to the surviving minutes of meetings, Mendel was still attending them regularly in 1883, until a worsening of his health shortly before his death prevented him from doing so (Orel 1970a).

Formally, Mendel was a member of all the specialized sections of the Agricultural Society, including that for forestry. But the ones he took an active part in were those for horticulture and apiculture. Shortly after being elected a member of the Central Committee of the Agricultural Society he was appointed to the commission for distributing financial support for the development of agricultural production. Its chairman was Count E. Belcredi, but in fact it was Mendel who took charge of the commission's work. This consisted mainly in supporting the rearing of cattle by buying breeding bulls from high-yielding dairy stock in order to improve yields in the villages. A close co-worker of Mendel's in this sphere was the veterinarian V. Čech, who from 1875 took over management of the agenda for distributing subsidies. Mendel continued to be a member of the commission. At that time they also began to monitor seed in order to improve yields of agricultural crops. In 1874 the District Agricultural Society in Olomouc, for instance, asked the commission for 200 guilders in order to buy a microscope for examining seed quality. When the application was considered, Mendel was in a dilemma as to what heading it should be placed under for purposes of the regulations which applied. The nearest category available was 'agricultural machinery', which he did not consider a suitable description. After considering the economics of the matter, Mendel concluded that a magnifying glass was quite adequate for the purpose.

Mendel was more interested in publishing reviews of new agricultural literature and evaluating the contributions submitted to the society journal

(Orel 1971*b*). When, in 1881, he was asked to perform critical reading of a new textbook on fruit-tree growing by Dr Lepkisch, he rejected it, considering it not up to the required standard.

In 1877 Mendel became a member of the commission which decided which books to purchase for the library of the museum administered by the Agricultural Society. He also took part in adjudication of publications submitted in response to essay prizes offered by the society. Another of his duties was the award of the society medal for merits in the development of agriculture, especially for distributing the seed of the best varieties of crops to farms which could be expected to propagate them and pass them on to others. According to the practice in other countries, Mendel also encouraged the holding of exhibitions of seeds and agricultural products in the Society's museum. He also frequently gave an expert opinion on the question of pests and infections of agricultural crops, and made use of the knowledge he had acquired during his entomological studies in Vienna. In 1878 he pointed out the danger presented by the Colorado beetle in western Europe, recommending that early action be taken to prevent its spread (Orel 1971*b*).

Today we can scarcely imagine high yields of agricultural plants without measures to ensure their health. In 1874 the committee of the Agricultural Society discussed a proposal by the horticultural section to set up an institute for the protection of plants from disease and pests. On Mendel's suggestion the proposal was also submitted for consideration to the committee of the Natural Science Society. Then, with his recommendation, it was passed on to the ministry of agriculture in Vienna; but no progress was made on the matter.

Mendel also applied his expertise in the field of agriculture on the commission of the provincial Diet responsible for fixing the level of agricultural taxation. Officialdom appreciated the fact that he knew local soil conditions well, particularly the soil quality in various localities. When, in 1872, the government introduced statistics of agricultural production, Mendel tried to obtain more financial support for the fulfilment of this task. Here, too, he showed his expert knowledge, emphasizing the exceptional value of such statistical records. He stressed that such surveys had to be extensive enough to ensure accuracy of the estimates they produced. He knew well enough from his plant-crossing experiments what it meant to keep statistics, and was anxious that the job should be done properly.

Mendel used to mark published reviews with an abbreviation of his name, GM, or M, or m. Under the letter M in 1877, a three-page article appeared entitled 'Against communism and socialism' (Anon. [M.] 1877). This article was examined in the Mendelianum by the author and Professor E. Lauprecht from Saxony, who compared the text in Brno with that in the agricultural journal in Lüneburg. Recently, Weiling (1991) drew attention to this article in connection with Mendel. The short introduction, written by the anonymous M, says that agriculturalists should find the topic interesting, though not as much as industrialists would. After a brief explanation of the ideas of socialism and communism in a historical context, attention is paid to the social conflicts between wealth on one side of society and poverty on the other. Special systems of education and social arrangements in the villages are recommended for creating favourable conditions for workers and preventing social conflicts.

Attention is also drawn to the education of village children in habits of thrift, which should be supported by the landowners. The building of homes for the workers with a piece of land attached is recommended for the new social programme that is considered as the precondition for contented and efficient workers. In addition to the basic school education, a new system of education up to the age of seventeen is recommended. This should teach the emptiness of the ideas of socialism and communism. According to this article, a neglected education encourages roughness of spirit in young people. This sentence is marked, and a footnote tells us that this roughness can be observed in the streets of Brno. The content of this article suggests the views of an experienced pedagogue in Brno, who was well acquainted with life in the villages. The anonymous M knew how important it was to save money. Abbot Mendel was not only an experienced teacher and an agricultural expert, but was later also the local representative of a bank. He may have recommended republication of the article in Brno.

The obituary of Mendel published in the Czech supplement to the Agricultural Society journal emphasized that 'All the deceased's studies took on a practical importance' (Matalová 1984). 'Fr. Gregor Mendel did not satisfy himself with lifeless words, but took an active hand in the agricultural affairs of Moravia at every opportunity, and always and in everything paid great attention to them.' This stemmed from the interest in agriculture he had acquired at his parents' home. Even before he was elected abbot, in 1861, an article in a supplement to the *Moravské noviny* said that: 'Professor Mendel of the Brno *Realschule*, a teacher of natural history and himself of an enquiring scientific bent, devotes time, money, and loving care to some question or other every year.' At the time he was especially interested in a pest of peas. The report in question also states that he walked around the fields 'to get to know its way of life, so as to be able to advise on how to deal with it'. This was at the time he was carrying out his *Pisum* experiments, and such a pest might well have ruined his results.

Attention can also be drawn to Mendel's and Napp's interest in natural sciences and research into agriculture. Napp, at the age of thirty-four, published, in cooperation with Diebl, an essay on two species of insects which had caused damage to plant production, encouraging investigation of their reproduction and behaviour (Napp and Diebl 1826). A summary of the results appeared two years later (Napp and Diebl 1828). At that time Napp was fully involved in the study of natural sciences for the promotion of plant and animal production on the monastery estates. As abbot he had to concentrate on other problems, but he supported research into natural sciences and agriculture in his post in the Agricultural Society. Mendel published his first paper at the age of thirty-one. The topic was insect damage in field production. He had been prepared for his investigation by his university training. Later Mendel was able to use his knowledge as a teacher and researcher. The only difference between the two was that Napp was in the position of an acknowledged expert and manager, and Mendel that of an expert and a researcher.

AMONG PLANT-BREEDERS

In their description of the history of the Pomological, Viticultural, and Horticultural Society which took over from the Pomological Association founded by André in 1816, Křívánek and Suchánek (1898), drew attention to Mendel's successes in the breeding of new fruit trees and fuchsias. The Society's journal recalled the initial success attained with breeding vegetables, which Mendel exhibited in Brno in 1859. This was at the time of his main experiments with *Pisum*. Even then he was linking his research to practical breeding. In his second letter to Nägeli, dated 18 April 1867, he gives the example of crossing a pea plant with differences in four pairs of traits (Correns 1905, pp. 200–1). He found in the hybrid progeny a form which was particularly outstanding for taste, and which remained constant in subsequent generations. From then on he grew it in the monastery's vegetable garden. This was in fact the first experimentally bred vegetable variety with the genotypes of the desired traits described. Later, in 1878, Mendel showed his new pea varieties to Eichling (1942), a specialist in seed-growing. This is dealt with in Chapter 7.

In 1863 there appeared among the ranks of the Pomological, Viticultural, and Horticultural Society a natural scientist and researcher who was to became a committee member five years later. In this sphere, too, Mendel was following in Napp's footsteps; however, he was not the chief organizer, as Napp had been, but once again more of an expert adviser. Like Napp, he was involved in holding regular exhibitions of fruit and flowers. He offered prizes for new varieties, and himself exhibited some of his own. In the year he was elected abbot, Mendel offered a prize for breeding new, frost-resistant varieties of climbing roses. One of the conditions was the use of hybridization. He also offered prizes for breeding new varieties of flowers and strawberries, and similar contests were held in subsequent years.

Interest in new varieties brought Mendel among Brno horticulturalists, who were beginning to acquire for themselves a reputation as flower-breeders which extended beyond the borders of their country. His closest associate was P. Olexík, director of the hospital on Pekařská Street, near the monastery, with whom he had already begun to cooperate in making meteorological observations. Later the two became friends, and worked together to breed flowers. Olexík built himself several greenhouses, and put flower-breeding on a business footing. Mendel's most frequent colleague was the ornamental flower-breeder J. Twrdy (1806–83). When, at the start of this century, Iltis began gathering information on Mendel, Twrdy's daughter Amálie stated in a letter to him that 'At the time the researcher was writing his now famous paper, he often met my father in connection with the work. Both were engaged in the same aim, though my father only to a limited extent. They were linked by an endeavour to use artificial crossing and artificial selection of plants to obtain far-reaching results. The researcher was a frequent guest in our garden, where he would speak with my father. They exchanged views on breeding. Out of friendship and appreciation the researcher gave my father a portrait of himself, though we no longer have it' (Mendelianum invent. no. 1738).

Twrdy had originally been the gardener at an aristocratic house in Brodek, about 40 km from Brno. In 1839 he won the prize for breeding a new variety of currant bush, the price having been offered three years previously by Abbot Napp and Professor Diebl as representatives of the Pomological Association. Later Twrdy moved to Brno, where he developed to the full his talent for breeding fuchsias and verbenas in particular (Vávra 1984). After 1870 he put up to ten new varieties on the market, selling them in several capital cities. At first he was in competition with a certain Belgian firm in the fuchsia-breeding business. Later he became Europe's leading exponent in the field. Libraries still have catalogues of the exhibitions where Twrdy published the names of his new varieties, and some of them contain coloured drawings on glossy paper. Twrdy named them after outstanding personalities and leading natural scientists. So, for instance, he used the names of the Czech historian Palacký or of natural scientists such as Galileo, Alexander von Humboldt, or Dr Fenzel, etc. Some species were named after their geographical origin, such as Moravia or Bruna. Twrdy also recalled the work of the Moravian thinker and pedagogue Comenius by naming one variety after his famous didactic work *Orbis pictus*. Others were given names like Concordia or Evolution. Twrdy was a practical gardener, and one can scarcely suppose that he invented such names himself. A frequent guest who, according to his daughter, was held in special esteem in their house, may have been involved in the choice of names. And in 1882 a new variety was named 'Prelate Mendel' (Fig. 6.4). According to a note by Twrdy, published by Iltis (1966, p. 210), it was a seedling of *Fuchsia monstrosa*, very large, pale-blue shading into violet, luxuriant, with regular structure, the sepals light, very beautiful, and blooming early. Mendel's enthusiasm for fuchsia-breeding was confirmed by Schubert (1899) in describing pomiculture in Moravia. He also pointed to the expansion of the fruit nursery at the monastery's farm in Nové Hvĕzdlice during Mendel's abbacy. Much attention was paid there to the artificial pollination of flowers of several different varieties, resulting in particular in new and successful varieties of drupes and fuchsias. The world fuchsia assortment still contains the names of varieties Twrdy bred, but only a fraction of them have survived. A leading fuchsia-breeder in Switzerland in the 1970s, E. Angst von der Leek, from Wil in the canton of Zürich, began to collect fuchsia varieties and records of them from the earliest times, and came to

Fig. 6.4 Gregor Mendel with fuchsia, taken from the group of the Augustinians.

the conclusion that the foremost breeder had been Twrdy, whom he considered to be a leading botanist from Brno. Von der Leek, wishing to find out more about Twrdy, was surprised to hear from the Mendelianum staff that the latter was a gardener who achieved great success in breeding flowers, in cooperation with Mendel (Vávra and Matalová 1983).

Twrdy soon became established as a leading figure among garden-plant breeders in Brno. It was he who suggested that the Pomicultural, Viticultural, and Horticultural Society discuss the question of plant breeding through hybridization at a meeting in 1862. The minutes of the meeting state: 'A love of flowers having been established, there still remains the desire for each one to grow his own, and if possible to breed them. Love of flowers will soon be extended to other plants, and interest in breeding will increase, especially when the joy of practical advantage is added to that of beauty, as is the case of fruit trees' (Orel 1975a). Among the other outstanding gardeners in Brno was Molisch, whose son Hans studied at the *Realschule*. He later became professor of the physiology of plants at Vienna University. Later H. Molisch (1934) also recalled Mendel's visits to his father's garden, and his father's interest in breeding flowers and vines.

Mendel's great love was fruit trees. He built on the experience with growing vines and fruit trees acquired at the monastery by the Augustinian A. Keller. According to the information of the monastery gardener he grew a variety of vine from seed he had brought from Florence in Italy. His interest in breeding vines is also borne out by a letter from M. Manner. He sends Mendel grape seeds as promised, brought from Graz, and assures him that the grapes from this vine are very tasty (Vávra and Orel 1976). There is other evidence of Mendel's breeding of fruit trees. He combined an interest he had shown from his early youth with scientific experience, in order to create higher-yielding varieties, which were exhibited anonymously in Brno. When one of them won a prize, it was not awarded, since the grower did not come forward, and it was carried forward to the next year. Mendel also exhibited his new varieties in Vienna. In 1883 he offered new varieties of apple and pear trees at a meeting of the Gardeners' Association in Hietzing, on the outskirts of Vienna, and was awarded a prize and the Association's medal.

After Mendel's death the monastery gardener J. Mareš stated that 500–600 pear, apple, and apricot seedlings were grown every year in the gardens (Iltis 1966, p. 209). According to documents from the period, the monastery had from the beginning of March a special fruit-tree gardener who carried out the necessary work according to Mendel's instructions. There are many pomological publications in the monastery library. Mendel made a detailed study of an eight-volume German book on pomology by Jahn *et al.* (1850–60), in which his notes are preserved. The flyleaf of the first volume carries a scheme of the crossing of various varieties of apples, and the flyleaf of the second a similar scheme for crossing pears. Mendel had twelve maternal apple varieties in his project and seventeen pollen varieties, and he considered which combinations were to be used to create a new variety (Orel and Vávra 1968). The description of these varieties indicates that he was trying to improve both yield and taste by combining selected traits of the parental forms. He may have had in mind the combination of the best taste qualities with resistance to climatic and soil conditions. One may give as an example the combination of French Grey Rennet with the sweet variety Holart. The maternal variety gave apples of the best quality only under warm conditions and with the best soil conditions. Under cooler conditions with poorer soil quality the apples lost their flavour and the woody parts were attacked by a canker. The apples of the pollen variety were sweet, but lacking in flavour. This was a late-blossoming variety which did not require particularly good soil. For this reason it was grown in cool and harsh conditions.

In other crossings Mendel was apparently trying to combine several traits of fruit quality. An example was the combination of the Canadian Rennet with white winter Kalvil. He might have considered reinforcing the desired taste traits by combination of the heredity units he had demonstrated in crossing *Pisum* and *Phaseolus*. He also ordered various varieties of pear tree from the nursery in Troje, near Prague. There is a surviving record of the evaluation of the time of ripening of the varieties under study, which is reminiscent of Mendel's quantitative evaluation of the qualitative traits in his plant-hybridization experiments. His interest in breeding pear trees is also known by pear-tree leaves labelled with their varieties (Vávra and Orel 1971). He probably crossed the variety Van Marum Flaschenbirne with Regentin. He also tried to improve the quality of the flesh of the winter variety Herzogin von Angoulême using Van Marum Flaschenbirne. At the time many other pomiculturists were trying to do the same. But unlike the others Mendel worked out his plans on the basis of theoretical assumptions.

Some of Mendel's schemes to create new varieties were in fact implemented. His varieties were grown in gardens until after the year 1920. According to the information of the monastery gardener, J. Hejl, three pear trees yielded fruit similar in colour to Forellenbirne, and were regarded as summer, autumn, and winter varieties. For a long time a variety of pear called Lemon Butter was also grown in the garden. It probably arose from a crossing of the varieties Herzogin von Angoulême and Regentin. Mendel's last information on plants also relates to pears and apples. In April 1883, in a letter to his nephews, he writes: 'Do me a favour; ask Alois Sturm [husband of Mendel's elder sister Veronika] to send me by you several grafts, one from Ginsbirne, two from the Quaglich pear, and three from the good apple tree from the pension garden' (Křiženecký 1965*b*, p. 126). At the end of his life Mendel became interested in using grafts of the varieties which his father had obtained from the parish priest, Schreiber. He recalled his father's efforts to improve fruit trees, and showed the same enthusiasm throughout his own life.

Mendel's improvement of fruit trees was also mentioned after the 'rediscovery' by A. Doupovec, whose mother worked at the monastery (Křiženecký 1965*b*, pp. 107–8). In his memoirs he recalls a remark by the gardener, Mareš: 'My dear fellow, our prelate, now that was a gardener! There isn't a gardener alive who could not learn from him.' And Mendel truly knew more than the gardeners of the day. He studied the newest fruit-growers' literature, and applied in pomicultural practice his theoretical findings relating to the heredity of traits, which he knew how to manipulate. He exhibited the results of his breeding work, gaining the respect of his contemporaries, who saw him as a breeder and a researcher.

APICULTURAL RESEARCH

In the early years of this century, reports began to appear concerning Mendel's activities in the field of apiculture. In his biography of the scientist, first published in German in 1924, Iltis (1966, pp. 212–20) pointed out the research character of Mendel's work with bees. Later H. Nachstein (1942), considering Mendel's research, briefly mentioned experiments with crossing white and coloured mice, adding that he can say as little about these as about Mendel's experiments with bees. Newly discovered documents now make it possible to describe his work with bees in more detail. It was of a research nature, and followed on from his plant-hybridization experiments. To a certain extent it was connected with Abbot Napp's efforts to form an organization of apiculturalists in order to develop new methods of beekeeping.

The first attempts to organize apiculture in Moravia date right back to 1818. André proposed the constitution of an association of beekeepers in Brno whose task it would be to organize the furtherance of apiculture in Moravia. But when he failed to find a suitable organizer for the association's work, he gave up the attempt. During the first half of the last century only individual beekeepers tried to raise the standards of apiculture, and no major progress was achieved. Then, half-way through the century, in connection with the latest findings in natural science, a new figure appeared on the European scene who was to set the apicultural ball rolling. At a congress of German-speaking apiculturalists held in 1853 J. Dzierzon (1811–1906), from Karlovice in Prussian Silesia, lectured on new techniques of beekeeping, and especially on new findings relating to the reproductive processes of bees. His mobile 'bar' allowed beekeepers to have easy access to hives, contributing to an improvement in the technology of beekeeping and increasing honey yields.

The interest shown by beekeepers in applying Dzierzon's new findings led to the setting up in 1854 of an Apicultural Association under the aegis of the Agricultural Society. Abbot Napp was the man behind the move. The first chairman of the association was school governor J. Hansmann, but he was a poor organizer. This showed itself in a stagnation of the activity of the new organization. After a new initiative by Napp, the Association got going again five years later. At the time it had 86 members. In 1860 the number of members had risen to 430, while in 1865 it was 689, and three years later the membership had reached a full 1200. This huge surge in the numbers of organized beekeepers was due to the work of a new type of apiculturalist, F. Živanský (1817–73). He came to Brno in 1860 as a pensioned-off army doctor. First of all he showed an interest in the activities of the Pomological and Horticultural Association, whose members got him interested in beekeeping. In 1861, Živanský's interest in the natural sciences led him to join the newly established Natural Science Society as a founder member. The beekeepers soon found him to be a capable organizer, and he was elected secretary of the Society. He devoted great energy and enthusiasm to organizational work, and his efforts soon bore fruit for beekeepers in general, resulting in a steep rise in the number of members of the Association.

By 1863 Živanský was propagating the idea of rational beekeeping among the members of the Association. He also emphasized the importance of providing pasture, drawing attention to the advantages of moving hives around to catch the various honey-bearing flowers in bloom. Živanský was aware that beekeeping was important not only from the point of view of honey production, but also for the fertilization of agricultural crops. The idea of crossing different races of bee caught his imagination. In July of the following year, Makowsky (1864*b*) displayed his hybrid bees at a meeting of the Natural Science Society.

An important milestone in the development of beekeeping in Brno was the congress of Austrian and German beekeepers held there on 12–14 September 1865, in which over three hundred leading apiculturalists took part. The year before Živanský had taken part in a beekeepers' congress at Gotha, where he had proposed holding the next one in Brno. The chairman of the congress was Abbot Napp, who offered the monastery's hospitality to leading participants from abroad. Among his guests were the leading bee-keeper of the day, Dzierzon from Silesia, and G. Dathe from Eystrup, near Hanover, who was known for crossing different races of bees in order to create a new race with a higher level of honey production. C. Zirkle (1951) drew attention to the fact that in the 1850s Dzierzon published a number of papers which reported, *inter alia*, that drones were hatched from un-fertilized eggs, but workers and queens came from eggs that had been impregnated; this caused a violent controversy at the time. In one of his experiments Dzierzon (1854) crossed German bees with Italian bees and found that the unmated hybrid queens produced German and Italian drones in equal numbers, a definite one-to-one ratio. This led to the sup-position that Mendel might have found inspiration here for his search for a ratio in the segregated traits found in his hybrid progeny.

At the Brno beekeepers' congress F.W. Vogel from Berlin reported that on crossing light-coloured Italian queens with black drones he observed non-purity as late as the third generation, and that three-quarters of the bees were black and a quarter light in colour. This was the segregation ratio of 3:1 which E. Lauprecht (1966) was to point out. It should be noted that at the time of the Brno beekeepers' congress in 1865 Mendel was teaching at school, and his name is not on the list of participants. Vogel's report cannot have escaped the attention of Abbot Napp, however, though we have no evidence to show that he brought it to Mendel's attention.

Following the Brno beekeepers' congress, which gave a major boost to efforts to improve beekeeping standards, Živanský began in 1867 to publish a beekeepers' journal, in Czech and German, called the *Honey bees of Brno*. It was to become the most widely-read German-language periodical con-cerning apiculture in the whole of Europe. The activities of the Apiculture Association expanded, and in 1868 Živanský formed an 'Association of Moravian Beekeepers', which by 1871 already had 1470 members. It was independent of the Agricultural Society. The beekeepers' association in neighbouring Bohemia, which had come into being in 1852, did not have half that number of members. The more active Moravian association attracted 121 members living in Bohemia. Among the members there were also beekeepers from Prussian Silesia, Hungary, Austria, and Bavaria. In 1870 Živanský accepted Mendel as a member, whom he had met at the Natural Science Society. This was the start of a short but very fruitful period of cooperation between Živanský the organizer and Mendel the researcher (Orel *et al.* 1965).

Mendel had come across beekeeping in his native village. His father kept bees in his garden in hives made of old tree-trunks. Beehives also stood in front of the village school for the instruction of the pupils. After the end of his *Pisum* experiments and the confirmation of their results in experiments with further species, Mendel sought a general application of his theory of heredity in the animal kingdom. He saw the keeping of a large swarm of bees as a suitable model for the experimental crossing of animals. We may ask to what extent Mendel's interest in bees was aroused when he came across Živanský. In 1870 he took part with the latter in a congress of beekeepers held in Kiel, and on the way there they called to see Dathe in Eystrup, to see for themselves his manner of crossing races of bees (Lauprecht 1966, 1982). The experience may have been useful to Mendel in his later apicultural activities. The memorial desk unveiled in 1969 by Lauprecht in the house where Dathe lived should serve to remind us of the contact between Mendel and this eminent beekeeper.

On his return from the congress, Mendel had what was at the time a remarkable beehouse built in the monastery garden (Fig. 6.5). It may be designated the first apicultural research institute in Central Europe. Mendel published nothing on his beekeeping experiments, but he reported on his activities at meetings of the Association. In the journal of the Association there were regular reports from which we can gain a picture of his apicultural episode. The first point to Mendel's experimental beekeeping after 1900 was Niessl, who recalled his visit to Abbot Mendel and the informa-

Fig. 6.5 Mendel's experimental beehouse.

tion of his friend Proskowetz (1902), according to whom: 'In the garden, at the foot of a steep slope, was an extensive swarm which was kept in an entirely methodical and scientific manner, and Mendel used it to study bastardization in bees.' But one has to go to the reports published in the journal of the Association to see the full extent of Mendel's apicultural experiments and experiences. After 1871 he took part in the monthly meetings. The journal was edited in Czech as *Včela brněská* and in German as *Honigbiene von Brünn* (in English *Honeybees of Brno*) in the years 1872–8. (In the following text the years of issue and the page-numbers of the Czech version are quoted.) In 1875 (p. 37) Mendel is recorded as saying: 'it is important for every beekeeper to carry out experiments, since this is the only way to achieve successful results'. He kept in his beehouse various races of bees which were discussed among European beekeepers in connection with the endeavour to breed a race with a higher yield of honey. He also kept Cyprian, Egyptian, and even South American bees, from which he could scarcely expect to obtain a direct advantage at Central European latitudes. These were in effect experimental animals. At a meeting in 1877 (p. 82) he states that 'Cyprian bees have certain advantages, and they would be useful in obtaining the so-called culture race, for crossing with other races, which the whole world is anxious to achieve.' He would appear to have been drawing a parallel with his experience in the field of plant-crossing to create new varieties.

We can only regret that the written notes on Mendel's research on bees have not survived. We only know that he gave every hive a number and carefully recorded the installation of queens, swarming, flight energy, mating flights, stinging, and the external morphological traits of bees. The quantitative evaluation of traits connected with production capacity is reminiscent of his evaluation of experimental plants. It is apparent from a report published in 1877 (p. 82) that he intended to breed a new race using the technique of crossing different races. In the same year he states that his bees were 'often flying for honey as far as the Red Hill, to Pisárky and even further; they fly a very long way, as far as the Imperial Road, and it is certainly highly demonstrative that a Cyprian drone from Old Brno fertilized a queen from Nový Lískovec' (pp. 100–5). The distance between the two places mentioned as the crow flies is about 5 km.

Mendel's apicultural research comes in a period when Moravian beekeepers were conducting trials with imported races of bee. The first foreign race was imported in 1862. Ten years later, Živanský admits in the journal (1872, p. 2) that this importation had not achieved any success, and that beekeepers had suffered by it. Two years later Živanský (1873, 1874) writes in the introduction to his textbook that reports of controlled mating of a certain queen with a selected drone 'inside the hive, in a jar or a net, or with the queen and the drone held by threads, etc.' are simply fabrications. But Živanský's criticism of attempts to perform controlled mating did not deter Mendel from bold experiments in crossing bee races. The Cyprian bees Mendel kept came from Count Kolovrat Krakowský, who kept them in a remote spot on Blaník Hill in Bohemia. He also undertook crossing of various races, and was aware that mating control is quite unreliable. Out of ten queens he observed eight or nine which had been fertilized by drones from their own race. Mendel knew the system of transfer of traits in plants, and he may have had similar ideas about crossing bees. But he could not have known that the queen mates at a great height with several drones, something which was not discovered until much later.

According to Iltis (1966, p. 202), Mendel's main interest in apiculture was to study the effect of crossing different races of honey-bees for obtaining data which would confirm his theory of heredity. To this end he had a special cage made, to permit mating of the queen by one particular kind of drone. Iltis (p. 215) published the drawings of this cage, which consisted of a wooden frame covered with fine woven cloth and measuring 4m × 4m × 4m (see Fig. 6.6). More detailed information has recently come to light relating to Mendel's research (Beránek and Orel 1988). This was described by F. Kühne (1881), a beekeeper from Hungary, who visited Mendel in 1879 and published his findings together with more detailed information contained in a letter from Mendel written the next year.

Kühne tells us that Mendel had 'invented a very ingenious mating apparatus to force the queen to mate with the required drone'. The young queen, separated by a net from the motherless colony, was let into the transparent cage along with the drones. But in spite of keen observation, Mendel did not record a single mating or even 'intimate approach which might lead to hope that they had mated'. After a period of vain waiting, Mendel found the queen dead in the extension, along with most of the drones, and in the hive below a young, strong queen, who had just begun to lay eggs. Mendel explained the failure of this experiment to Kühne as follows: 'The mating of the queen in the enclosed space was a failure because the drones were taken from the alighting boards of various hives, and knew the pleasant feeling of free flight. They were not therefore willing to undertake such an important act as mating in the confines of an enclosure. They may also have been a different age to the queen, who had perhaps left the queen cell only a short time before the experiment, which may also have contributed to its failure.' Mendel explained the occurrence of an egg-laying queen in the hive below as being due to the fact that the experiment was undertaken during swarming, and it was possible that a surplus queen had escaped death and taken refuge in a hive which had for a long time been waiting for an active queen, and was welcomed there. Later, when the bees already knew that there was a queen in the heart of the hive, they allowed her highness up in the extension to die of starvation. It is also possible that a queen, after a mating flight where she was successfully fertilized, mistook her hive, and the bees in the hive where she had mistakenly landed were only too glad to accept her instead of the unfertilized one in the extension. Mendel therefore tried to repeat the experiment using drones which had just left the cells, hoping he might succeed this time.

Kühne was curious to know how the next experiment had turned out, and he asked Mendel to keep him informed of his subsequent results. In his letter Mendel tells him: 'As regards the mating of the queen, this year also I did not obtain any result, though the experience I have acquired leads me to suppose that the experiment will eventually succeed. Last year the fault lay in the drones, this year in the queen, which, as I had the opportunity to observe, in the course of nine flights into the mating cage remained quite indifferent to her passionate lovers and repeatedly tried to escape from the cage. As an after-swarm queen she had already known free flight, whereas the drones had been put in the cage soon after they left the hives, and did not know what an open space was. The queen used last year was caught on an alighting board before taking off, and the drones were all caught on alighting boards. In the coming year 1881 I want to repeat the experiments with a quacking queen and drones which have no experience of free flight. I will also adapt the apparatus so that the queen and drones will be separated in two small beehives, whence they will be able to fly to the mating cage.'

Mendel knew that for the selection of more productive bees it was essential to control mating of queens with selected drones. He therefore tried out three different ways of arranging the hive and the cage. In the first system, shown in a drawing published by Iltis, Mendel placed the queen together with the drones in a netted frame in the middle of a long hive. The workers had access from both sides through the net. According to Kühne's report of 1881 Mendel placed the queen together with the drones in two similarly netted half-frames in a special extension. The workers also had access from both sides through the net, and from the hive below, which warmed the extension. The third system is mentioned in his letter to Kühne. The queen and drones were to be placed in two small hives. Only the workers from both would be able to reach the open air, while the queen and the drones separately would be able to fly only into the cage. According to this idea the queen would live in one hive and the drones in the other. The cage would apparently be half the size of that described by Iltis (1966, p. 216) as

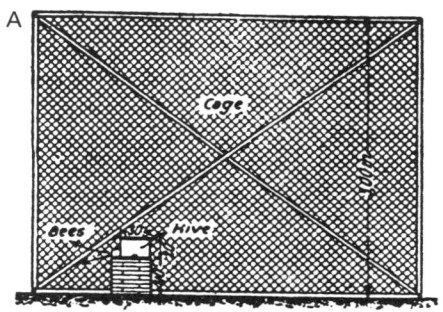

(*a*) Elevation of Hive and Fertilisa-
tion Cage.
(Scale about 1 : 80.)

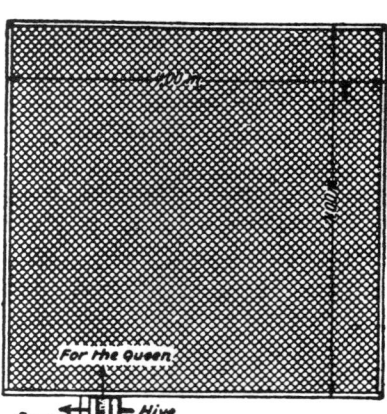

(*c*) Section of Hive and Ferti-
lisation Cage.
(Scale about 1 : 80.)

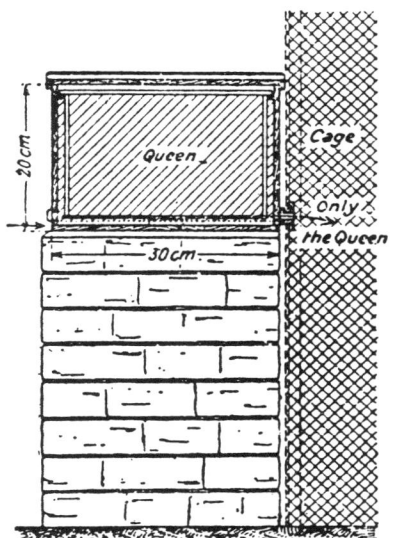

(*b*) Section through the Store-
Hive, in the Direction of the
Vertical Arrow, in *c*, or in the
Direction *a—b* in *e*.
(Scale about 1 : 10.)

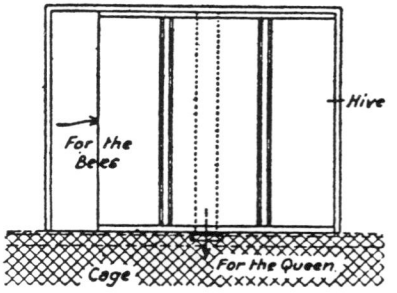

(*d*) Section of Hive alone.
(Scale about 1 : 10.)

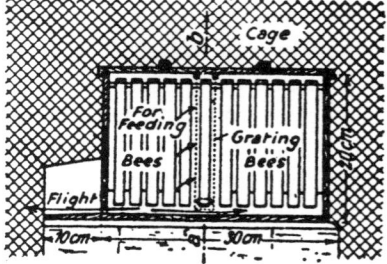

(*e*) Section through the Store-
Hive in the Direction of the
Horizontal Arrow in *c*.
(Scale about 1 : 10.)

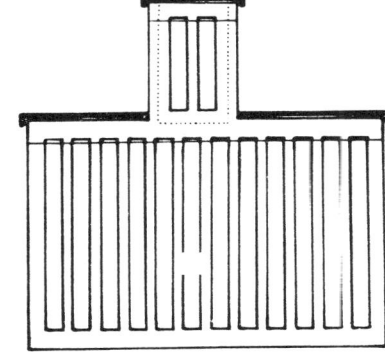
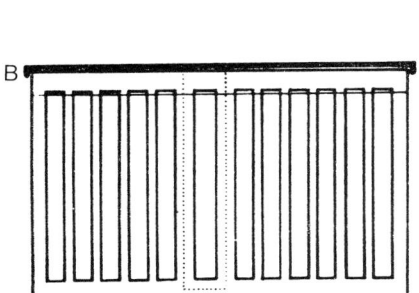

Fig. 6.6 (A) Reproduction of Mendel's drawings according to which Ludwig, carpenter and beekeeper, made a hive and mating cage (originally published in H. Iltis, 1924). (B) Alternative keeping of queen and drones according to Kuehne's report.

follows: 'There was a large cage with a wall of wire gauze, a cage about three metres high and with a floor space of sixteen square metres. To this was attached a store-hive, with two openings, one into the cage, for the queen, and one into the free air, for the worker bees.'

From the time of Mendel's experiments as described by Kühne much information has been collected on the mating of queen bees in confined spaces. Many experiments were carried out in the USA in the 1970s, but continued to be fruitless (Habro 1971). The methods, material, theoretical base, and results achieved by Mendel's well-thought-out experiments were in striking accord with modern conclusions.

In Mendel's experiments the workers were able to fly outside the cage into the open air, while the queen and the drones had access only to the cage. Mendel set out from the fact that the workers were unnecessary in the cage, since they do not prepare either the queen or the drones for their mating flight. Later research showed that they are hostile towards the mating pair. Mendel's assumption was therefore correct.

We do not know Mendel's views on the effect of light on bees in a cage. Kühne states in his report that it would be advantageous to build a pyramid-shaped cage where the bottom two-thirds would not let light through. The queen and the drones would find it easier to fly in the upper third, and mating was more likely. When such alternatives were later tried, it was shown that the queen and the drones flew into the walls and that mating did not take place.

In his letter Mendel states that he had observed nine flights of queen and drones, but we do not know over what period. The queen flies out several times a day, and the latest research has confirmed that bees which have not yet flown in the open air fly out more actively. When, in 1879, Mendel's experiment was not successful, he explained this as being due to the drones' already having known free flight, and thus not taking advantage of the possibility of mating in the cage. In 1880 he sought the reason for failure in the fact that the queen had known free flight. According to the content of the letter, Mendel planned a repeat of the experiment in 1881 with a queen which had not yet left the hive and drones which had not experienced free flight. He would be able to take a more detailed note of whether failure was due to the queen or the drones. He had made similar observations of the causes of failure to fertilize various species of *Hieracium*. On that occasion he found there was a difference in the ability of the maternal plants to be fertilized. Modern apicultural experiments have shown that drones in experimental fertilization in cages are attracted into the cage by the flying queen, but that the actual obstacle to mating also comes from the queen.

One can thus suppose that even in 1880 Mendel had an experimental cage and was performing the experiments mentioned by Iltis as the first variant. Iltis dated these experiments in his chapter on Mendel's declining years, mentioning the visit of J. Líznar, a meteorologist in Vienna, to the monastery in the summer of 1882. Mrs. Líznar, who accompanied him, noted the experimental cage, and Mendel gave her the following explanation: 'There are some drones and queens in that cage. The queen is choosing a proper husband, for it is just as unfortunate among bees as it is among human beings when a good woman is caught by a bad man' (Iltis 1966, p. 279).

Before Mendel began experimenting with the controlled mating of queen bees, Živanský (1874) had rejected reports that someone had succeeded in mating a queen with a selected drone. Mendel, however, paid no heed to this, and in his experiments sought a solution which was not available until artificial insemination was used in the second half of this century. But this technique was not usable for the central African species *Apis mellifica andansonii*, which, though a high producer, is very aggressive. According to Nowakowski and Morse (1971), controlled mating of this species was tried in the 1970s in large cages. In connection with this research a review of the world literature on attempts at controlled mating of bees in enclosed spaces was published (Habro 1971). It also recalls the pioneer work of Mendel.

Mendel gives the most frequent accounts of his experiments with the crossing of various races of bees in 1876. He kept two swarms of a Cyprian race, and emphasizes that he needed 'at least two years' experience' before he would be able to obtain more detailed information on them. Even then, however, he states that 'the two swarms kept last year with Cyprian queens are now the most resilient and are exceptionally fertile. This fertility is equally apparent in both swarms, and thus cannot be the work of chance.' This phenomenon is now called the hybrid effect, and Mendel had already observed it in crossing peas with tall and short stems. He kept various races of bees, and on the basis of different traits he was able to monitor the results of crossing. But it was essential that no one else kept the foreign race—this was the only way in which he could evaluate the hybrids with the Cyprian race. The other beekeepers mainly wanted to cross domestic bees with the Carinthian honey-bee, and wished to create a higher-yielding race (Iltis 1966, p. 219).

At the monthly meetings of the Apicultural Association Mendel also reported on his research into beekeeping techniques. His beehouse was exemplary, and he was most willing to show it to visitors. In 1877 the chairman of the Apicultural Association, Kment, said he had visited Mendel's beehouse and 'found the bees in the most beautiful condition a beekeeper might hope for. Swarm upon swarm. And a new rearing of queens, so that I should wish everyone to see this exemplary beekeeping, since I can express my firm conviction that everyone would take away something useful'. The Association's secretary, F. Bauer, was equally unstinting in his praise. Mendel showed him individual swarms. 'He explained various changes in the hives, and also different results, and also spoke in great detail of the Cyprian and Egyptian bees he had obtained. In the closed beehouse there were 36 swarms, most of them still in vertical hives divided down the middle and having two entrances, so that if one or another of the swarms is weak, or does not have enough honey, or has no queen, or in the case of artificial swarms, the quarter-inch partition can be horizontally removed, and if one of the swarms gets a queen, it can be combined with the other.' He emphasized that one strong swarm is created in this way, which finds it easy to overwinter. Stieber in 1877 also noticed that Mendel had a Dzierzon box 'for rearing queens, which, in view of the results we wish to achieve in the field of scientific knowledge, certainly should not be underestimated' (*Honey bees*, pp. 55–60).

In his technological experiments with bees we see Mendel the researcher, and also the experienced agricultural expert. At a meeting of the Association in 1877 (*Honeybees of Brno*, 1877, pp. 100–5) he instructed beekeepers as to the best choice of location for hives. In connection with the development of the weather, he states that 'the remainder of July will continue to be unfavourable. The earth has a plentiful supply of moisture, so the air is not too dry, and flowers are not producing much honey. On days which are too steamy the sugary substances from flowers grow so thick that the bees are unable to suck them out, even though the flowers exude sufficient quantities.' Mendel made use of his knowledge of botany to evaluate the nectar-producing sources of pasture for bees, especially in the period of nectar-gathering. He suggested that the bare south-western slopes of the Špilberk hill should be seeded with the simplest nectar-forming and pollen-producing plants. The reason he gave (1876, p. 68) was 'the chosen species are also suitable because of the richness of their flowers as pasture for bees, which is especially useful to beekeepers since flowering plants have been entirely pushed out of the fields by cultivation of beets'. In evaluating the main sources of nectar, he came to the conclusion that the best plants for the purpose in Brno were acacia, of which there were many on the Špilberk hill and in gardens. Similarly, he considered sainfoin, a large amount of which was sown on the nearby Red Hill, to be a useful source. According to Mendel limes were unreliable, since they had produced nectar only once in recent years. He recommended filling the gap after the blossoming of the fruit trees in those areas with early honey production by means of speculative feeding. By this he meant sugar solution. Mendel considered April to be the appropriate time of year as far as Brno was concerned. At the meetings chaired by Mendel consideration was also given to the possibility of obtaining nectar from spruce trees and to 'leaf honey'. He recommended growing plants which could form a good pasture for bees. He himself had experimented with sainfoin and the labiate *Leonorus cardiaca*, considering these species also suitable for the production of rope (1876, p. 98). Mendel evaluated the nectar production of plants according to the number of bees visiting them and the time taken by individual bees to suck the nectar from one flower. He also paid heed to the quality of the nectar, which he collected from the honey sacs of bees caught on flowers.

As an experienced beekeeper, Mendel emphasized (1877, pp. 100–5) the importance of strong swarms. He was convinced that 'the more bees there are in a hive, the more willing they are to work, and the more they produce'. He gives as an example of high yields the daily amount per swarm of 6.5 kg. He was also sure that strong swarms had a much better chance of overwintering successfully than weak ones. He studied the question of overwintering experimentally, and built a bee cave for the purpose on the slope behind his beehouse (Matalová and Kabelka 1982). The later development of beekeeping showed him to have been right. He also used his experience to make improvements to hives, simplifying them— a fact which was much appreciated by his contemporaries. In his beekeeping practice he had also come across the danger of an infection which was at the time known as 'foul brood'. In 1872–3 he lost his entire stock on account of this disease.

There were differing opinions among beekeepers as to the cause. Some supposed losses to be due to physiological weakening of the swarms, while others considered an infection to be involved. Mendel inclined to the latter view, and immediately destroyed affected swarms. At a meeting in 1875 (p. 132) he emphasized that no one should try to treat such bees, but must 'without mercy destroy the bees with sulphur'. He added that 'one's heart bleeds, but it is the only way to get rid of the evils'. He was sure such swarms were doomed anyway, and that any delay would only endanger the others. After fumigation with sulphur dioxide had killed the bees, he recommended scrubbing out the hives with a solution in water of the salt produced by the sulphur, which would 'destroy the infectious substance'. It was subsequently shown that a bacterial infection was the cause of such disasters. Mendel's approach was once more that of a pioneer, and his advice is useful to beekeepers to this day.

In 1879 a consignment of Pernambuco wood arrived in Brno from Brazil, for use in the textile industry. A swarm of stingless bees was found in the hollow of a trunk, and taken to the Natural Science Society. Through Professor Makowsky they were handed over to Mendel, who was by then a highly respected apicultural researcher. It is to him that the determination of the species as *Trigona lineata* is attributed. He was fascinated by this remarkable species, and began keeping them experimentally under local conditions, which were not at all suitable. *Trigona lineata* is one of the smallest of the stingless bees, differing from the honey-bee *Apis mellifica* not only in size and body structure, but also in the construction of its combs. The bees build horizontal combs with brood cells joined by a vertical pyramid. All cells are built with the same regularity of wax. Honey is deposited in wax cups, which are below the combs. In tropical regions there are several hundred kinds of Mellipods, the group to which the *Trigonae* belong. The first report of them was published by E. Drory (1872–4) in France. He kept several families experimentally in Bordeaux, investigating the possibility of their acclimatization to European conditions, but without success.

Nothing can be found in the pages of the Agricultural Society journal relating to these bees. It was not until much later that H. Molisch (1934, p. 126), who studied at the Brno *Realschule* at the time Mendel was a teacher there, stated that these bees had flown around the monastery gardens, and had produced honey. But they did not survive the winter, since they were not adapted to such low temperatures.

A zoological journal in Leipzig published two extensive studies describing Mendel's research with *Trigona lineata*. The author, Professor A. Tomaschek (1879), who taught natural history at the Brno Technical Institute, states in the first article that Mendel had special equipment made for keeping this species. He estimated that there were 300–400 bees in the hollow of the sandalwood when it arrived in Brno. He also made a microscopic analysis of pollen, proving it to be from *Haemotoxylon camechianum*. But he did not observe any drones, and supposed the queen had already been fertilized. She did, in fact, lay fertilized eggs. Tomaschek says the bees were handed over to 'an eminent expert'. He expected that expert to be able to carry out all manner of experiments on them.

The next year Tomaschek (1880) published the second part of his study, in which he states that by the end of November the bees had built nineteen honeycombs. The queen had laid forty thousand eggs. The development of embryos is also described. Tomaschek refers to experiments made by E. Drory with tropical bees in the French city of Bordeaux. These did not produce any results, as the bees soon died. Mendel kept his tropical bees in a hive which was inserted in a double-walled metal surround. Into this he poured warm water at a temperature of 40 °C every day. In this way he was able to carry on recording data right up to November. In conclusion, Tomaschek states the importance of research into such bees, and expresses his expectation that especially interesting data will be obtained in the winter, promising to publish them in a further study. But these data did not appear until 1885, and were published in Moscow in the form of a report by N. Zograf on a meeting of the Moscow Society of Acclimatization in the journal of the society *Zoologitcheskii sad i akklimatisatsya* (1885, **2**, 12–14), to which attention was drawn by Alpatov and Orel (1979).

At the time great attention was being paid in France and in Moscow to a study of the acclimatization of animals transferred into a new environment. A society for the acclimatization of animals was even founded in Moscow. Professor A. Butlerov was interested in bees, and tried to acclimatize various species to different conditions in Russia. He thus welcomed Tomaschek's paper, in the introduction to which there was a mention of stingless bees' having been brought to Brno, and having been entrusted to 'one of the most experienced beekeepers in Moravia'. The Moscow article also mentions that Mendel was able to keep the bees so long because in 1879 the autumn was especially mild. The bees even began to collect pollen and honey. Mendel supplemented their diet with sugar solution. In 120 days the bees made nineteen honeycombs. The first new bees were observed from 20 October. In conclusion the article states that owing to the efforts of the 'outstanding expert Mendel', the Brazilian bees lived in Brno up to the end of the next year. In France they had survived only up to the end of December.

According to Mendel's letter to Kühne (1881), he managed to keep *Trigona* up to May 1880 at least. He explained the successful overwintering on the one hand by heavy feeding throughout the whole period with a sugar solution and pollen preserved by the bees, and on the other hand by placing them in a heated room at a temperature of 17–18 °C during the winter. Mendel saw the first *Trigona* fly in spring as late as 25 May. at an air temperature of 20–22 °C. Then they began to carry out pollen and to build their combs, which they filled with nectar. For peak flying activity they required a daily temperature of around 25 °C. Mendel supposed it was not possible to consider acclimatizing this species under Moravian conditions. The required temperature of 25 °C occurs under local conditions on an average of fifty-four days a year. During this period the *Trigonae* would not be able to collect enough supplies to see them through the winter period. For comparison it can be said that the honey-bee (*Apis mellifica*) flies from the hive at an outside temperature of at least 14–16 °C, which occurs on average for 120 days a year. This is a sufficiently long period for them to lay down supplies, but truncates the period necessary for emptying the digestive tract.

The conclusion of the Moscow report on Mendel's acclimatization experiments with the bees was: 'We must acknowledge the experiments with the acclimatization of *Trigona* bees in Europe by him (Tomaschek) and Fr. Mendel as the most successful for fruitful science.' We might add to this assessment that Mendel's methodical approach in the apicultural experiments represents a continuity in research started with the hybridization of peas.

For his meritorious contribution to apiculture, Mendel was in 1877 made an honorary member of the association. In 1871 he was elected vice-chairman of the association, and his research work complemented the successful organizational work of Živanský. When Živanský died in 1874, the members of the association proposed Mendel as chairman. He declined, saying that 'private circumstances' would not allow him to accept. He was at the time coming into conflict with the authorities over the taxation of the monastery, and also, in the aftermath of the political stand he had taken, he was keeping a somewhat lower public profile. His nomination as an honorary member was the zenith of his recognition as a beekeeper. Suddenly, he ceased to attend meetings, and there were no more regular reports of his research activity in the pages of the Association's journal.

The last mention of Mendel's research in the journal of the Apicultural Association was in 1878. After that further information was published by Tomaschek, relating to the acclimatization of tropical bees mentioned above. Mendel was magnanimous enough to hand over his results to him, as he was later to give to Professor Líznar his data derived from research into fluctuations in ground-water levels.

Mendel began his beekeeping experiments twenty years later than those on plant hybridization. It was a new fascination for him, deriving partly from an effort to continue his research into trait transfer from generation to generation in animals, and on the other hand from the great interest bee-keepers took in progress in apiculture. At first he thought he would be able to cross races of bees with differing trait pairs, just as he had in plants. But he soon found that he must first research into the possibilities of controlling mating. This was at the time when he was working on the same problem in crossing experiments with *Hieracium*. After several years, however, he became convinced that under his new circumstances as abbot he was neglecting both the experiments with *Hieracium* and those with bees. But he did not lose interest in the question of trait transfer and the possibility of combining traits in either bees or *Hieracium*, as the newly-discovered documents relating to his beekeeping activities show.

THE ACKNOWLEDGED METEOROLOGIST

Of the many branches of natural science which interested Mendel, meteorology was the one which enjoyed his attention the longest. His obituary notice gave in first place after his name his membership of the Vienna Meteorological Society. No mention is made of his membership of the Natural Science Society, where he lectured on his plant-hybridization experiments. He was thus acknowledged foremost as a natural scientist among meteorologists. His expertise was acknowledged by the representatives of the Meteorological Society in Vienna, and in Brno and Moravia he was considered the greatest authority on meteorology. He must surely have been pleased with this fact. Every researcher likes his work to be appreciated, and Mendel was no exception. Iltis (1966, p. 278) states that before his death Mendel outlined the text of his obituary notice. If this is true, then we can see how significant he himself considered his activity in this area.

Interest in meteorology in Brno developed in the wake of pioneer work in the field in Prague. The latter is associated with the names of the great astronomers Tycho Brahe (1546–1601) and J. Kepler (1571–1630), who organized some of the first meteorological observations in Prague. After the Battle of the White Mountain (1620), which placed the Bohemian monarchy under Habsburg rule, scientific astronomy in Prague died out, and meteorological research declined. The first regular meteorological observations were made during the time of the astronomers J. Stepling (1716–78) and, in particular, J. Strnad (1747–99), who began regular records in 1775. These are among the longest-standing uninterrupted records in Europe. In the 1840s K. Kreil (1798–1862) began making regular geomagnetic observations in Prague. This outstanding organizer published meteorological writings, and made new meteorological instruments. Two younger natural scientists, K. Fritsch (1812–79) and C. Jelinek (1822–76), worked with him, and these two came into contact with Mendel.

Astronomical and meteorological observations were introduced in Moravia in the nineteenth century by Captain F. Knitlmayer (1750–1814), with whom C.C. André made friends in Brno. One of his subsequent colleagues was F.I. Hallaschka (1780–1847), who originally taught at the *Gymnasium* in Mikulov, in the south of Moravia, and later at the Philosophy Institute in Brno, and in 1814 was made Professor of Physics at Prague University. He was responsible for an increase in the number of meteorological observations made, and in 1816 André prepared the constitution of a Meteorological Association as part of the Agricultural Society. Meteorological observations were soon begun in Opava and Olomouc, and later in other towns too. In 1849 the Meteorological Association was incorporated into the activities of the Natural Science Section of the Agricultural Society.

Mendel met with meteorology even before coming to Brno, as has been explained in detail by Dubec and Orel (1980). Meteorological observations were made in Opava by F. Ens, a teacher at the *Gymnasium* who also published reports on them, and in Olomouc by the headmaster of the Philosophy Institute, C.R. Unchkrechtsberger. At the *Realschule* in Brno Mendel came into contact with Professor Zawadski, whose broad natural science interests also took in meteorology (in the previous post he had held, at the University of Lvov, he had actually taught the subject). In 1851 Napp also showed an interest in meteorology. Under his leadership the Agricultural Society sent a proposal to Vienna for the setting up of a meteorology institute in Brno. The result was the transfer of what was at the time extensive meteorological activity from Prague to Vienna. Leading meteorologists left Prague for the capital city of the Empire, where a new Central Institute for Meteorology and Earth Magnetism was set up.

Fritsch and Jelinek organized an Austrian Meteorological Society in Vienna in 1865. The list of founder members included the Augustinian F. Gabriel and the physician P. Olexík. Mendel was already working with the latter. When he became abbot, the Society's secretary Jelinek announced that Mendel had paid an additional 100 guilders and thus become a founder member.

Mendel began making meteorological observations at the monastery in 1857. At the time the Institute in Vienna nominated Olexík, who lived in Pekařská street, close to the Augustinian monastery, as its official observer. Mendel soon made friends with him, and during his absence handed over the results of his own observations following adjustment for the difference in height above sea level. From 1858 both men began to monitor ozone levels in the atmosphere. When in 1878 Olexík became seriously ill, Mendel took over his duties, and the monastery station became the official measuring point in Brno (Weiling 1984*b*).

Olexík and Mendel's observations are recorded at the Hydro-meteorological Institute in Brno. Compared with Olexík's somewhat disorganized data, Mendel's are outstanding for their clarity. Olexík sent in his data immediately after the end of the month. He supplemented them with details of the instruments used and adjustments to the data. From August to November 1883 the records of meteorological observations were rewritten by the young Augustinian L. Ledwina. The figures for December were sent with a considerable delay, not reaching Vienna until 26 January 1884, after Mendel's death.

Mendel recorded wind direction according to the flag on the tower of Špilberk Castle in Brno, and estimated wind force by the smoke coming from nearby chimneys. Rainfall was measured close to the prelate's garden. The ozone concentration was monitored according to a ten-degree scale by the coloration of an indicator paper impregnated with nitrous iodine and a solution of starch. The sites of these instruments were visited daily, and Mendel made a meticulous record of the data.

Information regarding the position of individual instruments for meteorological observations was recorded by Mendel's ex-pupil at the *Realschule*, J. Líznar, who later worked as an expert at the Meteorological Institute in Vienna. In 1881 he was Mendel's guest for six days, and got to know the instruments well. The thermometer and hygrometer were in a metal box screwed to the window frame on the first floor of the prelate's wing of the monastery, facing Pekařská street. The barometer was in the same place.

Mendel's first study, entitled 'Graphic and tabular overview of meteorological observations in Brno' was published in the *Proceedings* of the Natural Science Society in 1862 (Mendel 1863). A clear, large-format graph shows the data recorded. The Natural Science Society considered this publication so important that it paid for 500 special copies to be printed. It was published separately, along with the study by Makowský (1863) 'Flora of the Brno region'. Mendel illustrated the data recorded by Olexík in 1848–62. The fifteen-year mean of individual values was compared with the results for 1862, expressing the amplitude of fluctuations. In the upper part of the table the mean daily values of observations in the morning, at midday, and in the evening are given. The lower part shows the five-day averages of wind direction according to an eight-point compass rose, wind force in ten degrees, and cloud cover in five degrees. At the bottom are the five-day rainfall averages. Mendel closed his summary figures within fourteen days of the last observation, and was able to publish them within a very short time.

The most remarkable feature is the comparison of the fifteen-year averages with the annual observations. Mendel made use of statistical principles at a time when other meteorologists were only just beginning to consider the use of similar methods of evaluation. He had taken over the task in 1857 from Olexík. By chance he happened to be analysing the most difficult part of his plant-crossing experiments, which also involved statistics, at the same time. The question is sometimes asked: did he first think of using statistics in meteorology, or in his plant research? In reality he had learned the importance of large numbers of cases for evaluating natural phenomena while he was studying physics, and he was simply applying this knowledge to both meteorology and plant hybridization.

At a meeting of the Natural Science Society on 9 July 1862, it was decided to create in Moravia and Silesia a network of meteorological observation points, with regular publication of the results. Mendel's paper was the starting-point for the publication of results. In subsequent years summaries of meteorological observations were published from Těšín, Hukvaldy, Jihlava, Dačice, and Brno. Mendel (1863) included in the summary of observations for the previous year both annual means and extreme values. The only comment he made was on observations of the ozone content. In the following year he noted on the subject of precipitation that the low soil moisture content was connected with the major drought that year. Mendel published similar results for 1864, 1865, 1866, and 1869 (Mendel 1864, 1865, 1866b, 1870b). One can suppose from his notes for 1864 that he was at the time beginning to take an interest in levels of the water table. The results of observations in 1866 and 1867 were prepared for publication by I. Weiner, who studied at Vienna University with Mendel. The manner of their processing was the same as that used by Mendel. Mendel published the results of observations for the last time in 1870 (1870b). He was taking over other responsibilities, and he now confined his meteorological activities to making observations.

We can judge from the books Mendel bought that he was interested in new developments not only in meteorology, but also in astronomy. He made full use of his experience in 1870, when, on 13 October, an extraordinary whirlwind struck Brno. It was not long before, on 9 November, that the meeting of the Natural Science Society enjoyed a remarkable exposition of this exceptional meteorological phenomenon. In the following year Mendel (1871) published his ten-page report in the society journal. It is Mendel's most outstanding meteorological publication, exhibiting in addition to theoretical knowledge a remarkable talent for observation and a capacity for the quick formation of a hypothesis. The whirlwind swept past the monastery at two o'clock in the afternoon, and Mendel observed it from an open window. His description is supplemented by the evidence of other eye-witnesses. The first part describes the course of the whirlwind, and Mendel explains it as he would a physical phenomenon to a class of students. He speaks of a 'natural spectacle', with an indescribable roar, and of a 'hellish symphony, accompanied by the crash of window-panes and slates, which in some cases were flung through shattered windows to the other side of the room'. One slate even flew through the open door-way, across his desk, and into the next room. The moment the first wave of the whirlwind passed, Mendel goes on, 'my gaze soon uncovered the enemy. It was the first whirlwind I have seen other than in pictures and magazines.'

Another part of Mendel's exemplary account describes the shape of the oncoming whirlwind in the form of a diabolo; he estimates its size, direction of rotation, and speed as it travelled from Kamenný Vrch hill, through the Pisárky valley and the Old Brno brewery to Pekařská street, and up the south slope of the Špilberk hill to the railway station. He explains the origin of whirlwinds as a physicist through the meeting of two air streams of different dimensions and properties, arising shortly after the occurrence of a storm. The resulting formation in this case turned in a clockwise direction, which Mendel denotes as an exception to the rules of atmospheric currents in the northern hemisphere. He also considered the shape to be atypical.

The conclusion of Mendel's lecture shifts from a strict physical explanation to a lighter tone for the members of the Natural Science Society, from which we may quote, for instance: 'This brings to an end the discussion of our dangerous guest. But we must admit that, however we might have tried, we have got no further than an airy hypothesis, which is explained from airy material and on an airy basis.'

J. Munzar, a Brno meteorologist, studying Mendel's explanation of the origin and substance of the whirlwind, pointed out that Mendel's contemporary T. Reye (1872) failed in his own article about a whirlwind, dated 1872, to explain the underlying principles. It was not until much later that A. Wegener (1917) gave a scientific explanation of tornadoes, and did so in exactly the same way as Mendel, though without any reference to the latter. One might almost speak of the rediscovery of Mendel's fascinating publication on this subject, too.

In his meteorological activities, as in his plant-crossing experiments, Mendel had in mind the use of new scientific knowledge in practice. Even when writing his examination essay in physics in 1850 he had described the mechanical and chemical properties of air, stating in the end that more exact conclusions on the movement and intensity of winds would be acquired by the rapid transmission of data from meteorological observations in various places using the telegraph, which was at the time just being introduced into the Austro-Hungarian Empire. Later Mendel helped arrange the pioneer publication of weather forecasts by telegraph. In the old meteorological textbooks he studied, there was a description of prediction of the weather from long-term meteorological observations. In a textbook by Pilgram (1788) Mendel marked a mention of recorded data which failed to agree with experience.

In 1877 there were attempts by Moravian meteorologists to forecast the weather by means of data collected at various locations throughout Europe. The Meteorological Institute in Vienna began transmitting these observations by the then new-fangled telegraph. Members of the district agricultural societies in Moravská Třebová and Nový Jičín took up the challenge and introduced a system of weather forecasting to farmers during the harvest by means of couriers. This system met with difficulties, however, and the Society in Moravská Třebová went over to a system of optical signals. Using flags of various colours flown at vantagepoints in the countryside, they forecasted three types of weather between 2 July and 15 September: fair, unsettled, and rain. The project was well received by farmers. In October the Society made a written report to the provincial office in Brno regarding this successful experiment. The provincial office asked the committee of the Agricultural Society for their professional opinion, noting that the district society was applying for financial support from the government, which would be difficult to obtain. At the November meeting of the committee it was decided that the idea of making weather forecasts for farmers was a very good one. It was considered desirable to extend the activity. The governor's office was advised to ask the government for financing. The opinion of the committee was written and signed by Mendel. The original is to be found in the state regional archive in Brno. The governor supposed that it would be better if the Viennese institute sent a forecast to the Agricultural Society, which would then pass it on by telegraph to the district societies. At a meeting of the committee in February of the next year, however, Mendel pointed out that weather forecasts were issued for twenty-four to forty-eight hours, and therefore considered it more reasonable to transmit them direct from Vienna to the districts. The minutes of the meeting note that the matter is to be explained in the Society journal. Accordingly, on 28 February brief information on the subject appeared under the heading 'The significance of weather prognosis for agriculturalists' (Anon. [G.M.] 1878); in effect it is simply a reprinting of Mendel's report to the governor's office.

The question of weather forecasting was also discussed by the committee of the Natural Science Society. The minutes of the meeting contain a remark by the committee secretary Niessl: 'To what extent this will prove itself is to be seen (Orel 1969*a,b*).' At the beginning of 1878 the Ministry of Agriculture asked the Agricultural Society for the addresses of experts who were to be responsible for issuing weather forecasts in Moravia and Silesia. At the same time territorial divisions for individual experts were laid down. The subsequent organization of this system was undertaken by L. Jehle of Přerov, who went through the details of it with Mendel. In 1878 and 1879 the committee of the Agricultural Society discussed the organization of pioneer forecasts. In 1879 the Nový Jičín meteorologist J. Oborny published together with H. Voigt in Berlin the book *Meteorology and the telegraph in the service of agriculture* (Oborny and Voigt 1879). In the journal of the Agricultural Society a comment appeared which referred to Oborny as a pioneer of the introduction of weather forecasting in the Kravař region, and described him as a 'fitting apostle in this area'. The author of the article would appear to have been Mendel (Orel 1971*b*). In 1878 Mendel, from his position in the Agricultural Society, had recommended to the Minister of Agriculture that the successful practice of weather forecasting for farmers should be continued (Fig. 6.7), and he proposed that districts which took on this task should receive 150-guilder subsidies. The most outstanding of the forecasters, J. Jehle from Přerov, was in addition awarded the Society's silver medal on Mendel's suggestion.

Mendel's enthusiasm for weather forecasting is also apparent from his final publication 'The bases of weather forecasting', which appeared in the Society's journal in 1879 with the signature 'M' (1879). The introduction explains the physical composition of the atmosphere, whose elements inter- act to determine the way the weather develops. The author welcomes the new, scientific approach to issuing weather forecasts, and expresses his satisfaction that there will at last be an end to pseudoscientific forecasting based on the hundred-year calendar, phases of the moon, etc. At the same time he stresses that weather forecasting under Central European geo- graphical conditions is a very complex business. He is, however, convinced that by applying the great findings of 'sharp intellects' it will be possible to deduce 'from an apparently confused mass of facts' the underlying order. He points to the success which has been achieved in weather forecasting at sea, and anticipates 'equally good results' being obtained in the sphere of agriculture. The article also describes the instruments used for making meteorological observations, and emphasizes the importance of taking measurements simultaneously in several different places and of collating the information with the aid of the telegraph. The conclusion compares the

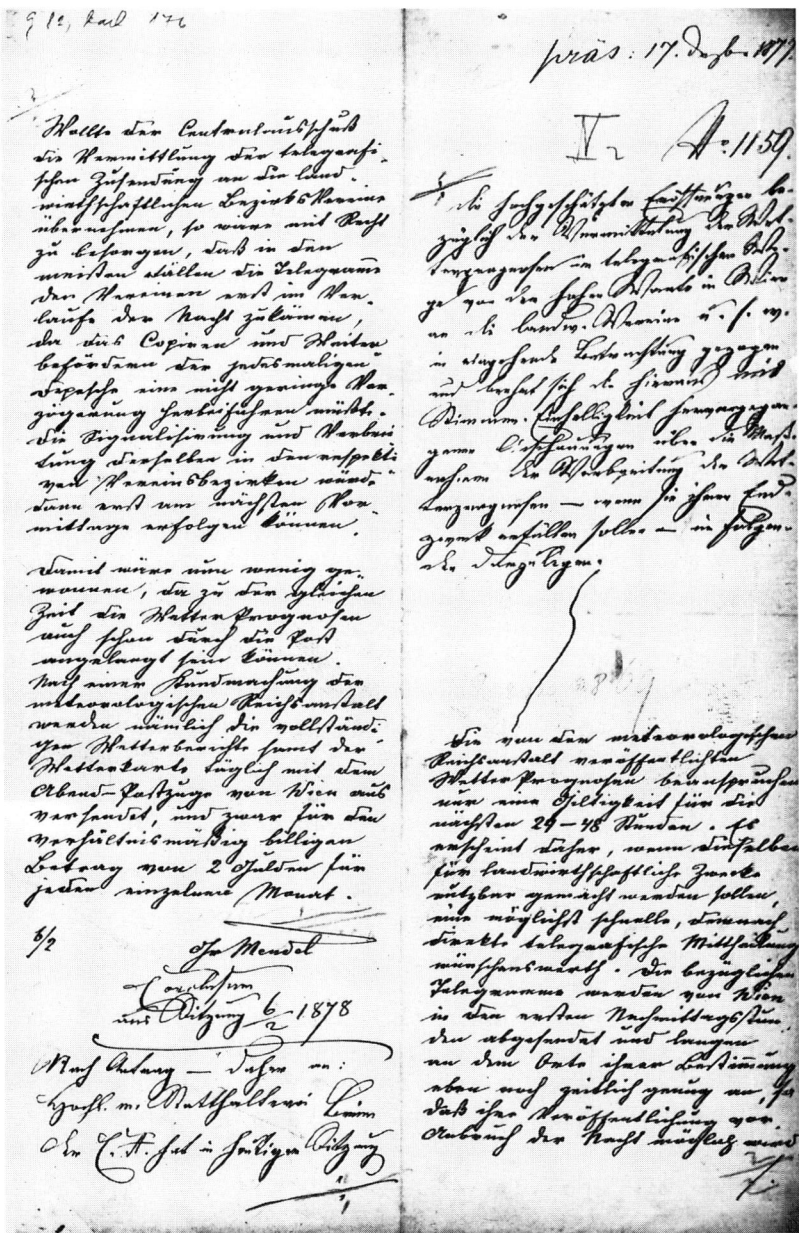

Fig. 6.7 The end part of the minutes from the meeting of the Central Committee on the Agricultural Society on 2 January 1878, with Mendel's recommendation to introduce the weather prognosis in Moravia.

reliability of forecasting according to the observations of about thirty European meteorological stations with the results of multilocational observations in the USA. The author states that the 'ever-practical' Americans are able to forecast the weather for the next twenty-four hours with almost 80 per cent accuracy.

The writer of the article shows a knowledge of European literature on the subject of meteorology, and an eminent interest in the publication of weather forecasts. Again, it could only have been Mendel. The same enthusiasm can be seen in his letter to Baron C. Barrata (published by Dubec and Orel (1980)), with whom he was in contact at committee meetings of the Agricultural Society. In 1877 the weather in autumn was cold and rainy, and Barrata was extremely worried about the autumn harvests at his Budišov and Třebíč farms. The two of them cooperated on the committee of the Agricultural Society in the application of new findings in agricultural practice. Mendel's letter states that the inclement weather had forced him to spend his spare time in meteorological research. According to previous records he supposes that there would be an improvement in the weather in the next five days. He adds that such forecasts are dangerous, but that Barrata will certainly understand how he intends these words. According to existing records of the weather at that same time, Mendel's forecast was proved correct within three days, so that his estimation of the way the weather conditions would develop was correct. He may have arrived at his conclusion on the basis of his knowledge and logical thinking. The final meteorological article on scientific forecasting again shows the originality of the scientist's thinking.

Towards the end of his life Mendel began to take an interest in an investigation into the appearance of sunspots. His astronomical telescope, which is on display at the Brno Mendelianum, testifies to the enthusiasm of a natural scientist for investigating new problems. In 1882, from January to the end of November, he carefully recorded the results of his observations in a three-centimetre circle. Niessl was professor of astronomy and geodesy, and he discussed these observations with Mendel. In those days the idea of a link between sunspots and the weather was something which had only very recently emerged. Only fragments of Mendel's records have survived, and were published by Iltis (1966, p. 224). We do not even know how far he continued along this line of research, or how he evaluated his results.

Mendel's health deteriorated to such an extent in the course of 1883 that he had to give up his meteorological observations. In a letter to Líznar, dated 20 December 1883, he provides evidence of how he concluded these activities, stating: 'Today I had to ask the management of the Institute to relieve me of all meteorological observations, for since May I have had heart trouble, and my condition is such that I am not able to read the instruments without someone's help.' The contents of the letter, published by Iltis (1966, p. 224) show the responsible approach of the observer, and also the friendly relations between teacher and pupil.

One might also consider Mendel's years of monitoring water-table levels to be part of his meteorological research. His work was based on an article by M. Pettenkofer (1862) pointing to the connection between epidemics and fluctuations in ground-water level (Weiling 1972, 1975b). From 1865 Mendel monitored the level in the monastery well, and in 1881 he handed his results over to Líznar. On the occasion of the fiftieth anniversary of the *Realschule* Líznar (1882) published these results in honour of his erstwhile teacher, by then a renowned scientist. Mendel did not reach any conclusion. His data show his drive to accumulate data on natural phenomena and to try to reveal a pattern from a large number of observations.

In the last year of his life the sickly abbot was reduced to hearing second hand how his flowers and fruit trees were getting on, and what was happening in the beehive. Meteorology remained an escape from the day-to-day chores of office and of public life.

At the outset of Mendel's extensive scientific career was the desire of a clever country lad to get an education. In the monastery surroundings he had chosen he was able to study and later to teach natural science, and to put some of his findings to work in research. His plant-hybridization experiments developed a new theory, and he applied their results in practice. To what extent can we say that he achieved his aim in life?

7

The man

In his curriculum vitae, at the age of twenty-eight, Mendel expressed his wish to devote himself to a study of natural science. He had by then lived for seven years in the seclusion of the monastery, where he was able to devote his spare time to a private study of the natural sciences which so fascinated him. He had also worked in a grammar-school environment in Znojmo, and tried his skills as a teacher. The year after his unsuccessful attempt in 1850 to pass his teacher's examination at the University of Vienna, he arrived at that revered institution to study the scientific subjects he was so fond of. This was an opportunity he could never have dreamed possible. In the university environment he came into contact with scholars who were developing new scientific ideas, and who encouraged their students to carry out their own research in the course of their teaching activities. In the Zoological and Botanical Society he lectured on his first research efforts and published his results.

On his return to Brno Mendel looked forward to being able to continue his research and thus make a contribution to progress in science. When he took up his post at the *Realschule*, he was joining a group of teachers who considered science a routine part of their job. He had the support of his superior at the monastery, Abbot Napp, and he spent a further fourteen years in the creative excitement of a natural science researcher who was uncovering new findings and was able to include them in his teaching. During this period he drew satisfaction not only from teaching, but also as a researcher whose scientific work met with a favourable reception among his contemporaries in Brno. They recalled Mendel's humble origins and his childhood during the feudal period, and remembered how he had lived in penury up to the age of twenty-one, when he entered the monastery. In fact, however, his position had never been as bad as it might have seemed. His father was a smallholder who, in spite of having to work for his feudal landlord, managed to build a new brick house and to keep his family. He was able to send his son to study at a *Gymnasium* in a distant regional town, but he did not have the wherewithal to pay the cost of his son's studies, board, and lodging. Johann was obliged to make ends meet by giving lessons to his less gifted fellow-students.

At the village school Mendel had come into contact with the schoolmaster and parish priest, who had taught the rudiments of natural history. He was able to form a picture of the importance progress in natural science had for peasants in their work, and how it could help give a family social security. He surely learned of the successful teaching career of his mother's uncle, A. Schwirtlich, who was very popular in the village. This may have inspired him not only to take an interest in study, but to consider becoming a teacher or entering the priesthood. The move to Lipník and Opava may also have brought with it a certain anxiety about how well he would get on at his new schools. In fact he was soon to find that he could hold his own with children from bourgeois families who had grown up in conditions much more favourable for study.

Mendel the student spent his holidays at home on his parents' smallholding, where he also learned about the work of the beekeeper and the fruit-grower. At the time his father obtained grafts of new varieties of fruit trees from Fr. Schreiber, and using these he produced improved fruit trees in his garden. Fr. Schreiber was at the time a great authority throughout the region on growing fruit trees, and he knew the importance of artificial pollination for obtaining new varieties. Young Mendel may have come into contact with him during the holidays and spoken of new findings in the field of natural science which could lead to increased crop yields. In Brno, towards the end of his life, Mendel recalled high-yielding varieties of apples and pears back home, and had grafts sent to him so the he could improve the fruit trees in the monastery garden.

Mendel's health was not always good during the period he spent away from home. For all children the best cure is a mother's loving care and when Mendel fell seriously ill in Opava he went home to his parents to recover his health. His teachers took into account his excellent results, and he was allowed to enter the next class without being examined at the end of the school year. In Olomouc he fell more seriously ill, had to convalesce at home for an extended period, and was obliged to interrupt his studies. When he moved to the Philosophy Institute the opportunities to eke out his meagre means were much more limited, a factor which may have contributed to his sudden illness, which today might be called stress syndrome. Could Mendel have been worried about whether he would be able to continue his studies and become a teacher or a priest?

At the age of twenty-one Mendel considered carefully the offer to enter the unknown environment of a monastery. It was a step of immense import, which would on the one hand remove all worries over his material needs, but on the other would commit him to spend the rest of his life in a closed community, with only limited contact with the outside world. He could also have taken up a post as an assistant teacher at a country school, with the prospect of later becoming a schoolmaster in his own right. But Mendel preferred to take up the offer made by his physics teacher, Franz, who explained to him the cultural and scientific milieu of the Augustinian monastery in Brno, where outstanding personalities such as Napp, Bratránek, and Klácel were already active. Franz's recommendation was decisive for Mendel's acceptance at the monastery. The boy had his favourite teacher's example to show the opportunities offered by combining spiritual service at the monastery with the study, and later teaching, of natural sciences.

A TEACHER WITHOUT UNIVERSITY QUALIFICATIONS

In Brno Mendel found conditions for his edification beyond his wildest dreams. He met educated monks who offered to help him with his studies. He discovered the large monastery library, and with it the realm of scholarship. He learned theology and philosophy, along with those areas of natural science connected with agriculture. All these were matters in which Abbot Napp shared an interest, and he surely welcomed his new monk's enthusiasm for learning. The master of novices, Fr. A. Keller, was an authority on plant improvement, which must have caught Mendel's attention. Fr. Klácel, in connection with the study of philosophy and natural science, was also performing hybridizing experiments with plants, which cannot have escaped the newcomer's notice. Agricultural science and natural history were taught in Brno by Professor Diebl, who was aware of the importance of natural science for the improvement of plant varieties. With the support of his fellow monks Gregor studied the natural sciences, concluding that in this field self-education was 'extremely difficult', leading 'only slowly' to its goal (Mendel 1850). He did not lose heart, however, but plodded on, until in 1848 a new obstacle was placed in his patch.

Because of a lack of priests, it was decided that Mendel should be among those ordained before they had completed their theology studies, and he received holy orders on 6 August 1847. As soon as his studies were completed, on 20 July the following year, Fr. Mendel was appointed a curate at the Old Brno parish attached to the monastery. It was a time of political upheaval, accompanied by revolutionary changes in public affairs, and Klácel, who had in 1843 been deprived of his teaching post, saw hope of being able to teach again. The vision of greater social justice during the deliberation of the Constitutional Assembly led Mendel to add his name to a petition written by Klácel demanding radical increases in the freedom of monks to teach at school and to cultivate science (Orel and Verbík 1984). At the time Mendel added to his signature the words 'candidate teacher'. We do not know how well he understood the possible ramifications of taking such a stand: most members of the monastic community declined to sign the petition. Perhaps it was more a question of rashness than of courage on the young monk's part. What in life may seem a bold decision is often no more than a hurriedly assumed attitude in an apparently hopeless situation. Mendel may also have been seeking a way out of his curate's job, where he was far from happy, a fact that was later to lead to an infirmity which called for extensive medical care (Sajner 1968).

The young curate was delivered from his misery in 1849. With the rapid growth of industry, there was a great demand for educated school-leavers, and the intake of pupils to *Gymnasia* was increased. Headmasters, faced with a shortage of teachers, were anxious to recruit new staff. Abbot Napp was aware of this situation, and he was still a very influential figure in the field of education. In December he was able to offer Mendel the job of supply teacher in the town of Znojmo. This constituted a major turnaround in the young priest's fortunes. He was no longer to be a curate, but a schoolteacher. By a new order of the Ministry of Culture and Education he was obliged to take a teacher's examination. He applied to be examined as a teacher of natural history, a subject he had hitherto studied only privately. His initial failure in the examination provided the stimulus for his university studies. At the age of thirty-one he returned to Brno as a natural scientist, with the prospect of earning his living as a teacher. He brought with him new ideas on the possibility of investigating unknown phenomena in nature. Similar notions had already been propounded by his fellow monks Klácel and Bratránek, but Mendel based his conception on a study of physics, which opened the way to the experimental investigation of problems. As a teacher at the *Realschule* he met the headmaster, Dr Auspitz, a strong personality who was introducing a new concept of natural science teaching, with the emphasis on solving practical problems. An example to all natural history teachers in Brno was the demoted university professor Zawadski, who introduced Mendel to natural scientists organized in the learned societies of Brno.

The young natural scientist who arrived to teach natural history and physics at the *Realschule*, although he had studied at university, had still not passed his teacher's examination. At the same time he set to work on a problem which was very topical for both natural scientists and agriculturalists. By 1855 he had revealed his new concept of the discreteness of plant traits, and he was aware that the problem he was tackling was connected with the unknown causality underlying the fertilization of the higher plants. It was then that he first realized the difficulties facing the researcher, who must prove and explain new theory to established experts. But he was all the more convinced that he had set out along a path that opened up new possibilities for scholarship. He prepared his experiments according to the principles he had learned in his physics lectures, and he knew that anyone repeating them must reach the same conclusion as he.

Failure at the second attempt in his teacher's examination was a great disappointment to Mendel, and led to a temporary deterioration in his health. But he soon recovered enough to continue his experiments. We do not know how Abbot Napp reacted to Mendel's failure. What is certain is that he did not withdraw his support from the young man, in respect of either his teaching post or his experiments. Mendel's headmaster, Dr Auspitz, also continued to report on his performance in glowing terms, and his teaching career was not interrupted, though he had to make do with a supply teacher's half pay. He never made a third attempt at the teacher's examination, and never became a fully qualified teacher. We can only ask ourselves whether his dream of becoming a teacher was fulfilled, or not.

NOT AN ESTABLISHED SCIENTIST

In 1856 Mendel was already experimenting with peas. In the same year Professor Zawadski lectured in Brno on a discussion which had taken place among the participants in a German natural scientists' and physicians' congress held in Vienna. Botanists evaluated hybridization in connection with research into the way cultured varieties of wheat had come into being. At the time Mendel was already aware that his research was of exceptional importance for the origin and development of organic forms. His previous dream of being a teacher was gradually transformed into a desire to become a researcher.

In 1857 Mendel took on the task of processing graphically the results of meteorological observations in Brno. He showed meteorologists the importance of statistical evaluation of the results of observations in various places;

it was this conception which was later to form the basis of his idea of scientific weather forecasting (Weiling 1970a).

The year 1859 saw the fall of the neo-absolutist government which had come to power in Vienna after 1850. It was a time when Mendel was entirely engrossed in his experiments with plant hybridization. When the natural scientists shook off their subordinate status in the Agricultural Society, Mendel immediately became a founder-member of their new society, whose aim was the promotion of pure science—the category into which his own research work fell. In 1865, when Professor Makowsky lectured on Darwin's theory, Mendel saw no contradiction between it and his own theory.

Published works often pose the question of how Mendel reacted to the fact that no one in Brno understood the implications of his theory. The audience at his lecture included natural scientists who were interested in research into hybrids in nature. But none of them had studied such varied natural scientific disciplines as Mendel, and none of them had his empirical knowledge of plant breeding. With every scientific discovery one can point to a certain period of assimilation into the system of the sciences involved. If the researcher is already well known, if he enjoys the support of established scientists, as did Darwin, this period of assimilation is shorter. But in Mendel's case the discoverer was a newcomer, and there was no acknowledged expert to throw his weight behind the new theory in the professional literature. One of those who might have understood his new concept of hybridization was J. Nave, who also understood fertilization in plants as a joining of one maternal cell and one pollen cell. He and Mendel had in common an interest in studying the mechanism of fertilization. In 1864, however, Nave was seriously ill, and died shortly before Mendel delivered his lectures. There were also plant breeders who worked together with Mendel. One of them was J. Twrdy, whose new plant varieties were well known even abroad (Vávra 1984). Twrdy was surely interested in Mendel's research, and saw in it a way of explaining how hybridization could be exploited to create new combinations of traits in plants. On the other hand, he was not equipped to understand the theoretical explanation of hybridization. So as far as Brno was concerned, Mendel remained a lonely figure, scientifically speaking. But he was not disappointed enough to abandon his dream of having his theory recognized.

After sending reprints of his *Pisum* paper to scholars outside Brno, Mendel must surely have expected a response to his work. One of these reprints was addressed to a professor of botany at the University of Innsbruck who was seven years his junior, Kerner von Marilaun, and whom he knew from Professor Unger's lectures. Kerner was also carrying out research into plant hybridization, and so Mendel expected a reply. But the Austrian was preoccupied with other aspects of plant hybridization, and did not react to Mendel's findings. A surviving reprint of the *Pisum* paper in Graz shows that even Professor Unger, who was by that time living in retirement in that city, paid no attention to it. The only one to react was Professor Nägeli, whose imagination was caught by Mendel's expressed intention to carry out his next series of experiments on various species of the genus *Hieracium*. Every scientist, on receiving the reprint of a paper, forms an impression of its content from the title and the opening sentences. Nägeli was not greatly impressed by Mendel's *Pisum* research. He thus paid little attention to the experiments of an author he considered a beginner in this field of research. But Mendel was not put off, and must have been fairly encouraged by the fact that such a well-known scientist even bothered to reply to him. Shortly after receiving Nägeli's letter he wrote to him again, giving further details of his experiments, and at the same time suggesting that the professor should repeat them. Mendel even sent labelled seeds for the purpose, and ventured the opinion that Nägeli must reach the same conclusions as he. This letter reveals Mendel as a self-assured experimenter who was offering his 'highly esteemed sir' a theoretical explanation of a problem Nägeli had dealt with speculatively in his own papers. Mendel also outlined the further research he would be undertaking to tackle problems that had come up in the course of his *Pisum* research.

In May 1868, Mendel wrote to Nägeli that his *Hieracium* experiments had yielded unclear results. He was unsure whether this research would be brought to a successful conclusion. But he expected the next year to bring new findings from crossing experiments using species of *Hieracium* sent by Nägeli. A year later, in his report on the *Hieracium* experiments, he mentioned (Mendel 1870a) Professor Nägeli's name in the context of the latter's interest in explaining the origin of new plant species. He did not, however, relinquish hope of solving in the course of further experiments the problem of the polymorphism of *Hieracium* traits, something which was a mystery even to Nägeli. It was a period during which the fulfilment of Mendel's dream of recognition by a prominent scientist, in the person of the professor from Munich, came close to realization. He was soon, however, to realize that his function as abbot was far from offering him more time for his research, as he had supposed, than his teaching post. In 1873 he was already aware that he was not going to be able to continue such demanding research, and hinted that he was about to bring it to a close. We are now unable to explain why he did not reply to Nägeli's last two letters. This brought to an end a remarkable correspondence between the university professor and the isolated researcher working in a monastery. It was also the end of Mendel's hopes of recognition by the scientific establishment of the day.

In 1878 Mendel received a visit from G.W. Eichling (1942), a representative of one of the large specialized plant-breeding companies operating in France. During a visit to Ernst Benary, the Nestor of European seed-growers, in Erfurt, he had been given Mendel's name. He wished to obtain details of Mendel's pea-hybridization experiments, and especially to learn the latest results of his research. Eichling published his impressions from his Brno visit more than half a century later. Abbot Mendel received him cordially, showing him round the gardens and the greenhouse, and also took him to see his beehives. In the gardens Mendel showed him several beds of green peas in full bearing, which he said he had 'reshaped in height, as well as in type of fruit, to serve his establishment to better advantage'. When Eichling wanted to know how he had achieved this, he was told: 'It is just a trick, but there is a long story connected with it, which would take too long to tell.' When Eichling pressed him, Mendel changed the subject, inviting his guest to inspect the hothouse, where he showed him beautiful ornamental plants. The only further mention he made of peas was of a variety which shelled out well, but was not high-yielding. He had crossed these with his tall local sugar-pod types and 'had tall shelling types, which were used at the monastery'. From this it can be seen that Mendel was still interested in breeding new varieties of peas for growing in the garden.

Eichling asked one of his customers, whose name he does not reveal, for his views on Mendel's research. He was told that in Brno Mendel was 'one of the best clerics', but 'not a soul believed his experiments were anything more than the maundering of a charming putterer'. Eichling visited Brno at the age of twenty-two, and his recollections were not committed to paper until sixty-four years afterwards, so that his statements regarding his conversations with and about Mendel can scarcely be considered reliable. But they may reflect the contemporary views of experts in plant-growing on Mendel's research. They knew of the monk's hybridizing experiments, but they did not understand what they were about.

IN CONFLICT WITH OFFICIALDOM

Thus the newly-elected Abbot Mendel was aware that he was also expected to take a role in public life, thus following in the footsteps of his predecessor Napp in this respect also. But the way of thinking of a teacher and researcher soon proved inadequate to the task of resolving the economic and organizational problems within the monastery and in a public sphere marked by a new political situation. The eighth letter to Nägeli, written on 3 July 1870, reveals how Mendel was aware of the difficulty of the task facing him, and the fact that he would be unable to continue his research as he would have wished. He describes the results of experiments which were opposed to the conclusions reached in his *Pisum* paper, and was experiencing the new excitement of a researcher seeking the answer to the mystery of trait polymorphism in hybrids. Earlier than this his letters to Nägeli had kept his friend informed of the results of experiments with other plant genera, and any natural scientist would have published them. In them he confirmed the validity of the theory of variable hybrids derived from his *Pisum* experiments. Later he tried to solve the much more complex problem of polymorphism, which was at the time coming to the forefront of scientists' interest. At first in his experiments with the crossing of various species of *Hieracium* he observed variability of traits in the hybrid generations instead of the uniformity he expected. When, in 1871, he found uniformity in some hybrids too, he suddenly interrupted his experiments. The only evidence of his continuing interest in solving the mystery is the second note fragment mentioned by Heimans (1970).

Experimental research finally led Mendel to the conviction that *Hieracium* hybrids were also variable. In 1873 he complained to Nägeli that his duties as head of the monastery were keeping him from his experiments with *Hieracium* and bees (Correns, 1905, p. 242). At the time he was no longer undertaking further plant-hybridization experiments. His experimental interest had shifted towards bees, and he worked on them until the end of his life. Weiling (1991) found that Mendel could have written his *Notitzblatt 1* in 1880, i.e. at the time when he was occupying himself with bees. Even then he returned to his data from the pea experiments, and it can be supposed that he saw in bees some connection with his preceding experiments into plant hybridization.

In 1874 the government in Vienna, formed by the Liberal Party, in an effort to find its way out of an economic crisis, proposed increasing the amounts paid by monasteries into the religious fund for covering the expenses of Church institutions. According to the estimate of the Brno monastery's property, the Augustinians should have paid 7330 guilders annually (Vybral 1971). Various State and Church organizations expressed their views on the new bill. Their representatives raised objections to its wording. But when the bill became law in 1875, Mendel was the only head of a monastery in the monarchy who refused to accept it, and he did so till the end of his life. On this account he earned the nickname 'the stubborn prelate'. The members of his community pointed out to him that the law had been introduced by a party he had publicly supported. This clearly amounted to criticism of their superior, whose political gesture at the start of the 1870s had also brought him into conflict with the Church authorities. He considered the law a gross injustice towards his monastery right from the start, and maintained his resistance with all his innate tenacity. When the State authorities refused to take account of his protests, he reacted in his own way. He refused to accept the validity of the law, even when the very best lawyers pointed out that his protests were in vain.

In vain, Mendel tried to convince the provincial authorities in Brno and the ministry in Vienna that the monastery could not be taxed, pointing to the fulfilment of responsible duties in public life. He offered to give the sum of 2000 guilders to the religious fund. This application to the provincial office was passed on to the ministry in Vienna, whose decision was unfavourable. Mendel was to obey the law. In 1876 the dispute came to a head to such an extent that administration of certain of the monastery's estates was forcibly taken over by the State, and the taxes were paid out of their profits. This was a great humiliation for the abbot. At this juncture Dr Auspitz once again figured in Mendel's life, with an offer he should have accepted: to change his views for the sake of the Liberal Party. In 1876 the new Mortgage Bank was set up in Brno, and Mendel was offered the post of vice-president, carrying with it a substantial annual salary (Vybral 1968). Mendel accepted the offer, and for a two-year period following the death of the bank's president he even held that post. But he still did not change his attitude to the law on monastery taxation. He went on preparing material for an appeal against the validity of the law, rejecting the advice of the best lawyers in Brno, Prague, and Vienna. Extensive manuscripts have survived from the time, showing just how much time and energy Mendel devoted to the dispute. In the midst of composing applications and making analyses he wrote on the back of a document relating to the tax question his views on the *Hieracium* experiments (Heimans 1970). These place the research in the context of that on *Pisum*, and we can thus see that even after his experiments had ended he returned in thought to the research programme he had devoted so much attention to over a period of almost twenty years.

His perseverance in research, coupled with enthusiasm for investigating the same problem over and over again, led Mendel to his discovery. In his research he was able to admit his own mistakes and change his tactics. But in his defence of the monastery's property rights he was too stubborn to change his standpoint regarding the validity of the law, and in the end he became isolated both in public life and within the monastery. In 1877 he approached the Ministry of Culture and Education in Vienna with yet another application, stating that because of his dispute he had suffered much in the two years he had been head of the monastery, and had 'gone grey and grown old before his time' (original MS in Mendelianum, invent. no. 612). Officials in Vienna and Brno had already given up all hope of convincing the stiff-necked prelate to change his attitude and comply with the law.

According to Iltis (1966, p. 268), Mendel's behaviour in the matter was considered pathological. In November 1879 he spent six days in Vienna and sought the support of lawyers for his actions. He later discussed with Dr Krása, a leading barrister in Prague, the possibility of submitting a complaint to the constitutional court which was being set up at the time. He was disappointed once again: Krása did not recommend this move. This was the greatest blow of all for the abbot. He began to be suspicious even of his brothers in Christ, a fact noted by Rambousek in a letter to Křížkovský of 8 May 1883. Rambousek comments that the abbot does not feel well and looks very poorly. 'On top of it all he is full of suspicion and sees himself surrounded by nothing but enemies, traitors, and intriguers' (Orel 1971c). Mendel supposed some members of the community to be on the side of the State authorities, agitating against him in the ministry.

After his election as abbot Mendel was anxious to fulfil his new role with the same success as his predecessor, and thus to win the recognition of his monks and of the public. But his political gesture on the occasion of the election to the provincial Diet and later his attitude to the new law relating to monastery taxation led him up a blind alley, where he remained until the end of his life. This put an end to his dream of becoming a highly esteemed public figure.

ESCAPE FROM PUBLIC LIFE

In his function as head of the monastery Mendel was able to pay more attention to his family. In 1870 he visited his sister Theresa and went on a botanical excursion with his nephew Alois on the hill overlooking his native village. On this occasion he also subjected the boy to a test of arithmetic and other elementary subjects, and was satisfied with the results (Iltis 1966, p. 244). Later he supported all three of his sister's sons during their studies at the Brno *Gymnasium*, and subsequently contributed to the needs of the eldest, Johann, while he was at the Brno Technical University, and of his younger brothers Alois and Ferdinand when they were studying medicine in Vienna. Mendel visited Hynčice for the last time in 1873, performing the marriage ceremony of his nephew Alois Sturm. Later he only maintained written contact with his relatives, or kept in touch with them through his nephews, who visited him regularly. During the time they spent in Brno they were in lodgings opposite the monastery, and were in contact with their uncle constantly. He liked to play chess with them in his free time, and took a delight in giving them difficult chess problems to solve (Iltis 1966, p. 243).

Fires were not an uncommon event in Hynčice, and they inflicted heavy financial losses on the inhabitants. Abbot Mendel suggested the establishment of a local fire brigade, and provided 1500 guilders for equipment. In a letter to his brother-in-law, Sturm, dated 20 June 1882, he mentions that the pumping equipment is ready in Brno, recommending that members of the fire brigade come to test it and take delivery of it (Kříženecký 1965*b*, p. 126). The grateful villagers made the abbot an honorary member of the fire brigade, which gave him great pleasure.

Mendel's relationship with his home village and the peasants' work are reflected in the motifs used to decorate the chapter hall of the monastery. When, in 1870, a whirlwind damaged the roof of the building, the abbot had to have it repaired at considerable expense. In 1875 Abbot Mendel had the ceiling of the great chapter hall decorated with paintings proposed by him. According to Richter (1943, pp. 121–3) during the later renovation of the monastery these paintings were removed. At that time Fr. Matoušek had photos taken of them 'to preserve this evidence for Mendel's spiritual direction' (*die Belege für Mendels Geistesrichtung*) and offered these photos to Richter for the illustration of his book. In the middle of the ceiling was a picture of St Augustine and his mother St Monica, the motif of the order. Paintings at the four corners depicted motifs reflecting Mendel's scientific interests. One showing fruit-tree grafting might have reminded him of his father and the man who advised him on fruit-tree improvement, Fr. Schreiber. In the background the silhouette of the village church can be seen, along with the hill overlooking the village. A second picture shows St Isidore, patron saint of agriculture, which may have reminded the abbot of his origins. A third shows an old and a new beehive, symbolizing Mendel's interest in beekeeping, which he also first came across in his native village. The last picture is of meteorological instruments, a globe, and maps, and may be understood as a motif of meteorology or of the natural sciences in general. In the decoration of the chapter hall, where Mendel received visitors, the abbot, becoming more and more a solitary figure, recreated the world of his childhood, along with that of science, his main interest in life. The connection between the decorations on his ceiling and Mendel's interest in scientific work (Fig. 7.1), along with his religious faith, escaped notice until pointed out by Matalová (1983). The tenth abbot of the monastery, Tomáš Martinec, after taking his office began to take responsibility for the

Fig. 7.1 Two of the ceiling frescos of Mendel's experimental plants. Pea flowers, leaves, and tendrils are shown in the bottom of the lower picture. Fuchsias are the central motif of the setting. (Photo by Bohuslav Havlíček.)

restoration of the whole area connected with the activities of his predecessor, Abbot Mendel, and in this context he also ordered the restoration of the ceiling decoration of the chapter hall.

At the same time as the chapter hall, Mendel had the ceiling of the monastery library decorated with paintings which still survive. Opposite each other are two abbots' coats-of-arms. That of Napp shows a vine, the improvement of which was a preoccupation of Mendel's predecessor. Mendel had a plough, symbol of the peasant's work, and a lily, symbol of purity, inserted in his coat-of-arms. In the corners and along the walls of the library are coloured paintings of fruit and flowers. Among them one can identify pea and bean plants, fuchsias, and others on which the scientist had experimented. The symbolism was dealt with by Matalová (1985).

Abbot F. Bařina was one of the last novices received into the monastery by Abbot Mendel. He often recalled Mendel, and mentioned his research. But he never said anything about the decoration of the ceiling. During his own term of office the ceiling of the chapter hall was whitewashed, and it is only thanks to the procurator of the monastery, A. Matoušek, that the photographs have survived.

From some fragmentary notes, which Zumkeller (1971) has explained as a preparatory draft for an Easter sermon, one can obtain an insight into Mendel's attitude to his priestly duties. He develops a dogmatic thesis using examples from the plant kingdom. Starting from a reading from the Gospel of St John, Mendel worked towards a number of ideas on the importance of faith. Using the statement that the risen Christ appeared, according to the Evangelists, in various forms, Mendel takes special note of the way in which he appeared to Mary Magdelen, in the guise of a gardener. This led him to the theme closest to his heart: he developed the homily of a gardener, seed, and plants, saying: 'The gardener plants seed or seedlings in prepared soil. The soil must exert a physical and chemical influence so that the seed or the plant can grow. Yet this is not sufficient. The warmth and light of the sun must be added, together with rain, in order that growth may result.' Mendel seeks to use this image from the world of nature to illustrate how 'the germ of supernatural life, sanctifying grace, is put into the soul of man,' and how, 'through the cooperation of man's goodwill and nourished by the supernatural food of the Eucharist, it develops and achieves perfection'. This was typically in line with the teaching of St Augustine.

In the final years of his life the solitary abbot experimented with bees, speaking about his work with beekeepers. A new area of research was his recording of sunspots, in order to improve the scientific methods of weather forecasting. A document which survives from this period gives a remarkable insight into the manner in which he sought a mathematical pattern in the occurrence of various names (Ferdinand 1966). In the archives of the Mendelianum there is a number of sheets of paper covered with surnames with the ending '-mann', e.g. Altmann, Borgmann, etc. The longest list survives not only as a fair copy, but in the original draft, in which additions are apparent, showing how Mendel subsequently arrived at new names of the type in question. One sheet of paper records names along with the sources Mendel used: the military year-book of 1877, the register of shippers and bankers, and the year-book of barristers, the last two being without date. On some sheets there are, in addition to names, numerical records similar to those in Mendel's hybridizing experiments.

According to Richter (1943, pp. 152–70) there had once existed more extensive material consisting of names collected by Mendel. Originals with groups of names according to their meaning, and especially the folder in which the lists were kept, are missing. The lost folder was covered with Mendel's numerical records and mathematical operations. It is only reproduced in Richter's book; the author, by his own admission, does not even dare to discuss the question of whether Mendel was involved in some sort of research, or whether this was merely a pastime, the 'spirited vagaries of a great savant'. Ferdinand (1966), according to various arrangements of the collected names and mathematical analysis, reached the conclusion that Mendel's idea was to find laws in the derivation of names. This was a type of research he could carry out in seclusion, and which brought him the satisfaction of seeking order in the apparently random occurrence of names, just as he had once sought significant proportions in the occurrence of plant traits.

At first the new abbot was visited by representatives of government institutions, such as the Lord Lieutenant, Count Vetter von der Lilie, councillors of the Lord Lieutenant's office (*Stadthalterei*), judges, schoolteachers, etc. They would play skittles or cards. As a result of Mendel's stubborn refusal to pay the monastery tax, the number of visitors declined, and the only guests he received were friends in the priesthood, education, and learned societies. In the summer of 1882 he was visited by the meteorologist J.Líznar and his wife; Líznar was interested in Mendel's meteorological observations. On that occasion he also examined the experimental construction for the mating of queen bees. According to Líznar (Iltis 1966, p. 274) Mendel was somewhat depressed at the time, but not so much on account of the tax dispute as of his failing health.

The ailing abbot may have recalled discussions with his older brother monks on scientific problems. His nephew Alois Schindler in a letter dated 1 January 1923 to H. Iltis, Mendel's biographer, wrote: 'One Sunday I went to see my uncle the Abbot. I noticed a photo album lying open there. I leafed through it and found at the end a photograph of a priest in the Augustinian habit, placed loosely inside. There was plenty of room left in the album, so I tried to put the photo into it. But my uncle told me to leave it, that the photograph did not belong in the album, since the priest had broken his vows. But he used to be a friend of mine, he added. This may have been a nostalgic recollection of the time when the teacher and researcher Mendel used to have lively discussions on natural scientific subjects with Klácel before he left for the USA. This affected his enthusiasm for research.'

The first information relating to Mendel's illness came to light in Křížkovský's letters in 1877 from the spa of Karlovy Vary (Karlsbad) (Eichler 1904, p. 102). He thanked the abbot for sending money, which he had not expected. He also complained of pains, reminding Mendel that he should not underestimate his own health problems, and should also undertake a spa cure. At the time Mendel had stopped attending the monthly meetings of the beekeepers' society, though his interest in beekeeping persisted. The reason was the ever more acute conflict over the taxation of the monastery, or perhaps his own state of health. Towards the end of his life Mendel even grew afraid of death. The Bishop of Brno is said to have given secret instructions to keep a discreet but close watch on the abbot (Iltis 1966, p. 267).

Mendel's last protest against the law imposing taxes on the monastery was in a note to the Lord Lieutenancy in May 1883 (Iltis 1966, p. 268). In June the provincial office, acting on the instructions of the Ministry of Culture and Education, turned to the Bishop of Brno to ask for his intervention in the dispute to prevent any further confiscation of monastery property. It was hinted that a considerable reduction in the tax payable might be considered. At the time Mendel was so ill that the economic affairs of the monastery were in the hands of the procurator, Fr. Ambrose Poye. When a note was received rejecting Mendel's protest, Poye did not show it to the abbot. He sent it back to the provincial office with a note to say that the abbot was seriously ill, enclosing confirmation from the physician, Dr Brenner, dated 30 June 1883, according to which Mendel had been examined for organic heart disease and general dropsy, 'conditions in which a perfect repose is essential, and all emotional disturbance must be avoided'. Later Fr. Poye told the ministry in Vienna that, owing to the abbot's illness, payment could not be laid before Mendel without gravely imperilling his health. While Mendel was still alive, Poye sent a protest to Vienna—but not against the validity of the law, rather against the amount of the contribution. Here he was already implying a way out of the dispute, which was found soon after Mendel's death by the newly installed abbot, Anselm Rambousek. The monastery was relieved of payment for the whole period 1880 to 1890.

In his last years Abbot Mendel lived a solitary life, and looked forward to each visit his nephews paid to him, which they always did when passing through Brno on their way to and from Vienna. On 4 April 1883, he wrote to them to bring him his grafts from fruit trees in his parents' garden on their way back to Vienna, wishing to grow them in the monastery garden (Kříženecký 1965b, pp. 126–7). In the summer of the same year he went to convalesce at the spa in Rožnov, in northern Moravia. His state of health had worsened to such an extent that in the autumn he was no longer able to go out in the gardens. By the end of the year his last hour was approaching, and Mendel knew it. On 20 December he wrote to his pupil from the *Realschule*, Líznar, that he was no longer able to make meteorological observations on account of serious illness, and took his leave of him as follows: 'Since we are not likely to meet again in this world, let me take this opportunity of wishing you farewell, and of invoking upon your head all the blessing of the meteorological deities' (Iltis 1966, p. 275).

On 26 December he wrote what was probably his last letter to Alois Schindler, his nephew (Kříženecký 1966b, p. 127). He entreated him to come to Brno for two days during the Christmas holidays and talk over 'an important professional matter'. Alois came, and on his uncle's own request gave his opinion as a professor of internal medicine as to the treatment called for by diseases of the kidneys and heart. But by that time the chronic kidney disease which had appeared a few years before was in an advanced stage. According to his nephews Mendel had already suffered from tachycardia for almost ten years, and at the end of 1883 his heart rate was up to 120 beats a minute. Mendel had become a heavy cigar smoker—in part from taste, in part because one of his doctors had recommended it as an aid to controlling obesity. If Iltis is to be believed on this point, he smoked up to twenty cigars a day, though they were 'not very strong' (Iltis 1966, p. 277). At the end of the year general dropsy set in owing to a failure of the heart and kidneys, so that Mendel could no longer get around. The end came from uraemia. Alois Schindler emphasized that his uncle awaited death stoically, as a natural necessity. But he was afraid of apparent death, and wanted a post-mortem examination to ascertain the nature of his illness (Iltis 1966, p. 278).

On 4 January 1884, his heart trouble grew so serious that his physician lost all hope of any improvement. The abbot died on 6 January, at about two o'clock in the morning. Sajner (1963) has made an attempt to explain his disease and the cause of his death. A post-mortem examination performed by Dr Brenner with the assistance of Mendel's nephew, Alois Schindler, showed the primary cause of death to have been chronic inflammation of the kidney (Bright's disease), accompanied by hypertrophy of the heart. On the day Mendel died the monastery issued a statement which, according to Iltis, was drafted by Mendel himself, as follows:

The Augustinian monastery of St Thomas at Altbrünn in Moravia respectfully and with profound regret informs the public of the death of the Right Reverend Abbot

Gregor Joh. Mendel

mitred prelate, companion of the Royal and Imperial Order of Franz Josef, emeritus chairman of the Moravian Mortgage Bank, member and one of the founders of the Austrian Meteorological Society, member of the Royal and Imperial Moravian and Silesian Agricultural Society, and various other learned and useful organizations, etc.

The funeral ceremony took place at the monastery church on 9 January, at nine in the morning, and the requiem was conducted by a man later to become a famous Moravian composer, Leoš Janáček. Among the mourners were representatives of the government authorities, of the Catholic clergy and other religions in Brno, of the institutions in which Mendel was active, schoolteachers, representatives from Hynčice, and many, many of the poor from around the monastery, to whom he had always been kind, and whom he had always tried to help. They followed the coffin to the Central Cemetery, where Mendel's remains are interred in the monastery tomb, close to the north-eastern corner of the enclosure.

One of those who sincerely regretted the abbot's death was Fr. Clemens Janetschek, who aptly expressed the sad tale of the deceased in the following verses (Iltis 1966, p. 280).

> Gentle, free-handed, kindly to one and all,
> Both brother and father to us brethren was he,
> Flowers he loved, and as a defender of the law he held out against injustice,
> Whereby at length worn out he died from a wound of the heart.

The contradictory position of the enthusiastic natural scientist and the responsible church functionary was well expressed by Abbot F. Bařina, who recorded Mendel's words relating to scientific work. They are often repeated in the literature, and though we may doubt the accuracy of the quoted words, they reveal the feelings of a man who spent his life as a natural scientist and researcher, while fulfilling the duties of a monk:

Though I have had to live through many bitter moments in my life, I must admit with gratitude that the beautiful and the good prevailed. My scientific work brought me much satisfaction, and I am sure it will soon be recognized by the whole world (Kříženecký 1966*b*, p. 6).

Mendel the scientist did not live to see his work appreciated. Published works on the subject refer to the overlooking of a paper from the provincial city of Brno, or its falling into obscurity, or speak of the new theory's being published before its time. But today we can ask the question: how was it that Mendel's work only attracted general attention among researchers after 1900?

8
Acceptance of the theory

Mendel revealed the very essence of heredity, showed its most complex and most fundamental laws, and introduced for its experimental investigation the simplest and most essential methods.

—B.L. ASTAUROV (1965).

The scientific legacies of outstanding natural scientists are repeatedly reassessed with the passing of time, and in the light of subsequent progress in science. The exceptional nature of Mendel's research was pointed out only by some of his contemporaries, but the true import of this theory was not brought home to natural scientists until 1900, when it was suddenly found to tie in very well with the latest findings on living systems.

At the start of the nineteenth century natural scientists still considered species to be stable, an idea which tallied with the static view of nature taken at the time. The proof of the existence of sex in plants led to an endeavour to explain the relative roles of the parent plants when new indi-

Fig. 8.1 The emblem of Abbot Mendel.

viduals came into being. New knowledge gradually transformed the static view of nature into a dynamic one, and the variability of traits of individual plants and animals became a focus of interest. At first natural scientists believed that the traits of an organism changed under the influence of outside effects, and that a process of graduation was under way in nature, controlled by as yet unknown forces. Later they began to ask themselves how new traits and even new species came into being.

Olby (1971) attributed the modern physiological viewpoint found in Mendel's research to the tradition of reductionist cytology advocated by Schleiden and supported by Unger. At that time chemical physiology was judged naive, and investigations of plant fertilization led to research into the morphological process. Mendel wished to explain the transmission of traits through the germ cells. At that time Herbert Spencer (1864) introduced his physiological units, which in different chemical compositions created different arrangements inside the cell and determined different morphological traits. This was the beginning of the search for units which determined heredity on a higher level of organization.

At the time of Mendel's lecture on the results of his experiments, in 1865, natural scientists' views of the question of variation and hybridization were already coloured by Darwin's Theory of Evolution—a situation quite different from that in 1854, when Mendel started his experiments. In 1859 Darwin provided a stimulus not only for research into the process of evolution, but also for a search for the determinants of heredity. The literature of the time contains more and more references to heredity. Cytologists were on their way to clarifying the structure of the cell, and later also the cell nucleus, and they too became interested in research into heredity. Interest in plant hybridization receded into the background, to re-emerge later in a different context. Today we are able to investigate the process of the acceptance of Mendel's theory among natural scientists at home and abroad.

IN THE EYES OF MENDEL'S CONTEMPORARIES IN BRNO

The view is still sometimes expressed in published works that Mendel's lectures in Brno in 1865 evoked no response. It was not until a hundred years later that a remarkable and extensive report from the daily paper *Tagesbote* was, on the instigation of J. Sajner, published in English translation by Gavin de Beer (1966). Following the first lecture, on 9 February, it was reported that Mendel had by artificial fertilization obtained plant hybrids which were of interest to botanists. It is said that Mendel pointed out that fertility of the crossbred or hybrid plants was proved, but that the hybrid trait did not remain constant and continuously tended to revert to the parental forms, a reversion which could also be accelerated by repeated pollination from the parental forms. Mendel is also said to have shown specimens of the relevant generations, according to which shared characters were transmitted reciprocally, but different characters led to the production of quite new characters. According to the report, what was particularly worthy of note was the numerical comparison in relation to the parental forms. The enthusiastic interest of the audience showed that the subject of the lecture was appreciated, and its delivery very acceptable.

The report on the second lecture, on 10 March, states that Mendel spoke about the reproduction of germ cells, fertilization, and the formation of seeds in general, and in particular about hybrids with reference to the experiments that he had carried out on *Pisum* with as much circumspection as success. Mention is also made of experiments which he intended to continue during the coming summer. He also announced that during recent years he had already carried out artificial fertilization experiments on many related named species of plants in order to obtain hybrids, and that his successful results had encouraged him not only to study hybridization further, but also to give an account of his results. After this lecture, according to the published report, Professor Niessl added that he had also observed by hybridization with the aid of a microscope in fungi, mosses, and algae and that further observations in this field would not only substantiate existing hypotheses, but also provide an interesting explanation.

Mendel's lectures were also reported, though not to the same extent, in the daily press in Olomouc, the province's second city. The written report in the Brno *Tagesbote* must have been the work of an expert who knew something about Mendel's experiments—possibly Niessl. The idea suggests that the paper's editor, Dr Auspitz, might have asked Mendel himself to submit a report on his own lecture, and then might simply have adapted it for printing. The meeting was attended by natural scientists who paid close attention to Mendel's account of his experiments. But their number did not include a scholar of the calibre of Thomas Huxley, who had at once thrown the weight of his authority behind Darwin's theory. Mendel's theory lacked such a patron.

In his lectures and in his paper Mendel challenged his audience and his readers to repeat his experiments. We may therefore suppose that he gave some of the members of the Natural Science Society reprints of his hybridization papers which have not survived, or are simply not known. However, none of Mendel's contemporaries had the conditions required to perform these experiments. Niessl, Makowsky, and Theimer, who had mainly studied hybrids in the wild, had quite different matters to attend to in pursuing their professions and were unable to undertake such demanding experiments.

In Brno further mention was made of Mendel's research in the obituary published in 1884, an account of which was given by Matalová (1984). Professor Niessl recalled Mendel at a meeting of the Natural Science Society held on the very day of his funeral, in the following words: 'The deceased was among the most enthusiastic and generous adherents of the Natural Science Society, supporting it financially and also with lively participation in the performance of its scientific tasks. He made use of every spare hour he was offered by his fortunate position exclusively for extensive natural science studies, which testifies clearly to his independent and special manner of reasoning. Here we may also place his observations of plant hybrids, which he grew in large numbers.' Niessl also pointed out Mendel's services to the furtherance of meteorology, concluding: 'He dictated the temperatures a few hours before his final agony. So he devoted himself to his last breath to his beloved sciences.' Niessl knew that Mendel's experiments were methodologically quite different from those he had learned about from studying the literature, but he was unable to grasp their far-reaching significance.

The longest obituary was published by the horticultural section of the Agricultural Society, where after a brief list of his extensive activities it was noted that he had been a member since 2 April 1863. It was also pointed out that in May 1859 Mendel had participated in an exhibition of vegetables, fruit, and flowers 'with a presentation of exquisite vegetables grown by him'. He had exhibited fruit-tree varieties he had improved himself in September 1883, when he won the grand gold medal of the gardeners' society of Hietzing, in Vienna, for a superb collection of drupes of his own production. The obituary notice continues: 'His first-class varieties of flowers, of which we must mention especially a full and beautiful fuchsia, were a successful achievement in domestic flower cultivation. How deeply the deceased was devoted to the plant kingdom was shown in his coat of arms, one field of which is occupied by a flowering plant, chosen as his symbol.' The notice end as follows: 'Many of his investigations in the field of natural science, especially of meteorology, continued after Dr Olexík's death, were published by him in the journal of the Natural Science Society. In fact his experiments with plant hybrids opened a new epoch. What he did will never be forgotten.'

The Agricultural Society in the published obituary connected Mendel's plant-hybridization research with his merits in the development of meteorology, apiculture, and pomiculture. In the overall evaluation, the exceptional significance of Mendel's work was lost. The most remarkable assessment of his research was that by the members of the Horticultural Section, who were aware of the significance of his research for the practical selection of new varieties. But they, too, failed to comprehend its far-reaching importance. At the January meeting of the Natural Science Society the next year Niessl (1903) stated that the reason for the underestimation of Mendel's work was connected with the 'spirit of the age at the time of his publication'. He adds: 'His work was well known, but ignored in the prejudice raised by the divergent and mutually exclusively views current at the time. From personal contact with Mendel of many years' standing I knew that he did not succumb to disappointment as a result of the fact that his botanical publications did not meet with immediate success at a time when for the explanation of the origin of new forms of plants the principles of the then generally acknowledged hypotheses of Darwin were almost exclusively decisive.' He did not mention Mendel's theory itself. During the same period E. Proskowetz, an acknowledged Moravian cereal breeder, also provided information (1902). He mentions experiments with the translation of *Ficaria*, in which Mendel tested Lamarck's theory of the effect of the environment on the inheritance of traits. In a talk on these experiments Mendel is said to have uttered the words, later much quoted, 'This much I already know, that in this way Nature does not introduce anything more in the formation of species; there must be something else involved.' Niessl also told Proskowetz that when speaking of his research Mendel repeatedly said 'my time will come'. Today we can only wonder why Niessl, even after 1900, said no more about Mendel's theory derived from the *Pisum* experiments. One can scarcely believe that Mendel did not speak to him about it.

But it must be taken into account that after 1900 Mendel's theory was little known, a matter which will be dealt with in more detail in the next section. Here one may recall the views of the prominent geneticist F.A.E. Crew, whom Kříženecký asked before the Mendel centennial in 1965 to report on the various views on Mendel's theory after 1900 in Britain. According to Crew (1966*b*), if English biologists had been shown Mendel's *Pisum* paper in 1900, 'it is most improbable that any of them would have appreciated its significance'. Niessl's 'spirit of the age' was still at work, not only at home, but also abroad.

IN THE VIEW OF CONTEMPORARIES ABROAD

Those of Mendel's contemporaries who might have reacted to his work include in particular the scholars who received reprints of his paper or who had the opportunity of reading it in the *Proceedings* of the Natural Science Society. We know that Mendel ordered forty reprints, of which the fate of only four was known up until 1965 (Kříženecký 1965*a*, pp. 19–20). At the time the *Proceedings* were published, Brno was occupied by the Prussian army, so Mendel did not get his reprints until the end of 1866. On the last day of the year he posted a reprint with a letter to Nägeli in Munich. This copy is now in the library of the Max Planck Institute of Biology in Tübingen. This reprint was to set off an extensive correspondence between Mendel and Nägeli. The next day, 1 January 1867, Mendel sent another copy, with an accompanying letter, to A.J. Kerner von Marilaum, who since 1860 had been professor of botany at Innsbruck University, and from 1878 was head of the botanical gardens in Vienna. In his letter Mendel noted Kerner's contribution to the determination of plant hybrids in nature, and considered it his duty to send him a reprint of his publication (Stubbe 1963, pp. 133–5). Kerner's reply to Mendel's letter has not survived, but it can be supposed that he was not particularly impressed by the monk's work. The reprint remained uncut, Mendel's experiments apparently seeming to the learned professor too remote from his main interest, that of explaining the origin of species. The third reprint Kříženecký mentions is in the library of Amsterdam University's Institute of Botany. All that is known about it is that in 1900 M.W. Beijerinck sent it to the prominent botanist Hugo de Vries (1848–1935). The fourth reprint Kříženecký speaks of is that in the collection of the Mendelianum, with the stamp of the Natural Science Society of Brno.

Later a fifth reprint was discovered in the library of the Institute of Botany at Graz University (Knoll 1967). It may be supposed that Mendel sent it to Professor Unger, who by then had retired. It also remained uncut. A sixth reprint was found in the library of the Augustinian monastery. A seventh was reported by Carlson (1973). It is kept in the Lilly Library of Indiana University in Bloomington, and comes from the legacy of C.V. Davenport, who visited Brno in 1910 and later in 1922, and seems to have received it as a present from the Old Brno monastery. An eighth reprint came to light at the National Institute of Genetics in Mishima, Japan. According to Nakazawa (1976) it was brought there from Brno in 1921 onwards, fetching a high price (Weiling 1984*a*).

After 1965 previously unknown citations of Mendel's publications predating the year 1900 were brought to light (Weiling 1966*b*, 1969, 1970*b*, 1971, 1984*a*). Citations of the papers on hybridization appeared in the German botanical journal *Flora* in 1866 and 1867, and in the *Proceedings* of the Viennese Academy of Science in 1871 and 1879. More attention was paid to the *Pisum* paper by H. Hoffmann (1869) in his book on the determination of species and variety. He wished to refute Darwin's Theory of Evolution, not believing in the importance of variations as a basis for the formation of new species. He stated with regard to *Pisum* hybrids that Mendel cultivated them from constant forms and observed them for six generations. He also described the technique of artificial fertilization, and in hybrids of various species of *Phaseolus* observed reduced fertility, or even sterility. Passing mention is made of Mendel in describing hybrids of *Aquilegia*, *Geum*, and *Lavatera*. In relation to the description of a *Geum* hybrid, Hoffmann commented that he did not find in Gärtner's monograph the statement citied by Mendel that these hybrid forms are constant. Mendel heard of this criticism, as has already been described in Chapter 5, pp. 169–74. Thus he must have been acquainted with Hoffmann's book, and had the opportunity to assess how that author had understood his theory. However, we know nothing further of this.

There are remarkable citations of Mendel's paper by students in Sweden and Russia. C.A. Blomberg in Uppsala had his attention drawn to the *Pisum* paper by Professor E. Fries, who was an honorary member of the Brno Natural Science Society. Blomberg (1872), in his thesis 'On hybrid formation', wrote that Mendel investigated curiosities which appeared in the hybrid progeny in various generations. In this connection he pointed to Mendel's observations of dominant and recessive traits in the hybrid progeny. He also mentioned the concept of constant hybrids. According to Blomberg, researchers fell into two groups. One of these groups believed that hybrids cannot remain uniform—in this group he mentions Kölreuter, Gärtner, Mendel, Naudin, Gordon, Wimmer, and Nägeli. These people are said to have 'played the most important role in this botanical field'. Their chief opponents were said to be Sageret, Wiegemann, and Wichura; and he also counted Darwin among this group.

Two years later, in Petersburg, I.F. Schmalhausen (1874) wrote in his thesis 'On plant hybrids' that he had received the *Pisum* paper after having completed his manuscript (Gaissinovitch 1966). In the German version of the paper he writes in the footnote that he finds it 'appropriate' to mention Mendel's paper 'for the method of the author and his way of expressing his achievements in formulae win one's attention, and call for further development'. Schmalhausen aptly states that Mendel's task was 'to estimate with mathematical accuracy the number of forms having originated from hybrid pollination and the quantitative ratio of the individuals of these forms'. He emphasized the finding that traits segregated in the hybrid progeny in numerical ratios. In the progeny of hybrids in which several traits were united, according to Schmalhausen, Mendel 'obtained a complete series, the number of which can be represented as originating from the combination of several series'. Schmalhausen added that 'by Mendel's observations and his mathematical considerations, among other things, constant members with new combinations of traits are always obtained'. Blomberg and Schmalhausen paid the same attention to Mendel's publication as to those of acknowledged scientists of the day. But their work also failed to elicit any response for a long time, and it was not until after 1910 that geneticists drew attention to it in the context of the search for Mendel's predecessors in the field of plant-hybridization research.

Interest in the study of Mendel's work was influenced most by W.O. Focke (1881), who in his book on plant hybrids mentioned not only Mendel's experiments with *Pisum* and *Hieracium*, but also those with *Phaseolus*. He considered the *Pisum* experiments to be quite similar to those of Knight, adding that Mendel believed he had found constant numerical relations between the different types of hybrids. He also mentioned briefly the fact that the forms of the parental plants appeared in the offspring of pea hybrids. Focke's well-set-out book on plant hybrids soon became the source book on the subject, arousing interest among his contemporaries, who later included Correns and Tschermak.

Mendel's papers are briefly quoted in the Royal Society's *Catalogue of scientific papers* (1864–73), published in 1879, at Vol. 4, p. 338, where mention is also made of his paper on *Bruchus pisi* (Mendel 1854). The *Hieracium* paper was cited by Neilreich (1871) and Henniger (1879). Both the *Hieracium* and the *Pisum* papers were cited by Besnard (1872). Mendel's name is also included by G.J. Romanes in the list of hybridists at the end of the entry on 'Hybridism' in the *Encyclopaedia Britannica* of 1881 (Romanes 1881–95). Olby (1985, p. 231) points out that Nägeli, too, whose relation to Mendel's publications is dealt with below, did not forget Mendel. In 1891, in the foreword to his monograph on European primulas, he mentioned Mendel's name in connection with *Hieracium* hybrids as one of the researchers who proved that the appearance of a hybrid plant is the same 'whichever of the two stem species was the fertilizing, and which the fertilized'. In his masterpiece, however, an 822 page monograph entitled *Mechanisch-physiologische Theorie der Abstammungslehre*, published in 1884, Nägeli did not mention Mendel's work. But in the following year Nägeli and Peter (1885) mention Mendel's *Hieracium* paper in a monograph on hawkweeds. A citation of the *Pisum* paper is made by L.H. Bailey (1892) in the bibliography later placed at the end of Chapter 2 of the second edition (1902) of his book *Plant breeding* (Bailey 1895). It is accepted that Bailey found the citation of Mendel's paper in Focke. MacRoberts (1984), however, found that he could also have learned of it from Jackson's (1881) *Guide to the literature of Botany*. Olby (1985, pp. 231–2) also pointed out a lecture by

R.A. Rolf on hybridization in London, in July 1899. Mendel's *Hieracium* hybrids are mentioned as evidence of the fact that hybridization in nature contributes to the wide range of forms in such a variable genus as *Hieracium*. Olby gives this as the 'first oral reference to Mendel in England'.

The last published information about Mendel's research before 1900 can be said to have been Carl Correns' (1864–1933) preliminary communication about xenia in *Zea mays*, a lecture given at the German Botanical Society in Berlin, on 29 December 1899. Correns (1899) criticized the view of T. Clarke for attributing to xenia colour changes in the seeds of *Matthiola*. According to Correns it is due to protein granules in the lower layer of the epidermis of the cotyledon. The conclusion was that the same state exists in the hybridization of yellow-and green-seeded races of peas, which 'Darwin and Mendel have already pointed out correctly'. But Correns did not cite Mendel's paper.

The most remarkable of Mendel's contemporaries was Nägeli, who had already dealt with the hybridization of plants in a thesis written in 1845. In the years of Mendel's experiments with peas, Nägeli (1865*a–d*, 1866*a–f*) published a series of speculative papers on the topic of plant hybridization. In his paper 'On hybrid production in the plant kingdom', Nägeli (1865*a*) writes that the origin of more or less stable varieties is not due to external factors, but to internal ones. He adds that in the progeny of self-fertilized hybrids 'three different varieties' appear, 'one which matches the original type, and two others which resemble more the progenitors', and that in succeeding generations these varieties have at the very least low constancy, and change easily. If the parental forms are close, according to Nägeli, there is a reversion of the hybrid form to one of the parental forms. In the following year, in his paper on hybrid formation, Nägeli (1866*a*) believed that the material participation of the paternal and maternal plants could not be the same, and thus a hybrid denoted AB was not the same as one denoted BA. In 1884 Nägeli altered his views on this matter, which became particularly apparent in his monograph 'Mechanical–physiological descent theory'. In the introduction he emphasizes that the problems of evolution can be investigated only by physiologists, who can make use of the findings of physics, chemistry, and mathematics. According to Nägeli, evolutionary theory was based on the model of animals, and most attention had been paid to the effect of the external environment. The process of internal changes in an organism was ignored. But in research into hybridization the main attention was paid to internal changes. Nägeli rejected Darwin's theory of pangenesis, and also Haeckel's theory of paragenesis of platidules, or the wave production of vital particles. He considered these to be speculative, and worked out his own theory of ideoplasm, the carrier of the hereditary *Anlage* in the cell, wishing to explain both heredity and evolution in this way. He considered the heredity *Anlage* to be solid and insoluble substances, with a special structure, and to be determinants of the origin of traits without being affected by the external environment, including the nutrition of the embryo. Nägeli drew a clear distinction between the trait and its determinants. On page 200, for instance, he wrote that 'The correct evaluation of the paternal or maternal contribution is a matter of *Anlage*.' In his view the estimation of the paternal and maternal contributions was a matter of probability theory.

Nägeli gives the striking example, which has already been noted by Weinstein (1958), of crossing a normal cat with an Angora cat. He points out that there is no blending of traits in the hybrids or their progeny; the racial traits remain pure. Nägeli was in effect accepting the idea of dominance and recessivity of traits. On page 213 he states that 'usually one of the colours of the flowers is *dominant*, the other coming out in the progeny of self-fertilized hybrids'. On page 209, Nägeli denotes the *Anlage* conditioning the origin of traits with large and small letters, and on the following page reaches the conclusion that the expression of a trait and its state of latency (*Latentbleiben*) 'is one of the more remarkable phenomena in the life of ideoplasm'. On page 227 Nägeli hints that the *Anlage* formed by various numbers of mycellae of various shapes and sizes are in the process of generation reduced to half the strength 'which they had in the parents'. The idea of *Anlage* as determinants of the transmission of parental traits to the progeny gradually led Nägeli (1884, pp. 273–4) to the idea of a 'law of heredity', which he considered analogous to the physical law of gravitation. He supposed variability to be 'largely inseparable from heredity', saying that 'heredity would be a revolution against the law of nature were variability to be removed from it'. In the end be achieved a narrower conception of heredity as the transfer of parental traits to the offspring, from which the problem of 'what is inherited' then followed.

F.B. Churchill (1979) pointed to the fact that five prominent German cytologists, M. Nussbaum, A. Weismann, E. Strassburger, O. Hertwig, and R.A. Köliker, in the years 1883 and 1885 developed the notion of the continuity of the germ structure, which significantly motivated the search for a theory of heredity. The most influential of these researchers was A. Weismann (1834–1914). He had the idea for his research from Nägeli, who in 1877 lectured on his theory of ideoplasm at a congress of German naturalists and physicians. Unlike Nägeli, Weismann sought the determinants of heredity in the germ of the cell.

In 1892 Weismann republished in German a series of articles, some of which had originally appeared in French and English. In this way he tried to show how his thoughts had developed on biological problems as important as sexual reproduction, heredity, and life and death, areas in which considerable progress had recently been made. In his introduction he expresses the conviction that progress would not come merely by the 'exact' method, and that these questions would remain unresolved only if measuring, weighing, and counting were employed.

Weismann's fifty-page lecture *On heredity* (Weismann 1892, pp. 75–121), delivered at Jena in 1883, tried not to deal with the whole question of the transfer of traits from parents to offspring, but only the then accepted theory of the inheritance of acquired characteristics. He emphasizes that an explanation had to be sought *in dem Stoff* (in matter), and thus in the substance of the germ cells. In Weismann's view a development of this idea would show Lamarck's and Darwin's explanations of changes in nature to be invalid.

In 1885 Weismann published a 105-page study called *Continuity of the germ plasm as the basis of theory of heredity*, in which he posed the basic question of how an individual can transfer the structure of his or her body to his or her progeny with such precision, and how a single cell can reproduce an image (*Portrait-Aehnlichkeit*) of the whole body, when it is made up of such different cells (Weismann 1892, pp. 191–302). He was aware that the answer would be very difficult to find, and that previous attempts had done no more than scratch the surface of the problem. He mentions briefly Darwin's provisional hypothesis of pangenesis, Häckel's (1876) theory of parigenesis of plastidules, a similar theory propounded by His (1874), and the more sophisticated theory of Brooks (1883). But none of these came up with a new approach to the problem, and Weismann proposed a return to Democritus' theory of heredity, according to which the seed contained particles of the entire body of both parents, which were brought to life by the body's power (*durch eine körperliche Kraft*), by means of which it recreated all the particles of the new organism. Weismann was only interested in the fundamental question: How can a single cell of the body contain in itself all the tendencies of heredity of the whole organism? He formed a hypothesis by which the germ cell (*Keimzelle*) was in fact an 'extract from the whole body', which arose directly from the germ cells of the parents. This led him to draw up his theory of the continuity of germ plasm. He knew, however, that it contradicted the then current theory of the inheritance of acquired characteristics.

It was from such considerations that Weismann was setting out when he distinguished between the somatic and germinal cells, and when he rejected the theory of the inheritance of acquired characteristics, and later when he formed his conception of inheritance based on the transmission of nuclear substance of a specific molecular constitution, called germ plasm. His idea was that this contained living elements which conditioned the appearance of particular traits of the body. They were active or passive, were not affected by the external environment, and on fertilization their number was reduced by half. In the course of their theorizing on this matter Weismann and the other cytologists, like Nägeli, did not cite Mendel's work.

THE DAWN OF A NEW AGE

New ideas arising from cytology led Hugo de Vries to seek units of heredity in the structure of germ cells, which would explain the variability of heredity and sudden changes in the traits of plants. It was during the period when varying degrees of emphasis were being placed on variations acquired during the life of an organism. Different theories also appeared to explain both heredity and evolution. As a student at Leiden University, de Vries became an enthusiastic adherent of Darwin's Theory of Evolution. After successful and intensive research on the physiology of agricultural plants, he extended the scope of his research interests into the fields of variability, heredity, and the origin of species. Like Weismann, he soon came to the conclusion that the continuous variations caused by the influence of environment during an individual's life could not be inherited, and that there was a heritable basis for selection.

This led de Vries (1889) to his theory of intracellular pangenesis. He rejected major parts of Darwin's hypothesis of gemmules, and changed the names of the carriers of heredity to pangenes. He also rejected Weismann's theory of the constancy of germ plasm, supposing it would mean in effect the splitting of species. According to de Vries (1889, p. 133), living protoplasm consists of pangenes, and these and only these form living elements. All the traits of an organism are, according to this theory, represented in the nucleus of the cell. By de Vries's account (p. 134) the pangenes 'are invisibly small, but have a quite different arrangement to chemical molecules, and each of them is composed of their infinite number; it must grow, reproduce, and on division of cells must be divided among all or nearly all the cells of the organism'. In Olby's opinion (1971), de Vries, like Weismann and Nägeli, failed to explain heredity because he never considered the possibility of only a small number of units per character. At the same time Olby wrote that Mendel 'postulated a liberation from the association of differentiating elements determining the traits, but made no stipulation as to how many such elements are present for any one character pair'.

De Vries distinguished two types of variability: fluctuating, conditioned by a varying number of pangenes and the effect of the external environment, especially nutrition, and species-forming, conditioned by the sudden appearance of new pangenes. After publishing the theory of pangenesis, de Vries devoted his attention to research on the occurrence of monstrosities, believing that hybrids would show a relatively high frequency of new monstrosities. Hybridization became a tool in his research. Towards the end of the century he was convinced that inherited qualities were independent units. At the International Conference on Hybridization in London, de Vries (1899) reported his observation of the crossing of hairy and hairless species of *Lychnis*. In 1893 hybrids had been hairy, but the next year de Vries described in the progeny 99 hairy and 54 hairless plants. In his lecture he stated erroneously that 'only about three-fourths were hairy, the rest hairless'. Meijer (1985) pointed out this complicated calculation error. The segregation ratio was in fact 2:1.

In 1893 de Vries was experimenting with crossing *Papaver somniferum* Mephisto, a dark-coloured type, with the lighter-coloured type Daneberg, and in the hybrid progeny he gave percentage occurrences of 77.5 per cent dominant parental traits and 22.5 per cent recessive ones. Many years later Peter W. van der Pas (1976) pointed out that the Botanical Institute of the University of Amsterdam still possesses plates on which de Vries had recorded the data from his experimental investigations into the relative proportions of active and talent traits among the hybrid progeny.

These data were not published by de Vries (1903) until he brought out his masterpiece *Die Mutationstheorie* (1901–3). L. Darden (1985) published another plate from the same room of the institute, on which it can be seen that in 1895 the researcher also crossed starchy and sugary varieties of *Zea mays*, and observed the proportions of these traits. De Vries (1899) published these results before the data from the experiments with *Papaver*.

In Meijer's (1985) view, de Vries did not prepare this poster until 1900, when he studied Mendel's *Pisum* paper. He had known about it previously from the bibliography of H.L. Bailey (1892), who probably took over the citation of the *Pisum* paper from Focke's book. Edwardson (1962) explained how Bailey drew the attention of de Vries to Mendel's paper. In his letter to Bailey de Vries wrote: 'Many years ago you had the kindness to send me your article on "Cross-breeding and hybridization" of 1892; and I hope it will interest you to know that it was by means of your bibliography therein that I learnt some years afterwards of the existence of Mendel's papers, which now are coming to so high credit.' Much later Tschermak (1958, p. 47) showed that when he visited Amsterdam in 1898 de Vries already knew Mendel's laws of heredity in peas, and was verifying the pattern of the inheritance of traits in various plant species after bastardization in the first and second generation of the offspring of hybrids.

After a study of the available documentary evidence relating to de Vries' role in the rediscovery of Mendel's work in 1900, Meijer (1985) reached the conclusion that he knew of Mendel's work before 1900, but 'failed to understand its full extent'. He did not make a detailed study of it until after 1900, when M.W. Beijerinck sent him a reprint of the *Pisum* paper, as was explained by Stomps (1954). Then for the first time, according to Meijer, 'this event must have disturbed him greatly' 'and after studying the paper he was prepared to absorb the idea of numerical segregation ratios. He therefore quickly published selected data from his own previous experiments, and only then referred to a segregation ratio of 3:1, using Mendel's terms dominant and recessive.' According to Meijer (1986) the appreciation of Mendel's theory by de Vries lasted only from January to August 1900; later 'Mendelism became subordinated to his theory of mutation'.

In his previous publications on plant hybridization, de Vries also described intermediate hybrid forms, but in 1900 he referred only to the effect of dominance, like Mendel, to which fact attention was drawn by Correns (1900*b*). In his first paper, published in French, de Vries (1900*a*) did not cite Mendel, and Correns suspected that he wished to cover up Mendel's priority in discovering the segregation ratio of 3:1. In his second paper, written in German, de Vries (1900*b*) did cite Mendel. In the introduction he mentioned that his experiments were based on his theory of pangenesis, according to which the whole character of the plant was composed of certain units connected with a material carrier. In his research he set out from an investigation of variability, mutability, and observation of bastards. He pointed to the complete change in the field of research into hybridization, making it essential to see species as being composed of independent factors. According to this the law of segregation was valid only for true hybrids. He deliberately published collected experimental data showing a proportion of 22 to 28 per cent recessive traits in the hybrids' progeny. These were the results of crossings of several plant species resulting in a segregation ratio of 3:1. De Vries added briefly that he had obtained a segregation ratio of 1:1 from backcrossing experiments and trait recombination on crossing plants differing in two pairs of traits. His conclusion was as follows: 'From these and other numerous experiments I draw the conclusion that the law of the segregation of hybrids as discovered by Mendel for peas finds very general application in the plant kingdom, and that has a basic significance for the study of the units of which the species characters are composed.'

In his third paper, published in French, de Vries (1900*c*) analysed in more detail the relation of his results to his theory of intercelullar pangenesis and his conclusion was: 'This law is not new. It was stated more than thirty years ago, for a particular case (the garden pea). Gregor Mendel formulated it in a memoir entitled *Versuche über Planzaenhybriden* in the *Proceedings* of the Brünner Society. Mendel has here shown the results not only for monohybrids, but also for dihybrids.' Here de Vries's statement was inaccurate. Mendel also gave the results of trihybrid experiments.

At the beginning of 1900 de Vries supposed that his publication confirming Mendel's law of segregation would reinforce his theory of pangenesis. Soon, however, he changed his view. At the time he was preparing for publication his life's work *Theory of mutation*, which subordinated Mendel's theory. This was aptly expressed by de Vries (1901–3, p. 329), as follows: 'Mendel's laws are valid for characteristics of varieties, whereas the characteristics of species give rise upon hybridization to constant hybrid characters.' At the time he was concentrating on research on the model plant *Oenothera lamarckiana*, and thought he was explaining the origin of new species by mutation.

Immediately after receiving the reprint of the French paper by de Vries, Correns (1900*b*) published a paper entitled 'G. Mendel's law concerning the behaviour of progeny of variable hybrids' In this very title he was acknowledging Mendel's priority. According to Roberts (1929, p. 358) it was his teacher, Nägeli, who told Correns about Mendel's research 'only of the *Hieracium* investigations'. In his first response to Roberts, Correns stated that he had read Mendel's *Pisum* paper a few weeks after October 1899, so that after reading the reprint of de Vries's paper he was already prepared to write his paper as quickly as possible. He was originally attracted to the problem of xenia in plant breeding. On crossing varieties of various seed colour he observed segregation of colour, and pointed out that the colour of the individual kernels was that of the embryo. The results of the first experiments of Correns (1899) were published under the title 'Investigation of xenia with *Zea*'. He described 'very interesting, but also very complex relations', but was only able to conclude that 'basically the behaviour is the same as that found when yellow and green pea seeds are bastardized, as has been correctly pointed out by Darwin and Mendel'. But Correns did not cite Mendel's paper. In 1900 he concluded, like de Vries, that he thought he had found 'something new' (Correns 1900*b*). He discovered, however, that 'Abbot Gregor Mendel had, during the sixties, not only obtained the same results through extensive experiments with peas, which lasted for many years, as did de Vries and I, but had also given exactly the same explanation, as far as that was possible in 1866.' In the next line Correns added that: 'Mendel's paper is among the best that have ever been written about hybrids, in spite of some objections.'

Correns described his experiments with crossing peas differing in seed colour over six generations. He was the first to emphasize Mendel's distinguishing of traits and their determinants, *Anlage*, and he introduced the explanation of Mendel's theory in the form of the determination of each trait by means of two hereditary units, which Bateson (1902) called allomorphos, later alleles. In generalizing Mendel's theory, Correns was the first to state not only the numerical ratio of 3:1, but also the ratio of 9:3:3:1 for experiments crossing organisms differing in two traits pairs. His conclusion was: 'This I call Mendel's Principle; it also includes "a loi de disjonction" of de Vries. Everything else may be derived from it.' In his final statement Correns objected to this principle as follow: (1) that in many pairs of traits there is not always one dominant trait; and (2) that Mendel's principle of segregation cannot be generally valid.

In the same year Correns (1900*c*) published a paper on *Levkojen* (snowflake) hybrids, referring to the preliminary character of his previous rediscovery papers. Here he defined two Mendel principles, and proved the linkage of characters, considering this to be an exception to the characters described by Mendel. In a footnote Correns adds that Mendel himself described grey, grey-brown, or leather-brown seed-coat coloration in connection with violet flower colour and reddish markings at the leaf axils as a dominant character. Foreseeing such exceptions, Correns soon discovered further important genetic phenomena, making him one of the important cofounders of Mendelism.

The third person usually numbered among the 'rediscoverers' of Mendel's work in the year 1990 is Erich von Tschermak (1871–1962). In the same botanical journal where de Vries and Correns had published their findings Tschermak (1900*a*) published a seven-page abbreviated version of his thesis. He began his experiments in 1898, trying to explain the effect of crossing plants upon the vigour of the offspring. Later Tschermak (1900*b*) published his thesis under the same title, ninety pages in extent. In his memoirs (Tschermak 1958, pp. 53–5), he states that he had studied Mendel's paper along with other literature relating to hybridization in the winter of 1899–1900. A book by Focke (1881) led him to study Mendel's papers. After reading the papers by de Vries and Correns he prepared as quickly as possible a shortened version of his paper, in which he confirmed 'the Mendel doctrine' as follows: 'The simultaneous discovery of Mendel by Correns, de Vries, and myself is especially pleasing to me. I, too, thought in the second year of the experiments that I had discovered something new.'

In crossing pea varieties whose seed differed in their shape, or colour, or both, he concluded that 'the characteristics traits of individual varieties in relation to their heredity do not appear in the same value'. In this connection he pointed to Mendel's explanation of the segregation of dominant and recessive traits, adding that Mendel's 'premiss of regular unequal quantivalency (*Ungleichwertigkeit*) of traits for heredity is fully confirmed in my experiments with *Pisum sativum*, and in the observations of Körnicke, Correns, and de Vries in *Zea mays*, and in de Vries's interspecific crosses, and is shown to be of the utmost importance in the science of heredity in general'. The relation of dominance and recessivity was understood by Tschermak as the 'unequal quantivalency of various traits'. In his abridged publication he states that 'the number of bearers of the dominating or prevailing trait is in proportion to that of the bearers of the recessive approximately as 3:1'. He adds briefly that the combination of two dominating or recessive traits in one parental form brings the same behaviour in the production of the seed of blending as in the case of isolated traits.

In his thesis Tschermak (1900*b*) also mentions the numbers of segregating seeds in the progeny of blendings. For seed colour the figures were 1854:660, i.e. a ratio of 2:8:1.0, and for seed shaper 884:288, i.e. a ratio of 3.1:1.0. His thesis also contains data on experiments crossing peas differing in seed shape and colour, which resulted in a segregation ratio 8.3:3.0 and 2.6:1.0. In the shortened version of Tschermak's paper he also mentions the crossing of blending plants with the parental form with a dominant and a recessive trait, and in the second case notes that here there is an increase in the bearers of the recessive trait compared with the case of the progeny of the self-fertilized blendings (*Mischlinge*). In the thesis, Tschermak (1900*b*, p. 80) states a proportion of traits of 1.2:1.0 for seed form and 1.75:1.0 for seed colour. Curt Stern (1966, p. xi) drew attention to this confused explanation of back-crosses, and in particular to Tschermak's words 'the influence of the trait *yellow* on the seeds of the hybrid was reduced by 57 per cent, that of the trait *smooth* by 43 per cent'. According to Stern, Tschermak had 'not realized that back-crosses should give a ratio of 1:1:'. Olby (1985, p. 123) explains this on the grounds that Tschermak's interpretation of paired traits and their segregation as their 'differing valency' was based on the fact that he regarded the Mendelian conception of germinal segregation 'with suspicion proper to a devoted nineteenth-century positivist'.

In stating the proportion of traits these experiments, Tschermak (1900*b*, p. 80) writes that the behaviour of plants arising from both types of seeds must be further studied, 'especially from the point of view of the question of whether the recessive traits of seed formed here remain constant in the progeny'. He adds: 'Such experiment with the pollination of castrated blendings of varieties with the pollen of one of the parental varieties as well as the investigation of the quantivalence of traits which follow from this have not, in my view, yet been published.' In fact such experiments were published by Mendel, and he repeatedly suggested that his contemporaries or his readers repeat them. For this reason we can agree with Stern (1966) that 'Tschermak's designation as a rediscoverer of Mendel has only limited validity.' If M. Kottler (1979) and M. Campbell (1980) showed that de Vries and Correns had not reached their conclusions independently of Mendel, then this is all the more true of Tschermak, who in his memoirs defended himself against 'arranging in order' (*Staffellung*) in the rediscovery of Mendel's work. Tschermak's publication also differs from those of Correns and de Vries in that he did not use the term hybrid or bastard, instead using the term blending (*Mischling*). De Vries and Correns in their publications in 1900 dealt with the problems of hybridization and the transfer of traits of plants from generation to generation, and did not use the term heredity. But Tschermak linked the question of crossing with heredity, and had his own concept of heredity potency.

Dunn (1965*b*, p. 49) and Křiženecký (1965*a*, p. 15) wished to demonstrate that W. Bateson should also be counted among the rediscoverers of Mendel's paper, alongside de Vries. Bateson (1894) first demonstrated the discontinuous character of variation in the plant and animal kingdoms, in contrast to Darwin's concept of continuous variation. From 1896 he had been experimenting with the crossing of plants and poultry, and considered cross-breeding a method of investigation of a particular case of evolution and of determination of which variations are discontinuous and which continuous. In his lecture delivered to the Royal Horticultural Society in London, Bateson (1899) wished to investigate the problem of heredity of discontinuous traits 'on a scale sufficiently large to give statistical results'. He supposed that 'true and precise experiments in these fields' had never been made. Later Bateson (1901) drew attention to Mendel's paper in a lecture published under the title 'Problems of heredity as a subject of horticultural investigation'. Beatrice Bateson (1928) states in her memoirs that her husband had already prepared his lecture, and it was only on the train down to London that he 'read Mendel's actual paper'. She says that Bateson immediately altered the text of his lecture. Olby (1987) was the first to explain that in his lecture Bateson explained only the law of segregation according to de Vries (1900*a*). It was only in the version prepared for printing that he included a citation of Mendel's paper. Soon afterwards Bateson and Saunders (1902*a*) published their results of experiments with crossing poultry, and thus for the first time proved the validity of Mendel's theory in the animal kingdom. Bateson's junior colleague M. Pease (1958) seems, however, to have overestimated his affirmation that Bateson himself would have discovered Mendelism had he started his experiments a year earlier. In fact before 1900 Bateson was not thinking in terms of numerical segregation ratios, and it was only after he changed his experimental methods following his study of the *Pisum* paper that he reached the conclusions which he published in 1902, thus making an outstanding contribution to the development of Mendelism. Cock (1973) analysed the influence of Mendel on Bateson, and came to the conclusion that 'the enormous stimulus given to Bateson's work by the rediscovery of Mendel is reflected in the greatly increased number of chicks produced in 1901: over three times as many as in the previous year.'

An analysis of the publications by the 'rediscovery' of Mendel's work in 1900 showed that none of them had got as far as the idea of pairs of discrete traits, which are transmitted by determinants in the germ cells and segregate out in the hybrid progeny according to mathematical series. Fisher (1936) thus rightly described the story of the sensational rediscovery of Mendel's work in the traditional teaching of the history of biology as a tale. A more detailed analysis of the publications relating to Mendel's work from around 1900 shows that there was a process of incorporation of Mendel's innovative ideas into the thinking of naturalists, who were investigating the problems arising from the development of knowledge in the late nineteenth century. The chief problems concerned new findings relating to the reproduction of animals and plants in connection with development of cytology and with progress in ideas on evolution in connection with Darwin's theory.

De Vries, Correns, and Tschermak were studying different problems, and most probably they began their experiments without knowing of Mendel's paper. After reading the paper they definitely revised the methods used in their further experiments, and above all interpreted the results taking into account Mendel's approach to research in plant hybridization so far as they had perceived it. In 1900 'rediscoverers of Mendel' were introduced into the literature, and as late as the 1920s the historian of science, Roberts (1929), began to elucidate their role in the process of the acceptance of Mendel's theory. Recently Corcos and Monaghan (1987a,b) and Monaghan and Corcos (1986) made a fresh examination of the interpretation of Mendel's theory by de Vries, Correns, and Tschermak, concluding that each of them accepted only one part of Mendel's explanation. At the same time Monaghan and Corcos (1987) wished to prove that the explanation of Mendel's observations and of his methods of analysis and representations 'were new and not easily understood or easily connectable to the usual patterns of thought about hybridization'.

De Vries and Bateson were concerned with investigating the character of the variability of animal and plant traits in connection with the process of evolution. De Vries subordinated Mendel's theory to his own theory of mutation, with which he wished to contribute to the development of genetics after 1900. Bateson saw in Mendel's theory confirmation of his concept of discontinuous variation. He can be credited with the publication in 1902 of an English translation of the *Pisum* and *Hieracium* papers, and his research according to Mendel's methods contributed significantly to the development of the early science of heredity. But when the chromosome theory emerged, Bateson refused to accept it, and in England he made a stand against the development of this new direction of research on heredity (Olby 1987). Correns was the one who was informed in most detail about Mendel's research, and in 1905 he made available Mendel's letters to Nägeli, with information on Mendel's experiments on other plant species. He founded the German school of genetics, and made a significant contribution of his own to the development of the science. Tschermak did outstanding work in the field of the application of Mendel's theory to breeding new plant varieties. His efforts also aroused the interest of Brno natural scientists in Mendel the researcher, and in the search for documentary evidence relating to the personality of the founder of genetics, who was soon to achieve posthumous fame.

Between 1865 and 1900 there had been major changes in research into living systems, and thus also in the thinking of naturalists. In 1865, in the words of Margaret Campbell (1985), everything about Mendel's theory was implausible. Naturalists up to 1900 did not conceptualize their results in the investigation of plant hybridization as numerical ratios, so their explanations were different. Mendel's interpretation of the role of probability in the production of germ cells was also implausible, as was trait combination and the theoretical categories as discrete traits. Their determinants seemed to be in the Platonic transition, which was in contradiction with current evolutionary thinking, influenced mostly by Darwin. The idea of material determinants seemed deterministic to some, for example Morgan (1903). Mendel, in his second lecture, first predicted the results of planned backcrossing experiments and then gave experimental results confirming the *a priori* elaborated theory. This was an original manner of thinking in biology, but unintelligible to his listeners and to those who were later to read his published paper. Campbell deduces from this that the plausibility of the new theory was more important in the process of rendering the new theory acceptable than the experimental proof of the theory's soundness.

As regards the plausibility of the new theory, Mendel himself did little to promote it in later years. We may well ask why he did not publish the results of experiments with other plant species, about which he reported in detail in his correspondence with Nägeli. In effect each of his letters was a scientific communication which might have been published after slight editing. In his second letter he suggested to Nägeli that he repeat the experiments, even providing the seeds for him to do so. The seed packets were labelled with Mendel's symbols, and the idea was for Nägeli to perform the experiments and verify Mendel's predicted results. Nägeli did not take up the challenge, and Mendel does not mention the matter in subsequent letters. But he continued to ply Nägeli with new evidence of the validity of his theory. We do not know when or how Mendel got to know of the citation of his work by H. Hoffmann (1869), along with an erroneous explanation of the constancy of a *Geum* hybrid. Mendel must have had every reason to publish the results of further experiments, and prove the wider validity of his theory. In particular, the results obtained in further experiments with *Hieracium* would have obtained a better response. But he was silent, and his theory had to wait until 1900, sixteen years after his death, to achieve recognition.

THE ORIGIN OF GENETICS

At the first international genetics conference in London in 1906, Bateson stated that ideas on hybridization and plant breeding had already developed into a new science, which he suggested should be called genetics. In his view the name 'sufficiently indicates that our labours are devoted to the elucidation of the phenomena of heredity and variation' (Saunders 1906, p. 91). Before that, Bateson's friend Weldon (1902) had, setting out from a biometric investigation of heredity using statistical methods, rejected Mendelian heredity, seeing it as being contradictory to the law of ancestral heredity introduced by Francis Galton. This induced Bateson (1902) to publish his *Mendel's principles of heredity. A defence.* In 1900 Bateson believed that the new theory, then being called Mendelism, could resolve some of the problems inherent in the Darwinian theory. Later he rejected the efficacy of natural selection as an evolutionary agent, and became a critic of Darwin's theory of evolution, in the belief that Mendelism, considered a theory of heredity, in conjunction with the theory of mutation, would also give an explanation of the process of evolution. Olby (1987) showed that later on Bateson, with his concentration on Mendelism, created an obstacle to the further development of genetics in Great Britain. Towards the end of his life he even told his son Gregory, who was named after Mendel, that 'it was a mistake' to have devoted his life to Mendelism, that it was 'a blind alley which would not throw any light on the differentiation of species, nor on evolution in general' (Koestler 1971, p. 121).

The year 1900 was beyond doubt a milestone in the development of experimental biology, into which genetics was just beginning to penetrate. In the course of the years to come biologists began to make use of Mendel's method to carry out research into the inheritance not only of morphological traits, but also of physiological and later mental ones. It was only then that the process of acceptance of Mendel's ideas began to take its place in the sphere of research into the heredity of plants, animals, and even man. In this period de Vries (1901–3) published his mutation theory, and soon researchers tried to find out more about the relations between heredity, mutation, and the theory of evolution. In this connection one can point out a forgotten evaluation of the later perception of Mendel's theory, written by D.J. Scourfield (1911) in 'A sketch of Mendel's life and work', delivered as an address to the Mendel Society in England, on 6 June 1910. In his introduction he emphasizes that the rediscovery of Mendel's work by de Vries, Correns, and Tschermak 'was not the discovery of a new scientific fact, but the unearthing of a little paper by one Gregor Mendel, which had been published as far back as the year 1866'. Scourfield's address continued to the following effect: 'We can imagine their astonishment as they read that old paper, to find that it actually contained the clue for the puzzle they had themselves been struggling with in the course of their work. They lost no time in making known their discovery, and thus was inaugurated what, from the point of view of the study of heredity, may justly be termed the Mendelian era.'

In the present day, as we know by experience, the name Mendel and words such as *Mendelism, Mendel's laws,* and *Mendelian facts* are consistently forcing themselves upon our notice, sometimes in the most unlikely places. And yet, in spite of all that has been said and done around the name of Mendel during the past decade, it is doubtful whether very many even of those who take a real interest in biological matters have a clear idea of who Mendel was and what he did. In his contribution Scourfield wished to clarify both Mendel's life and his research. He based his statements mainly on the data which were available in English-langauge publications, but was unable to add anything to the interpretation of Mendel's research as such.

Kříženecký (1965a, p. 18), in preparing to republish the *Pisum* paper along with twenty-seven papers from the rediscovery era, emphasized that soon after 1900 Mendel's discovery and cytological findings 'gave an admirably coherent picture'. External phenomena of Mendelian heredity were thus 'explained in their internal mechanism', and 'the union of Mendelism and knowledge about the chromosome apparatus, which took place in the years 1903–04, formed the real solid basis of modern genetics. It holds true, therefore, if Wilkie (1962) declares that only the years 1903–4 can properly be considered the birth-date of modern genetics.'

Later some geneticists who investigated Mendel's research attributed to him a knowledge of certain scientific findings which came to light only when genetics developed further in the twentieth century. Others, on the other hand, reduced Mendel's achievements to the laws of segregation and recombination of traits. These contradictory perceptions of Mendel's theory have formed the subject of many studies by geneticists and historians of science which Orel and Hartl (1994) recently elucidated. Crew (1966a), quoted beneath the title of his book *The foundation of genetics* the line first used by John Buchan of Oliver Cromwell: 'A great man lays upon posterity the duty of understanding him.' Crew's book contributed to our understanding of Mendel the researcher, and of his influence on the development of genetics. At the time he was writing his book he was corresponding with J. Kříženecký, who was then working on the plans for setting up the Mendelianum in Brno as a place of research into the history of genetics. Before that Kříženecký had published hundreds of papers in the field of experimental biology, but still found time to go back to studying Mendel's work, publishing a number of historical papers on the subject, culminating in his book *Fundamenta genetica*, published on the occasion of the centenary of Mendel's lectures in 1865 on his experiments into plant hybridization. He also took part in the evaluation of Mendel's research in Brno in 1922, with the participation of geneticists from abroad, and later in his experimental research returned to Mendel's work, and, as he stressed in 1964, found in it ever new stimuli for his own research.

One is tempted to seek an answer to the question: How was Mendel's theory perceived in Brno, where Mendel's contemporaries still lived after the turn of the century, in the context of the development of genetics abroad?

9

Mendel remembered

The centenary of the birth of Gregor Mendel demands of his native country and of international science a debt of honour, that due homage be paid to the discoverer's memory, in the place where he lived and worked.
—From the declaration of the international committee on the occasion of the 1922 Brno memorial.

In 1902 the secretary of the Natural Science Society in Brno, Niessl, informed members of the society of Professor Tschermak's application for permission to republish Mendel's publications on plant hybridization in Oswald's series *Klassiker der exakten Wissenschaften*. In the same year Alois Schindler (1902) published a lecture containing more biographical details on his uncle, and also some information on his research. He mentioned Professor Bateson, who, he said, put Mendel's law 'in line with the theory of atoms, forming the basis of chemistry'. Schindler's booklet led Proskowetz (1902) to publish recollections of Mendel related to him by Niessl.

Alois Schindler's publication also induced Bateson to seek out more facts relating to the scientist who fascinated him so much. At the first International Conference on Genetics in London in 1906, William Bateson delivered an enthusiastic evaluation of Mendel's achievements, noting that seven years previously, at a conference of the Royal Horticultural Society in London, he had still been doubtful 'whether Science had anything to contribute to Practice, looked at from the trader's point of view'. He went on: 'But seven years have gone by since then, and I now know that we have something. The scientific and practical have gone to form a perfect and fertile hybrid. I think segregation will occur, and that Science will ultimately separate from Practice; but that date is remote, and it is quite enough for us to rest on the absolute fact that for many years to come Science and Practice will go hand-in-hand and assist each other. There will come a time when Science will have learned all it can from Practice, and possibly there

Fig. 9.1 Gregor Mendel, painted after his death and preserved in the Mendelianum.

will come a time when Practice will have learned all it can from Science, and, as in the profession of electricity, Practice will develop into Science. The practical electrician of today is at the same time the scientific electrician, but in horticulture I expect a century must elapse before the same complete union of Science and Practice will be achieved.' (Saunders 1906, p. 60).

Bateson was the first naturalist who realized the far-reaching significance of Mendel's theory from the points of view of both the development of the science of heredity and the application of the results of the new theory. It can be supposed that he spoke with the same enthusiasm in December 1904 in Brno, when he first visited the place where Mendel carried out his research. Among members of the Natural Science Society, this aroused a new interest in their recently famous former member. And in the same year a native of Brno, J. Wiesner (1901) published in Vienna an article on Mendel, and E.R. Proskowetz (1902) published his memoirs of Mendel's activities as related to him by Niessl. But Iltis (1911) relates that even much later on there were very few people in Brno who had any real grasp of what Mendel's work was about.

At a meeting in January 1905 the Brno naturalists heard Niessl's (1906) report on Bateson's visit to Brno in December of the previous year, which he denoted 'a pilgrimage to the city in which Abbot Gregor Mendel carried out the experiments and preliminary studies for his work, published in the *Proceedings* of our Society many years ago, on the laws of inheritance in the plant kingdom'. Bateson, according to Niessl, was 'a spirited supporter of this doctrine, and holds Mendel's memory in great piety'. After that naturalists in Brno organized lectures explaining the significance of Mendel's research in the development of the natural sciences, and for the application of new findings in agriculture and medicine. This was the start of efforts by Brno natural scientists to grasp the significance of the scientific achievements of an almost forgotten member of their society.

During his first visit to Brno Bateson also visited the monastery, thus stimulating interest among the Augustinians in their former abbot, who had since 1900 become an extremely important figure in world science. The young conventual, Fr. Anselm Matoušek, after his ordination in 1906 devoted all his spare time to searching for manuscripts and recollections among Mendel's surviving contemporaries (Orel and Marvanová, 1966).

THE 1910 MONUMENT TO MENDEL

In June 1906 the committee of the Natural Science Society put forward a proposal to obtain funding for a Mendel monument in Brno. These efforts were greatly supported by E. Tschermak, who was at the time a professor at the agricultural university in Vienna, and enjoyed international authority as the third rediscoverer of Mendel's work. The nine-member committee of the Natural Science Society was joined by the Augustinian abbot, F. Bařina, and the parish priest of the monastery parish, Fr. C. Janeček. Both these men had been received into the monastery by Abbot Mendel. In 1907 Professor Tschermak and H. Iltis, later to become known as Mendel's biographer, published an appeal to biologists of the world to contribute to a fund for the erection of a monument to Mendel in Brno. In the appeal it is said that Mendel arrived at a knowledge of the laws governing nature in the area of 'heredity and bastardization'. The text states that 'The discovery and the actual determination of the laws of hybridization opened up and indeed made possible a new, extraordinarily fruitful era for experimental research into the heredity of individual characters, for the systematics of plants and animals, and not least for the microbiology of the process of generation and for practical breeding.'

The appeal, published in German, Czech, French, and English, found support among many scientists abroad. The organizers anticipated that the monument would 'speak to future generations of the excellent and exceedingly modest researcher and his contribution to the biology of all lands'. Within a relatively short time the organizing committee obtained 5000 crowns from foreign natural scientists. More money was obtained from the municipality of Brno, the Ministry of Culture and Education in Vienna, and from some of the leading personalities of the day. Abbot Bařina, for instance, contributed 3300 crowns, and Duke Joseph von Liechtenstein 1000 crowns. The total cost was 64 000 crowns. The committee asked five Moravian sculptors to submit ideas, and that of Theodor Charlemont from Znojmo won the day. Out of four proposed sites for the statue, the square outside the monastery was chosen, and when the statue was unveiled it was renamed Mendel Square.

The statue was officially unveiled on 2 October 1910 (Fig. 9.2). Before the ceremony, visitors from abroad were invited to view an exhibition of documents and mementoes of Mendel's life and his scientific work, presented by Iltis (1911) with the full support of Abbot Bařina. The ceremonial gathering in the square, which was decorated for the occasion, was attended by representatives of the municipality and of public bodies, government officials from Vienna, and thousands of ordinary citizens. The published report of the meeting particularly noted the presence of Professor Tschermak, from Vienna, Professor Bateson from Great Britain, Dr Hagedorn from France, Professor J.P. Lotsy from Holland, Dr H. Nielsson-Ehle from Sweden, and Professor E. Baur from Germany. In his introduction, Iltis described Mendel as a person and as a scientist, and then Tschermak evaluated his work from the viewpoint of international science. In his view Mendel's great work consisted in 'constructing a rational, indeed mathematical problem, and an exact method of research into heredity'. The chairman of the local organizing committee, Baron S. Haupt, handed the memorial over to the mayor as 'one of the most beautiful ornaments of our city'.

At an evening meeting of two hundred leading participants, the representative of a French plant-breeding firm, Philippe de Vilmorin, emphasized that thanks to Mendel's research, 'breeding practice is moving out of the age of empiricism and on to a logical basis; it is ceasing to be an art, and becoming a science'. He was repeating the words spoken by Bateson at the first international genetics conference in London in 1906. In conclusion, H. Molisch, professor of the physiology of plants at the University of Vienna, who was a pupil of Mendel's at the Brno *Realschule*, noted that Mendel had been a regular visitor to the greenhouses of his father, who was a Brno gardener. He also pointed to the fact that the monument in Brno was the first to be unveiled to Mendel in the whole of the

Fig. 9.2 The Mendel monument unveiled in Brno in 1910.
(Sculptor, Th. Charlemont.)

Austro-Hungarian Empire, adding that the capital, Vienna, had many monuments to statesmen, generals, and artists, but none to a natural scientist.

So 1910 saw the appearance in Mendel Square in Brno of a memorial with the inscription: 'To the natural scientist Fr. Gregor Mendel, dedicated by friends of science in the year 1910' (Fig. 9.3). But the general public still knew very little about Mendel. Iltis mentions a chance conversation between local citizens standing in front of a shop window in which the scientist's portrait hung. When one asked the other who this man had been, the other replied: 'Don't you know: the city got an inheritance from him?' (Iltis 1911, p. 342). This goes to show the public's understanding of heredity. Representatives of the Natural Science Society resolved to publish in the

Fig. 9.3 Plaque of Mendel issued by the Augustinian monastery on the occasion of the erection of the Mendel monument in 1910. (Sculptor, Th. Charlemont.)

forty-ninth volume of the Society's *Proceedings* both Mendel's plant hybridization papers and his meteorological paper on the whirlwind. Iltis also included in the *Proceedings* his report on the organizational preparations for building the monument, and an account of the official unveiling ceremony. Cock (1982) describes Bateson's impression of this.

The first paper published is thirty-two pages long, by P. Kammerer (1911*a*), in which Mendel's principles are explained in agreement with speculation on the heredity of acquired characters. Paul Kammerer (1880–1926) did not take part in the unveiling ceremony, but lectured on the subject at an extraordinary meeting of the Brno Society on 14 March of that year. He mentions in an introduction that his explanation clarifies the relation between Mendelism, Lamarckism, and Darwinism. Reference is made to the publications of contemporaries who were investigating deviations from the simple segregation of parental characters in the hybrid progeny. According to Kammerer, morphologically defined units of heredity cannot determine directly the expression of characters. He preferred the explanation that each property or ability was materially conditioned through 'chemism of the germ'. He thus explained changes in characters under the influence of internal factors such as enzymes or various external factors. To support his theory he referred to examples in published works of influencing segregated recessive traits of a quality different from the dominant traits. In conclusion, Kammerer supposed that 'Mendel's science is not contradictory, as we have heard from several sources, to the view that characters of single individuals acquired during their lifetime are transferred to the progeny; indeed, it rather brings this theory, on which we are working at the moment, to its most beautiful perfection.' Here Kammerer was in fact changing Mendel's theory of particular heredity, which E. Mayr (1982, p. 687) called 'hard' heredity, quite substantially, turning it into the neo-Lamarckian concept of 'soft' heredity.

In a further twelve contributions, leading scientists such as H. Nielsson-Ehle and L. Cuenot pointed to the results of the more complex phenomena of heredity, mostly arising from the interaction of genes, and research into quantitative characters of animals and plants; Hurst (1911), in his contribution, paid similar attention to man. According to a remarkable contribution by C.C. Shull (1911) the influence of Mendelism was already being felt not only in the academic world, but also in agriculture and medicine and in the new discipline of eugenics, and 'in certain cases sheds a clear light where before was rather darkness'. The conclusion was that 'The influence of Mendelism in modifying the Darwinian conception of evolution is already considerable, and the reflex of this influence must, sooner or later, be put more in the region of philosophy'.

A fifty-three-page paper by a representative of German developmental mechanism, W. Roux (1911) was of quite different content and conception. He set out from Weismann's theory of *Keimplasma* and sought the explanation of heredity and evolution together in the chemical assimilation capacity of organisms. He considered his theory to be borne out by Kammerer's 'highly meritorious experiments'. He did not mention Mendel's theory, but accepted in its full extent Kammerer's explanation of the heredity of acquired characteristics. In the same year Kammerer (1911b) published his lecture to breeders in Berlin, given on 24 February, where he pointed to the breeding of improved animal and plant races not only by crossing, but also by the suitable use of external conditions to create new characters. His published lecture, fifty-two pages long, argued the inheritance of acquired characteristics from the results of his own experiments and those of other authors, citing sixty-seven sources. Mendel is mentioned only in connection with the segregation ratio of 3:1 obtained in Kammerer's crossing of midwife toads (*Alytes obstetricans*). According to Kammerer, this corresponded 'according to the discoverer' to the Mendelian segregation scheme or scheme of prevalence. This was a term used only by Tschermak (1900a,b) in connection with the quantivalence of traits, as was mentioned in Chapter 8, p 287. One has the impression from Kammerer's publications that he was acquainted with Mendel's paper only through Tschermak's publications, according to which he developed his ideas, without considering Mendel's explanation in the *Pisum* paper, in particular his rigorous attitude to the significance of the published data.

The Mendel memorial in Brno was initiated by members of the Natural Science Society, whose history has been described by Iltis (1912). The unveiling of Mendel's statue in 1910 inspired interest in developing the theory for purposes of plant breeding in the Institute of Plant Production which had been set up in Brno in 1899, and also in the Higher Horticultural School established in Lednice (Eisgrub) in southern Moravia, according to the proposal of E. Fenzl, Professor of Botany at Vienna in 1894. His nephew, later Professor E. von Tschermak, suggested in 1912 the building in Lednice of a new institute for plant breeding, together with research into the development of Mendel's theory for horticultural plants. With the support of Duke Johann of Liechtenstein the institute was indeed built and called the Mendeleum. Its history has been described in detail by Recht (1976). The institute's activity is historically connected with that of the *herbarium vivum* in Lednice, mentioned in Chapter 2, p. 17. In 1918 the Mendeleum was taken over by Franz von Frimmel (1888–1957), who worked there until 1945 and achieved outstanding results in both research and the breeding of new plant varieties.

The erection of a monument also increased interest in the use of Mendel's theory in animal breeding. A pioneer in this area was J. Taufer (1869–1941), who had already tried to organize the breeding of farm animals in Moravia according to the achievements in neighbouring German states. He realized how important it was to develop Mendel's theory in animal breeding, and stressed that 'Mendel's laws lead us away from unplanned and hopeless crossing, based on mere empiricism without any real scientific basis and logical connections, to the one correct path of consanguineous mating within the domestic breeds, without importing foreign blood' (Taufer 1914). In conclusion he stated that 'those who still underestimate the practical importance of Mendel's laws in animal breeding, and that is the majority, are like sailors who reject the compass because a captain's ignorance of the sea and its danger spots sometimes lead to shipwreck.' In 1919 Taufer, as professor of animal breeding at the newly established Agricultural University in Brno, became the first Moravian member of the new German Genetics Society, which was at the time the only society spreading genetics teaching in Central Europe. Two years later he became head of the new Research Institute of Animal Husbandry in Brno. His attempt to develop Mendel's theory in animal breeding brought him, through V. Růžička (1882–1924), Professor of Biology at the medical faculty in Prague, into contact with Kammerer, an encounter which had a major effect on the way in which Mendel's theory was interpreted in Brno.

THE PERCEPTION OF MENDEL'S THEORY IN BRNO IN 1922

After 1910 scholars in many lands developed the science of heredity in connection with cytology and the theory of mutation. They soon showed that chromosomes constitute the physical basis of heredity. Research into heredity in populations and into mutations gradually led to an explanation of the relation between the theories of evolution and heredity. Various theories, which were often contradictory, appeared on the scene in relation to both Mendel's ideals and the interpretation of the units of heredity. This was abundantly clear from the events held in Brno to commemorate the 1922 centenary of Mendel's birth.

In 1921 the members of the German Society for Science and the Arts asked H. Iltis to make preparations in collaboration with the Natural Science Society for a celebration of the hundredth anniversary of Mendel's birth. The proposal was accepted, and when the organizing committee announced that a memorial gathering would be held in Brno, they soon received applications from 125 prospective participants from fourteen countries, among them C. Correns and E. von Tschermak. A congress of German naturalists and physicians to be held in Leipzig was also planned for 1922, and later it was announced that the second congress of the German Genetic Society would be held in Vienna. Representatives of learned societies and universities and universities in Brno reached an agreement with the representatives of the above bodies to coordinate the dates of all three conferences. In the wake of the First World War biologists were anxious to support the notion of international cooperation in the new branch of science Mendel had founded. The government of the young Czechoslovak state provided financial support for the scientific meeting, which was attended by many more participants from abroad than the unveiling of Mendel's statue in 1910. On the eve of the conference, on Saturday 22 September, they had the opportunity to see an exhibition of documents connected with Mendel's studies and research which had been found by Iltis when preparing his biography of the discoverer. On Sunday morning the abbot of the Augustinian monastery, Fr. Bařina, invited participants to tour the monastery, and to view in one of the rooms documents relating to Mendel's life and work. This was to form the basis of the Brno Mendel Museum. Participants in the congress were also able to see Mendel's bee-hives and the experimental garden, where a sandstone monument was erected with inscriptions in Czech, German, French, and English, stating: 'Prelate Gregor Mendel performed here the experiments for his law.' The dates 1822–1922 were attached (Fig. 9.4). There was also a plaque by V.E. Staff, of which several copies were made in plaster and bronze.

The memorial meeting is described in the *Proceedings*, edited by Iltis (1923*a*). It was opened in front of Mendel's statue in the presence of the guests from abroad. On the invitation of the committee of the Natural Science Society the ceremony was conducted by Professor A. Rzehak, who had been a pupil of Mendel's at the *Realschule*. He pointed to the import-ance of international cooperation in the evolution of the science of genetics, emphasizing that: 'Even ideas and scientific theories are inherited—therefore let the spirit of brotherhood between scientific workers be domi-nant, expressed in Mendel's terms.' On behalf of T.G. Masaryk, president of the Czechoslovak Republic, the gathering was addressed by B. Němec (1873–1966), professor at the Charles University. He quoted the poet Novalis, saying: 'We carry the burden of our forebears, just as we receive good things from them, so that in fact we live in both the past and the future.' Professor Němec was convinced that the poet was in this way expressing his perception of heredity. But according to Mendel's theory, he said, we were obliged to preserve and multiply for the future above all that which was good. 'If we link in ourselves the past and the future without our own endeavours, and we are not to blame for our own specificity, then the result should be a deep-rooted mutual tolerance of all the qualities of individuals, nations, and races.' Němec also pointed to our joy in the diver-sity of nature, where every bird sings in its own way, and flowers bloom in a thousand different shapes and colours. Why, then, should we not take pleasure in the great diversity of people, their qualities, and their languages:

Fig. 9.4 The sandstone monument erected in Mendel's experimental garden in 1922 by the Augustine monastery.

the correct use of Mendelism could, in Němec's view, lead mankind at last to greater tolerance, to true humanity, and to real peace.

The representative of the German geneticists, Professor E. Baur, considered the short period of progress in genetics since 1900, and stressed that 'for four decades Mendel's discoveries remained unnoticed. When the time was ripe, they were rediscovered, and it took further decades for biologists to begin to understand the far-reaching implications. When the monument was unveiled in 1910, Mendel's theory was still not generally accepted. It was still necessary to fight against many prejudices. Since then ten years of misery and tears had elapsed, but ten years which were also rich in scientific work, and now biologists throughout the world were united in recognizing Gregor Mendel as a truly great pioneer of science, who was at first known only to a narrow circle of scientists. Now the structure of Mendel's research was in the front line of all biological research. It could be declared that no other method of working had brought us any nearer to the most profound problems of life. Mendel's discoveries are not only pioneering ones for theoretical research, but are also of primary significance for practical selection of plants and animals, for medicine, for population policy and for racial hygiene.'

Iltis (1923b), wrote a fifteen-page report on the meeting, giving a brief résumé of the contributions of other participants in front of the memorial, and later in the banqueting hall of the Grand Hotel. The head of the Carnegie Institute in the USA, C.B. Davenport, greeted the gathering on behalf of the geneticists of the United States, where the science had already made great progress. In his opinion the large number of Mendel institutes, breeding stations, and eugenic institutes which had come into being in America demonstrated better than mere words the influence that this 'quiet' Brno scholar had upon American science. Correns was unable to attend because of sickness, and in a letter to the participants he mentioned how, thanks to Mendel, decisive progress had been made in genetics. He was sure that the evolution of the science started by Mendel was only just beginning. How was genetics going to develop over the next century? What a pity no one there would survive to see the 200th anniversary of Mendel's birth, he said. Of the Russian geneticists, Professor Rimsky-Korsakov of Leningrad sent a letter, stating: 'As everywhere else, in our land too the name of Gregor Mendel is a symbol of scientific ideas and fruitful opportunities in the sphere of research into heredity. I hope that the way of internationalism in the future will also lead Soviet scientists to join the rest of the world.' At the time the Soviet state was just beginning to rebuild its scientific base. This message to Brno in 1922 indicated great interest in genetics in that country. Ten years later there were already major scientific centres in Moscow and Leningrad, which stood in the forefront of international science.

The *Proceedings* of the Brno meeting contain twenty-three contributions, mostly from visiting authors. They relate to genetic research in plants and animals, and the use of genetics in breeding practice, medicine, and physiology. Five contributions are of a rather different nature. R. Goldschmidt and S. Minami (1923) reported on the heredity of secondary sexual characteristics. V. Haecker (1923) pointed to the complex determination of some characteristics, recommending that they be investigated through the historical-development analysis of characteristics or phenogenetics; he thus introduced a new term and at the same time a trend in research into genetics. He was sure that the way to carry out such research had been shown by Mendel's theory. A.C. and A.L. Hagedorn (1923), experienced plant breeders, explained the differences in the way genetics was taught to natural scientists and farmers. Using concrete examples they showed the way in which genetics could be used in practice. In conclusion, they showed how Mendel's work is grasped by those practical men who have succeeded in penetrating it without any interpretation by geneticists. For example, in 1912, W. Arkwright, an English poultry-breeder, had written that he had read the main pamphlets and books on Mendelism, and reached the conclusion that it could never be of practical use. Later he realized that no writer had ever tried to extract from it any application to the breeding of livestock.

He was led 'to consult the work written by Mendel himself'. His conclusion was that Mendel had 'written very little', but that 'his words were clear and very refreshing after the bewildering verbiage of his followers'. This was the view of an experienced animal-breeder regarding Mendel's work, in connection with the development of contemporary genetics in textbooks by biologists. K. Tjebbes (1923) digressed from the programme of evaluating Mendel's work to consider communism in society. In his view Mendel's theory was also capable of explaining sociopolitical problems. From this point of view a communist society could be created only if the majority of that society lived under similar social conditions, and thought and acted in a similar manner out of conviction. But this situation could not arise in society in the long term, and the author therefore considered communism a form of society which was impossible.

H. Przibram (1923) pointed to a problem which was topical at the time. He was the head of the Experimental Biology Institute in Vienna, where Kammerer had worked from 1902 to 1918. In his introduction Przibram noted that the greatest enemy of progress in science was insistence on the validity of new theories, which tended to become dogma. The example he gave was the theory of natural selection, in which, Przibram claimed, Darwin had adopted Lamarck's theory of the origin of species. But if Darwin had known of Mendel's classic experiments he would certainly have revised much of his own work, especially the sections in which he states that in the progeny of hybrids no individuals occur with exactly the same traits as the parental forms. Przibram added that it was equally impossible to adopt as a generally applicable rule the segregation of traits in the hybrid progeny. It would be in the spirit of Mendel himself to avoid overestimating this rule, and to take up a critical stance, in view of the new theory of mutation, which Przibram considered the starting-point for an explanation of the process of evolution. He cited the prominent geneticist R. Goldschmidt, who at the previous meeting in Leipzig had explained the origin of new species by a cumulation of minor mutations according the theory of the Morgan school, as opposed to a leap through one mutation, which was what biologists had previously assumed according to the mistaken theory of de Vries.

Przibram's (1923) view was that geneticists were in this way reverting to Darwin's idea of small changes which gradually led to a change in species. He mentioned in support of this idea Kammerer's experiments with a species of *Proteus*, in which he showed how an eye had developed from a rudimentary formation under the influence of artificial lighting. The conclusion was that new species in nature arise from modifications and not from mutations. According to this school of thought the task of Mendelism was to investigate the conditions under which the heredity of traits changed through the action of the environment, before segregating in line with Mendel's theory. Here we see the direct influence of Kammerer on his superior, who in 1926 was still accepting in full the latter's explanation of the inheritance of acquired characteristics.

Kammerer published the results of his experiments from 1907 onwards, attempting to show the inheritance of acquired characteristics not only in the case of morphological and physiological traits, but also in the behaviour of animals. Koestler (1971, p. 41) stated that in 1923 Kammerer explained, with the passing of time, that in the early days of his research he was not setting out to prove Lamarckian inheritance. On the contrary, he was 'under the spell of Weismannism and Mendelism, which both agree that acquired characters are not inherited'.

Only when he observed purposeful adaptations of appearance and behaviour did it occur to him to test whether these changes might be inherited. Later Kammerer (1914) tried to show the heredity of the characteristics of man acquired in the process of education and upbringing. He considered this conception of heredity to be 'progressive', and was convinced that it was based 'on modern bastardization research, on Mendelism'. The opponents of the idea of the inheritance of acquired characteristics, such as Weismann, Hertwig, and Johannsen, he dubbed 'antigeneticists', unable to understand the role of genetics in evolution.

At the gathering of geneticists in Brno in 1922, differences in interpretation of the theories of evolution and heredity became clear; at a subsequent meeting in Prague, they stood out even more starkly.

THE PHANTOM OF THE INHERITANCE OF ACQUIRED CHARACTERISTICS

The hundredth anniversary of Mendel's birth was also commemorated on 19 October 1922 by the Czech Eugenic Society in Prague, which also published the *Proceedings* (Brožek *et al.* 1925). At the opening ceremony the chairman of the society, L. Haškovec (1925), emphasized that the rules or laws of heredity were basically applicable not only to biology, but also to the psychiatry and pathology of man. He emphasized that their significance extended to 'our individual and social life'. Haškovec supposed that it was not only a question of the heredity of corporeal traits and the factors governing them, but also of 'all psychic functions and properties, both normal and pathological'. He recalled his speech on the occasion of the fourth congress of Czech physicians and natural scientists in 1908, where he had emphasized that 'on the basis of the laws of heredity, especially psychological heredity, we proclaim the right of all nations to preserve national types and characters, the national specificity and peculiarity, and national tongues; to respect and preserve these is the duty of a cultured nation'. He added: 'One cannot overestimate the scope of those pillars of genetic science raised by Mendel, which is why our eugenic society has taken the opportunity today, rightly and properly to revive and recognize the significance of the life of Gregor Mendel, which peered with unprecedented acuity of vision into the problem of heredity, the peak of achievement of biological science, at a time when the official representatives of that science were still a long way away from this new light.'

Professor Němec, mentioned above, gave a detailed evaluation of Mendel's theory. He explained the delayed recognition of Mendel's work, and its relation to the theory of evolution. According to him Mendelism was often placed in the context of the theory of evolution in general, and Darwinism in particular. On the other hand, it had to be admitted that in one respect Mendelism spoke in favour of the original views of Darwin, as regards individual varieties within a Linnaean species. As Linnaean species comprised a great number of minor species which mutually intercross without any difficulty, considerable variations in progeny arise, thus providing material for natural selection. He also mentioned differing views of the action of external conditions on changes in the species, and attributed to Weismann 'great merit as regards Mendelism, for he was against the older uncritical views on the inheritance of acquired characters'.

Professor Růžička (1925), a Prague geneticist, presented his own version of the older uncritical views. He pointed to the critical period in the development of genetics which was under way, with the theory of preformism 'which he associates with Mendel's idea of genetic units' standing face to face with that of epigenetics, which was 'about to be constructed'. In this time of discord, Růžička saw the great edifice of the evolutionist nucleus rising up, an edifice of which Mendel had involuntarily laid the foundations. Růžička was convinced that Mendelism had not shown the qualitative diversity of species to be due solely to various combinations of the same genes. His conclusion was that here historicism clashed with Mendelism, since a cumulation designed to explain the gradual evolution of organisms into more complex ones presupposes the co-action of external factors, or, in other words, propounds the inheritance of acquired characteristics, which Mendelism in essence denies, though the role of heredity in the sense of the science of descent should be 'the maintenance and transmission of what has been accumulated, acquired'. Růžička was simply repeating the views which he had set out in his lectures as professor of biology at the medical faculty, an appointment he took up in 1908 (Růžička 1914). His institute did not offer the conditions for laboratory research on heredity. He had previously studied histology and cytology under Professor O. Hertwig (1849–1922), and under the influence of German developmental mechanics and the teaching of biochemistry sought an explanation of the basis of heredity in the chemical substances of cell structures. He adopted the views of Kammerer, who was trying to prove their validity experimentally.

Růžička's pupil Kříženecký (1896–1964) tried at this meeting to find a compromise between the proponents of evolution and heredity. In his lecture 'The heredity of acquired characteristics and the significance of Mendelism for the theory of evolution', Kříženecký (1925) set out in detail ideas he had published in short articles in 1914, as explained by Orel and Matalová (1990). He states: 'The knowledge which Mendel's research and discoveries produced, and which were multiplied and built by later research into a whole system of so-called Mendelism, at first sight argue against all evolutionary thinking.' Kříženecký supposed the basis of heredity in Mendel's conception to be a passing on of qualities from generation to

generation. 'Instead of external qualities', he states, 'internal pre-
dispositions, or genes, were put forward. The concept of heredity as a maintenance of these qualities or traits was imperfect, primitive, and not very
scientific.' 'One might propose a new concept of the gene', says Kříženecký,
'in its true evolutionary–physiological or evolutionary–mechanical action'.

Kříženecký expected that evolutionary analysis of traits and an evolutionary–mechanical view of genes would lead biology out of the cul-de-sac into
which the contemporary 'Mendelism' had taken it, creating a conflict with
the science of evolution. His conclusion was that mutations were acquired
hereditary traits arising through gradual adaptation to the environment in
the course of generations. He therefore considered research into the effect of
the environment on the occurrence of mutations to be very important,
referring to the ideas of V. Haecker (1923), who distinguished evolutionarily simple traits, which he supposed were governed by Mendel's laws, from
evolutionarily complex traits, which had to be investigated. Kammerer was
trying to reconcile Lamarckism and Mendelism, and under his influence
Kříženecký believed the results of his experiments, published in leading scientific journals. He thought the Mendelian laws which were generally
applied at the time could not be considered universal, 'even the principle of
the independence of genes, their purity and segregation', and suggested
that the basic concepts of Mendelism should be revived, without seeing this
as contradictory to Mendel's own theory.

Kammerer's publications affected the interpretation of Mendel's theory not
only in Brno, but throughout the world. Koestler (1971) has described the
main details of Kammerer's life and his efforts to fulfil his ambitions in various countries, especially in Britain and the USA. Geneticists in these two
countries and later in the USSR were solidly opposed to the idea of the inheritance of acquired characteristics. Koestler (1971) knew nothing of
Kammerer's contact with Kříženecký in Brno, nor that in 1921 the latter
had offered him an important post in the newly set up Research Institute of
Animal Husbandry in Brno. In a letter dated 7 February 1922, Kříženecký
begged Kammerer to accept the post, since he was the best qualified to make
a contribution to progress in the field of evolutionarily oriented biology, a
direction from which Kříženecký expected innovations in animal-breeding
methods. Kříženecký's views at the time show clearly how geneticists were
seeking a way out of the impasse presented by the juxtaposition of heredity
and evolution. It was not until much later that, in the words of Mayr and
Provine (1980), there occurred an 'intellectual transformation, through
which the legacies of Mendel and Darwin were fused into a new biology'.

Kammerer's book *The inheritance of acquired characteristics*, published in
the USA in 1924, was denoted an epoch-making work, for which the

author was acclaimed as a 'second Darwin' (Koestler 1971, p. 93). T.H. Morgan was adamant that his name should not be used in connection with Kammerer's lectures in the USA. J.B. Watson, on the other hand, founder of the behaviourist school, wrote of Kammerer in glowing terms. The denouement came in 1926, when the American zoologist G.K. Noble showed that the discoloration of the nuptial pads of the midwife toad, the cornerstone of Kammerer's proof of the inheritance of acquired characteristics, was due not to natural causes, but to the injection of Indian ink beneath the skin of experimental animals. The tragic aftermath was Kammerer's suicide on 23 September 1926.

The year 1924 saw a Russian translation of the book by Kammerer published in the USA. The views expressed in it caught the imagination of the Minister of Culture and Education, A.V. Lunacharsky, touching as they did on the question of acquired characteristics in the process of education and teaching. In May 1926 Kammerer was offered the post of head of a laboratory in the USSR to carry out research designed to establish the validity of Lamarckism. Kammerer accepted, and was about to leave Vienna for Moscow. In a letter to Moscow shortly before his suicide, he states that although he had 'no share in the forgery', he no longer felt qualified to accept the post (Koestler 1971, p. 118). Leading geneticists in Moscow and Leningrad rejected Kammerer's views as firmly as Bateson and Morgan. But there was a group of biologists and medical practitioners in Moscow who, under the spell of Marxist ideology, saw in Kammerer's research into the inheritance of acquired characteristics the possibility of fulfilling their revolutionary ideals. If people could inherit what they were taught, then under a state-controlled system of education it would be possible to breed a new race of human beings imbued with the principles of social justice. Even after the 1926 midwife toad fiasco, so tempting was the prize that many of the proponents of this genetically-based system of improving mankind continued to research into the inheritance of acquired characteristics, thus paving the way for the scientific and human tragedy of Lysenkoism (Gaissinovitch 1988).

W. Bateson died shortly before Noble proved that Kammerer's results were falsified, deprived of the final vindication of his stance against that school of thought, which dated back to Kammerer's very first publications. However, his pupil and younger friend, N.I. Vavilov (1887–1943) who studied genetics and plant-breeding with Bateson at the John Innes Horticultural Institute in 1912–14, had to deal with the after-effects of Kammererism. He had been appointed head of the Institute of Applied Botany in Leningrad in 1921. In 1929 Vavilov organized in Leningrad a conference of geneticists and breeders, to which eminent guests from abroad were invited. Among these was R. Goldschmidt, who at the time held a post at the genetics institute in Berlin. The huge attendance of geneticists from the Soviet Union showed how extensive genetic research was in the USSR. Later Goldschmidt (1949) recalled the showing at the Leningrad conference of the film 'Salamandra', propagating the idea of inheritance of acquired characteristics. The script was by Minister Lunacharsky. In effect, Kammerer was hailed as a pioneer in the field of genetics who had succeeded in changing the colour of an experimental animal by the action of the environment. It was suggested that the 'class enemy' had successfully organized against this researcher, who had all the qualities of a true communist, a smear campaign that presented him as a confidence trickster, thus driving him to suicide.

At the time Vavilov himself was committed to organizing genetic research and the breeding of cultivated plants, and was not interested in propaganda or the resuscitation of long-discredited scientific theories. He also found the time to make a detailed study of Mendel's classic paper, and wrote an introduction to a new Russian translation, describing Mendel's theory as one of the most splendid chapters of modern biology. At the same time Vavilov asserts that biologists still do not fully understand Mendel's theory, though expressing a conviction that 'the dialectics of the facts will lead them all to the laws discovered by Mendel' (Vavilov 1935).

In the 1930s agricultural production fell dramatically in the Soviet Union as a result of the enforced collectivization of peasant farmers. Vavilov was criticized, as a representative of agricultural science, for the fact that scientists were not delivering the goods in the area of agricultural productivity. In an atmosphere of considerable social tension the pathologically ambitious agronomist T.D. Lysenko (1898–1974) took a critical stance towards Vavilov, propounding the new teaching of 'Michurin biology'. This was founded on the idea of the inheritance of acquired characteristics, which promised rapid increases in the yields of agricultural crop plants (Medvedev 1969).

Vavilov was removed from office as president of the agricultural academy, and in December 1936 became the target of the attacks not only by Lysenko and his followers, but also by Marxist ideologists, led by I.I. Present. Genetics was to develop 'purely on the basis of Darwinism'. The interpretation of Darwin's theory, however, was not to be that of the nineteenth century, but one which presented the great biologist as one of the stalwarts of Marxist philosophy. These demagogic outbursts were denounced by geneticists, led by Vavilov and N.I. Kolcov (1872–1940), who was at the time head of the prominent Institute of Experimental Biology, where genetic research played an important role. They were supported by the American geneticist H.J. Mueller (1890–1967), who had been in Moscow since 1932 (Medvedev 1969; Soyfer 1989). At the height of the attacks on Vavilov the University of Agriculture in Brno awarded him an honorary doctorate, along with E. Tschermak in Vienna. But Vavilov was no longer free to travel out of the Soviet Union, and could not receive the presentation. He wrote to the university that he was unable to fulfil his ambition to visit Brno, 'the home of modern genetics' (Orel 1988). From then on he was persecuted by the secret police, until in 1940 he was finally arrested, to die in prison in 1943 from the effects of starvation and unbelievable privation.

In connection with the change of views on the significance of Mendel's theory in the 1930s, new ideas were also put forward in Brno as to how Mendel's legacy might be developed. In 1928–9 Kříženecký studied genetics in the USA on a Rockefeller fellowship. There he acquainted himself with the latest developments in genetics, and with the biological synthesis of the theories of genetics and evolution. In 1930 he and the professor of biology in Brno, J. Bělehrádek, proposed the establishment of a Gregor Mendel genetics institute at the Masaryk University in Brno. A Mendel Museum was to form part of it. The abbot of the Augustinian monastery supported the project. The economic depression and the threat of war prevented this idea from coming to fruition; but Kříženecký returned to it when the war was over. In 1948 he gained the support of international bodies for the project (Kříženecký 1974). But a new danger was emerging, whose far-reaching effects could not be foreseen.

In February 1948, the Communist Party took power in Czechoslovakia under threat of force, and the country fell under the control of Moscow. The new political trend was accompanied by the imposition of Marxism, which was soon to mean the subordination of science to that ideology. In August of the same year a conference took place in Moscow under the title 'On the state of the biological sciences', where Lysenko, with the overt support of Stalin, confirmed his dominance of biology (Zirkle 1949). Lysenko declared genetics to be a bourgeois false science, which he deprecatingly referred to as Mendelism, Weismannism, and Morganism. This was to be replaced by 'class Michurin biology', representing the dogmatic teaching of inheritance of acquired characteristics. The published *Proceedings* of this conference show that at the session in question S.I. Alichanyan pointed to the previous primitive experiments on the inheritance of acquired characteristics, mentioning Kříženecký's name along with that of Kammerer (Alichanyan 1948). When the Czech translation of the proceedings appeared, Kříženecký explained to his students that this was a misconception, that he had never conducted any such experiments, though in his twenties he had believed in Kammerer's experiments. Now, in 1948,

Kříženecký rejected Lysenko's speculation as being unscientific, and defended Mendel's theory and genetics in general (Plesnik and Orel 1949–50). Lysenkoists in Czechoslovakia denounced Kříženecký as a reactionary Mendelist, and so at the peak of his scientific activities, when he was poised to create a Mendel institute in Brno, he was forced to leave the university, and became the main tragic victim of Lysenkoism in Czechoslovakia (Orel 1992).

THE REINSTATEMENT OF MENDELISM AFTER 1950

The golden jubilee of genetics was celebrated on the occasion of the fiftieth anniversary of the 'rediscovery', in September 1950, at Ohio State University. Geneticists from the Soviet Union and the other states where genetics had been erased from the list of sciences were precluded from attending. Mendel's homeland was symbolically represented by H. Iltis, who had left Brno on the eve of the Second World War, and lived in the USA. C.L. Dunn (1951a,b) published the *Proceedings* of this gathering, and states in his introduction that: 'Because the potential contributions of genetics to human welfare in the next fifty years are so great, the continuation of freedom and cooperation among geneticists of all countries of the world may well be the hope which provides force and direction for the greater effort and devotion which the future will require of all scientists.' Goldschmidt (1951) showed how genetics had, in the short lifespan of fifty years it had thus far enjoyed, had a remarkable effect on almost every branch of biological science. Iltis (1951) tried to unfold a characteristic picture, to sketch some historical miniatures from Mendel's life, and to draw some conclusions from these facts concerning Mendel's philosophy and his legacy to us. Prominent geneticists from various countries evaluated the spread of genetic research to biochemistry, microbiology, immunology, cell biology, and cancer research. In the final contribution, Sir Julian Huxley (1951) thanked the Genetics Society of America for organizing the golden jubilee of genetics, concluding that 'In the fifty years since Mendel's Laws were dramatically rediscovered, genetics has been transformed from a drooping incertitude to a rigorous and many-sided discipline, the only branch of biology in which induction and deduction, theory and experiment, observation and comparison have come to interlock, in the sort of way they have done for many years in physics.'

In April of that jubilee year of genetics, 1950, the Augustinian monastery in Brno was forced by the secret police to close down and the monks were arrested. The buildings were handed over to industrial organizations which neglected them. Subsequently the greenhouse built when Mendel began his experiments was knocked down. In 1959 an official of the regional national committee even ordered Mendel's statue to be removed from the square, and for many years it was committed to the yard of the monastery. The only thing which somehow survived the period when genetics was banned and Mendel was denounced as a representative of the Church was the name of Mendel Square. The documents from the Mendel Museum were placed in the trunks of the Moravian Museum, in the safekeeping of natural scientists who believed that the time must soon come when they could be returned to the place where the scientist carried out his plant-hybridization experiments.

After 1950 genetic research began to develop quickly on the cellular and the molecular level, and Lysenkoists could only look on in amazement at the remarkable progress resulting from close international cooperation. With the support of the Soviet Union's all-important atomic scientists, Lysenko was criticized most effectively by a prominent geneticist of the 1930s, newly returned from a labour camp, N.V. Timoféev-Ressovsky (1900–81). When in 1957, N.S. Khrushchev criticized the unfavourable results of chaotic crossings of varieties of cultivated plants, Křiženecký (1957) saw this as the beginning of the return of genetics to Mendel's native land, and published a critical study of the development of the science of animal-breeding, pointing to the pernicious consequences of the proscription of genetics, and concluding as follows: 'for Lysenko to deny the Mendelian conception is like rejecting the Law of Gravity.' Shortly after this he was attacked in the press as a representative of reactionary Mendelism, and was subsequently arrested. He was released eighteen months later, his health broken. But Křiženecký did not lose hope of Mendelism's soon being reinstated at universities in Czechoslovakia, and began gathering documents relating to Mendel in connection with the approaching centenary of the publication of the *Pisum* paper, which geneticists were planning to commemorate in 1965. In 1962 he was given the task of undertaking the initial preparatory work directed towards setting up a Mendel museum in Brno (Křiženecký 1963a)(Figs. 9.5 and 9.6). His intention was supported abroad by his friend, the animal geneticist Hutt (1962). The project for the museum was adapted by Křiženecký so that it compassed an institute for the history of genetics. He was conscious that in this way he would be able to realize only the historical part of the plans he had formed in the 1930s. He was by then in poor health, and it took all the energy he could muster to prepare source books on Mendel, his scientific work, and the 1900 rediscovery for printing. Recalling the 1950s, when Mendel's statue was removed from Mendel Square, he noted that the scientist, 'a discoverer

Fig. 9.5 Mendel's experimental garden, renovated in 1965 on the occasion of the opening of the Mendelianum.

whose significance in twentieth-century biology ranks alongside that of Darwin' had been 'buried in Brno for a second time' Kříženecký 1974). From the beginning of 1964 Kříženecký worked intensively on preparations for an international symposium in Brno to commemorate the hundredth anniversary of the publication of Mendel's discovery. But in that year he died of cancer, and was denied the possibility of taking part in the opening of the Mendel Museum by the Moravian Museum the following year.

In August 1965, the Czechoslovak Academy of Sciences, together with the Moravian museum, the Genetics Section of the International Union of Biological Sciences, the International Atomic Energy Agency, and the Council for International Organizations of Medical Sciences, organized a Gregor Mendel Memorial Symposium in Brno (Sosna 1966) (Fig. 9.7). Geneticists from all continents descended on Brno to pay homage to the founder of genetics in the place where he lived and worked. By then Lysenko's teaching had finally lost its hegemonic position in Mendel's homeland, and the participants were able to visit the Mendel memorial and see documents of the man and his work.

Opening the symposium, F. Šorm (Sosna 1966, p. xvii), president of the Czech Academy of Science, expressed his admiration for 'the creative

317

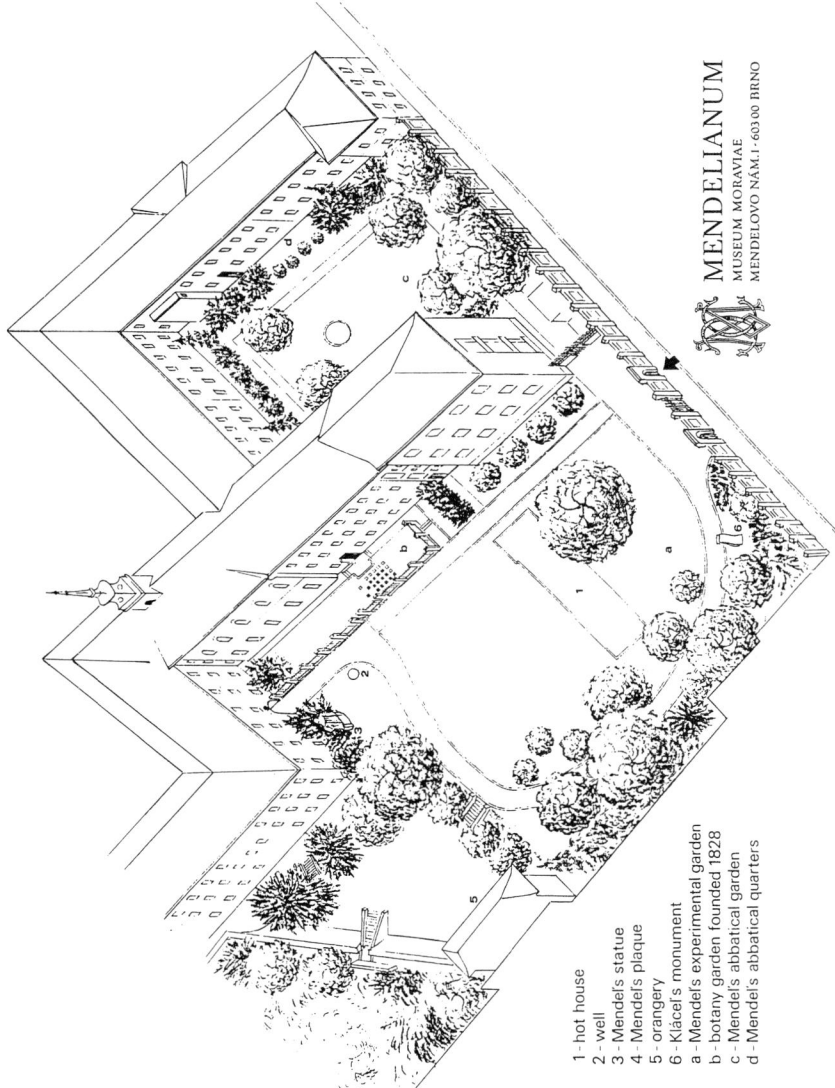

1 - hot house
2 - well
3 - Mendel's statue
4 - Mendel's plaque
5 - orangery
6 - Klácel's monument
a - Mendel's experimental garden
b - botany garden founded 1828
c - Mendel's abbatical garden
d - Mendel's abbatical quarters

Fig. 9.6 Scheme of the renovated Monastery area dedicated to Mendel's scientific achievements.

genius, experimental skill, and scientific integrity of Mendel *inter alia* from the viewpoint of the current almost explosive growth of the life sciences'. The prominent Russian geneticist and dissident, N.V. Timoféev-Ressowsky was only allowed to send his paper elucidating Mendel's original method-ological contribution to the research of heredity, which he later recapitu-lated in *Folian Mendeliana* (Timoféev-Ressowsky 1981). The participants in the symposium in Brno sent a message to Nobel prize-winner H.J. Muller, who had made an important contribution to the development of genetics in the USA, and later also in the USSR. He evaluated Mendel's research as follows:

The science that we at present term *genetics*, that has its first clear start in the brilliant work of Gregor Mendel, contains the main clue to the means by which life arose out of non-living material, to the nature of the threads that have woven evolution, and to the way that man must follow when he transcends himself. Mendel has been twice resurrected, but man will be resurrected repeatedly and even continuously. This rebirth will proceed by way of two reciprocally reinforcing methods: the manipulation and improvement of the physical and cultural envir-onment in the interest of humanity, and the raising of man's inner genetic con-stitution. In this work of self-creation all of mankind will participate and cooperate. Science is many-faceted, but unified. In pursuing science and its applications the many-faceted family of man will also attain a higher union. In this great adventure genetics, started by Mendel, will play a central role (Sosna 1966, p. xxiii).

The American historian of science, C. Zirkle (1966), noted in his lecture in Brno some oddities in the delay before Mendel was recognized both before and after 1900. He explained that contemporary natural scientists had not understood Mendel's innovative way of thinking, adding 'Those of us whose teachers matured before 1900 can remember what difficulties they had in re-orienting themselves, and how much they had to unlearn before they could accept Mendelism. Some never could! Even today a very few anti-Mendelians still exist, but they have little influence in world science. The memorial symposium is striking evidence that Mendel has come into his own.'

In 1965 the Mendel Museum returned to the now abolished monastery where Mendel had worked as priest, teacher, and abbot, and in his spare time had carried out research. This research gave rise to a science which is now regarded as the core science of biology. From 1965 the Mendel Museum has hosted scholars from many countries who wished to see where it all began, the place they regards as the Mecca of geneticists. The Mendel medals issued in different countries exhibited in the Mendelianum illustrate the high appreciation accorded to Mendel by scientists internationally (Obermajer 1974, 1985).

In 1990, after forty years' absence, the members of the Order of St Augustine were able to return to their monastery. The tenth abbot, Fr. Tomáš Martinec, is now trying, along with the General of the Order, Miguel Angel Orcasitas, to set up an institute for the study of Mendel's legacy to modern science, with special emphasis on ethical issues arising from the latest developments in genetics. Father Mendel has finally come home.

Fig. 9.7 Opening of the Mendel Memorial Symposium in Brno in 1965. *Standing*: Professor B. Němec, President of the Organizing Committee, Prague. *Seated. from left*: Professor C.H. Waddington, President of the International Union of Biological Sciences; B. Sekla, representative of geneticists in Czechoslovakia; F. Šorm, President of the Czechoslovak Academy of Sciences; S.J. Geerts, President of the Genetic Section of the International Union of Biological Sciences.

10
Epilogue

Robert H. Haynes (1989), in his presidential address to the XVIth International Genetic Congress in Toronto, pointed out the fact that genetics had achieved progress in 'veritable quantum jumps' since the rediscovery of Mendel's ideas at the turn of this century. His conclusion was that at present physics is still the most fundamental science, but genetics 'the most pivotal', having become 'the primary focus of contemporary natural philosophy'. This book explains how Mendel's research brought together ideas from various scientific disciplines. He tackled his research problem using the most important knowledge from physics, which was decisive in making possible his 'quantum jumps'.

Recently the molecular geneticist Alison Smith (1988) explained at the John Innes Institute in England the molecular basis of the element responsible for one of the traits Mendel described in pea seeds. In cloning the gene determining the wrinkled form of ripe seeds it was found that the structure of the starch depended on the presence or absence of one of two forms of an enzyme called the starch-branching enzyme I (SBEI). A seed which lacks the enzyme has a higher sugar content than one that has the enzyme. This causes the accumulation of more water, so that it swells to a large size in the early stages of its development and shrinks as it dries out. This shrinkage gives it its wrinkled appearance. Bhattacharyya *et al.* (1990) cloned the fragment of DNA that encodes the enzyme from the pea Mendel would have used, and explained that it lies at the *r* locus. The messenger RNA transcript of the SBEI gene in the wrinkled seeds produced only about one-tenth of the amount of transcript that round seeds did. The extra genetic material in the sequence coding for this gene was an inserted fragment of about 800 base pairs, very similar to certain transposable sequences of DNA in other plants, such as maize. This insertion causes the wrinkled phenotype of pea seeds and prevents the expression of the gene determining the round form. P. Brown (1990), evaluating this research, concluded aptly: 'Mendel observed that peas were smooth or wrinkly. Now we know why.' According to

Fincham (1990) students of genetics tend to think that they have already had enough Mendelism at school, and they are understandably eager to hurry on to the most recent advances in molecular genetics and gene cloning. The paper by the group at the John Innes Institute (Brown 1990) could be a boon to teachers looking 'to help them bridge the gap between the old and the new'.

In the twenty-five years since Stern's evaluation of Mendel's research, geneticists have penetrated further into the depths of nature, taking major steps forward in investigating all levels of organization of living systems, and now making use of their latest findings in the area of genetic engineering. In the light of the questions, theories, and discoveries of their predecessors they can appreciate the present level of knowledge, and, again in the words of Robert Haynes, they can 'understand why they are doing what they are doing'.

In the last quarter of the century many more studies and speculations have appeared in journals and books on Mendel and his scientific work than in the preceding century. Historians, philosophers, and sociologists of science now take an interest in Mendel when studying how scientific progress comes about. As always, opinions differ when it comes to an evaluation of the various aspects of Mendel's scientific work. The present book collates information on Mendel's origin, his studies, and his career as a teacher and monk, and finally as abbot. In the years to come there will certainly be new views of Mendel's personality and his work. New information may come to light, or new and more detailed ways of assessing the evidence surrounding the way he went about his experimentation and the creation of his theory. But it is my opinion that, even in the future, geneticists and biologists will study Mendel's papers on plant hybridization and the dramatic history of his research, not only in order to understand better the origin and early development of genetics, but also to learn under what modest conditions a scientist who can be called an innovator in the life sciences arrived at such a major discovery.

In 1984, Pope John Paul II, commemorating the centenary of Mendel's death, remarked that Mendel 'would not have fallen under the condemnation of St Augustine, according to whom many people have inclined more to admire the facts than to search for their reasons'. In fact, Mendel did both. In this connection the Pope assumed that in his deliberation on the research problem Mendel would also have contemplated the following statement of St Augustine (*Enarration in ps.*, 144, 13, 1876):

The beauty of the earth is like a dumb voice rising from it. You observe it and look on its beauty, fertility, and sources. You observe how the seed begins to germinate and gives a completely new thing which was sown. You take note of all of it, and it is as though you were asking in your mind why it is so. Filled with wonder, you search on, going to the root of things, and finally you discover a great beauty and a magnificent strength. Since this strength cannot come from itself, you think at once that he, the Creator, must have given it, so what you have discovered in the created is the voice of his declaration, and leads you to the glorification of God.

I thank Fr. Paulus Sládek OSA, from Zwiessel in Germany, for drawing my attention to the evaluation of Mendel's achievements by Pope John Paul II and for sending me the copy of the quoted statement by St Augustine.

References

Abbreviations

Folia Mendeliana
Published yearly by the Moravian Museum in Brno since 1966. Issues 1–9 appeared as special volumes. Issues 10–20 were published as a supplement to *Acta Musei Moraviae, sc. nat.* with the pagination of these volumes. The sections of the *Folia Mendeliana* were also reprinted separately. *Folia Mendeliana* Vol. 21 and further volumes are being published again as special volumes with their own pagination.

Mittheilungen, Brünn
Mittheilungen der k.k. Mährisch–Schlesischen Gesellschaft zur Beförderung des Ackerbaues, der Natur- und Landeskunde in Brünn.

Agricola, G.A. (1716). *Versuch der Universal Vermehrung aller Bäume, Stauden und Blumen–Gewächse.* Regensburg.

Alichanyan, S.I. (1948). In Stoletov *et al.* (1948), p. 240.

Alpatov, V. and Orel, V. (1979). An 1885 Moscow report on Mendel's research into acclimatization of tropical bees. *Folia Mendeliana*, **14**, 237–42.

André, C.C. (1795). *Der Zoologe, oder Compendiöse Bibliothek des wissenswürdigen aus der Thiergeschichte und allgemeinen Naturkunde.* Eisenach und Halle.

André, C.C. (1802). Footnote to the essay of Schreiber (1801).

André, C.C. (1812). Anerbieten, Gutbesitzern auf dem kürzesten Wege zu höchsten Veredlung ihrer Schafherden behülflich zu seyn. *Oekonomische Neuigkeiten und Verhandlungen*, Prague, **24**, 181–3.

André, C.C. (1815). Rede bey der ersten Eröffnung der vereinigten Gesellschaft des Ackerbaues, der Natur- und Landeskunde. In *Erster Schematismus der k.k. Mähr. schles. Gesellschaft zur Beförderung des Ackerbaues, der Natur- und Landeskunde* (ed. C.C. André), pp. 93–111. Brünn.

André, C.C. (1818). Note on Anon. (1818), p. 303.

André, R. (1816). *Anleitung zur Veredelung des Schafviehs. Nach Grundsätzen, die sich auf Natur und Erfahrung stützen.* Calvé, Prague.

Anon. (1818). Wirksamkeit der Ackerbaugesellsachaft in Brünn. *Oekonomische Neuigkeiten und Verhandlungen*, Prague, **38**, 297–304.

Anon. (1834). Wie muss man die Getreidearten veredeln. *Mittheilungen, Brünn*, **42**, 333–4.

Anon. [G.] (1837). Ueber die durch kreuzende Befruchtung bewirkte Veränderung in der Farbe der Erbsen. *Allgemeine Deutsche Gartenzeitung, Bayern*, **15**, 213–14.

Anon. [V-g-1.] (1854). Der Erbsenkäfer *(Bruchus pisi)*. *Mittheilungen, Brünn*, **13**, 100–1.

Anon. [M.] (1877). Gegen Kommunismus und Sozialismus. *Mittheilungen. Brünn.* **1**, 6–8.

Anon. [G.M.] (1878). Die Bedeutung der Wetter-Prognose für Landwirther. *Mittheilungen, Brünn*, **49**, 385. Mendel's authorship is explained by Orel (1969*a,b*).

Anon. [M.] (1879). Die Grundlage der Wetterprognosen. *Mittheilungen, Brünn*, **13**, 29–31. Mendel's authorship is explained by Orel (1969*a,b*).

Anon. (1989). Auch Galileo hat gelogen. *Kosmos das Naturmagazin*, **5**, 78–82.

Apelt, E.F. (1854). *Die Theorie der Induktion*. Engelmann, Leipzig.

Astaurov, B.L. (1965). K stoletiu otkrytija osnovnych zakonomernostej nasledstvennosti. In *Gregor Mendel. Opyty nad rastitelnymi gibridami* (ed. A.E. Gaissinovitch), pp. 107–17. Nauka, Moscow.

Astaurov, B.L. (1966). Mendel's laws and some selected pages from the establishment of genetics. In *G. Mendel memorial symposium 1865–1965* (ed. M. Sosna), pp. 57–62. Academia, Prague.

Ayala, F. (1988). P. Medawar: limits of science. *History and Philosophy of Life Science*, **10**, 129–36.

Bacon, F. (1620). *Novum organum*. Reprinted in *Artificial selection and development of evolutionary theory* (ed. C.J. Bajema), Benchmark papers in systematic and evolutionary biology, Vol. 4, pp. 11–13. Hutchinson Ross, Stroudsburg, 1982.

Bailey, L.H. (1892). *Cross-breeding and hybridizing*. The rural library series, Vol. 1, pp. 1–44, Rural Publishing Co., New York.

Bailey, L.H. (1895). *Plant breeding*. Macmillan, New York. Reprinted in 1902.

Balcárek, P. (1977). Die ökonomische Bedeutung der Veredlung des Schafviehs in Mähren in der ersten Hälfte des 19. Jahrhunderts. *Folia Mendeliana*, **12**, 223–8.

Barry, M. (1840). Researches in embryology. *Philosophical Transactions of the Royal Society of London*, **130**, 529–93. Quoted by Wood (1989–90).

Bartenstein, E. (1837). Ueber Vererbung in der Schafzucht. *Mittheilungen, Brünn*, **50**, 393–5.

Bateson, B. (1928). *William Bateson. F.R.S., naturalist. His essays and addresses*. Cambridge University Press.

Bateson, W. (1894). *Materials for the study of variation*. Macmillan, London.

Bateson, W. (1899). Hybridisation and cross-breeding as a method of scientific investigation. *Journal of the Royal Horticultural Society*, **24**, 59–66. Reprinted in Kříženecký (1965*a*), pp. 217–25.

Bateson, W. (1901). Problems of heredity as a subject for horticultural investigation. *Journal of the Royal Horticultural Society*, **25**, 54–61. Reprinted in Kříženecký (1965*a*), pp. 226–34.

Bateson, W. (1902). *Mendel's principles of heredity. A defence*. Cambridge University Press, London.

Bateson, W. (1908). *The methods and scope of genetics*. Cambridge University Press.

Bateson, W. (1909). *Mendel's principle of heredity*. Cambridge University Press.

Bateson, W. and Saunders, E.R. (1902*a*). Part II. Poultry. In *Reports to the evolution committee of the Royal Society of London* (ed. W. Bateson), Vol. I, pp. 87–124. Reprinted in Kříženecký (1965*a*), pp. 120–57.

Bateson, W. and Saunders, E.R. (1902*b*). Part III. The facts of heredity in the light of Mendel's discovery. In *Reports to the evolution committee of the Royal Society of London* (ed. W. Bateson), Vol. I, pp. 125–60. Reprinted in Křiženecký (1965*a*), pp. 242–75.

Baumgartner, A. (1823). *Die Naturlehre nach ihrem gegenwärtigen Zustande mit Rücksicht auf mathematische Begründung*. Gerold, Vienna.

Baumgartner, A. and Ettingshausen, A. (1842). *Die Naturlehre nach ihrem gegenwärtigen Zustande mit Rücksicht auf mathematische Begründung*, (7th edn). Gerold, Vienna.

Beadle, G.W. (1967). Mendelism, 1965. In *Heritage from Mendel* (ed. R.A. Brink and E.D. Styles), pp. 335–50. University of Wisconsin Press, Madison.

de Beer, G. (1964, 1966). Mendel, Darwin, and Fisher (1865–1965). *Notes and Records of the Royal Society of London*, **19**, 192–225. Addendum published in the same journal in 1966, **21**, 64–71. Both reports on Mendel's lectures from the journal *Neuigkeiten, Brünn*, Nos. **40** and **69** were published in English by Gavin de Beer in the Addendum. Reprinted by Olby (1985), pp. 220–1.

de Beer, G. (1965). Other men's shoulders. *Annals of Science*, **20**, 303–22.

de Beer, G. (1966). Genetics: the centre of science. *Proceedings of the Royal Society of London* **B.CIXVI**, 154–66.

Benedík, J. (1971). A contribution to the traits combination in Mendel's paper. *Folia Mendeliana*, **6**, 61–7.

Bennett, J.H. (ed.) (1965). *Experiments in plant hybridisation—Gregor Mendel*. Oliver and Boyd, Edinburgh.

Beránek, V. and Orel, V. (1988). New documents pertaining to Mendel's experiments with bees. *Folia Mendeliana*, **23**, 5–20.

Besnard, A. (1872). Alphabetische Uebersicht der speciellen Literatur des 'Genus *Hieracium* L.'. *Flora*, **55**, 390–4.

Bhattacharyya, M.K., Smith, A.M., Noel Ellis, T.H., Hedley, C., and Martin, C. (1990). The wrinkled-seed character of pea described by Mendel is caused by a transposon-like insertion in a gene encoding starch-branching enzyme. *Cell*, **60**, 115–22.

Blixt, S. (1975). Why didn't Gregor Mendel find linkage? *Nature*, **256**, 206.

Blomberg, A. (1872). Om Hybridbildning hos de fanerogama vaxterna. Dissertation, Stockholm University.

Blumenbach, J.F. (1771). *Über den Bildungstrieb und das Zeugungsgeschäft*. Göttingen. (The 3rd ed appeared in 1791.)

Blumenbach, J.F. (1780). *Handbuch der Naturgeschichte*. Göttingen. (The 12th ed was published in 1830.)

Boole, G. (1847). *The mathematical analysis of logic, being an essay towards a calculus of deductive reasoning*. Cambridge.

Bowler, P.J. (1989). *The Mendelian revolution. The emergence of hereditarian concepts in modern science and society*. Johns Hopkins University Press, Baltimore, MD.

Brannigan, A. (1979). The reification of Mendel. *Social Studies of Science*, **9**, 423–54.

Brannigan, A. (1981). *The social basis of scientific discovery*. Cambridge University Press.

Bratranek, F.T. (1853). *Beiträge zu einer Aesthetik der Pflanzenwelt.* Brockhaus, Leipzig.

Bratranek, F.Th. (1874). Goethe's naturwissenschaftliche Bedeutung. In *Goethe's naturwissenschaftliche Correspondenz (1812–1832)* (ed. T. Bratranek), pp. XXXV–LXXXIX. Brockhaus, Leipzig.

Broad, W. and Wade, N. (1984). *Betrug und Täuschung in der Wissenschaft.* Birkhäuser, Basle.

Brooks, W.K. (1883). *The laws of heredity.* Baltimore, MD.

Brown, P. (1990). Molecular biology solves the riddle of Mendel's wrinkly peas. *New Scientist,* **2**.

Brožek, A., Haškovec, L., and Růžička, V. (eds) (1925). *Memorial-volume in honor of the 100th birthday of J.G. Mendel.* Borový, Prague.

Burger, J. (1819). *Lehrbuch der Landwirtschaft.* Vienna.

Burkhardt, R.W. (1977). *The spirit of system: Lamarck and evolutionary biology.* Harvard University Press, Cambridge, MA.

Callender, L.A. (1988). Gregor Mendel: an opponent with descent of modification. *History of Science,* **26**, 41–75.

Campbell, B. (1980). Did de Vries discover the law of segregation independently? *Annals of Science,* **37**, 639–55.

Campbell, M. (1985). *A century since Mendel.* Illert publications, Adelaide.

Carlson, E.F. (1973). A Mendel reprint in the United States. *Folia Mendeliana,* **8**, 255–8.

Catcheside, D.G. (1966). Sir William Macley memorial lecture, 1966: The centenary of Mendel. *Proceedings of the Linnean Society of New South Wales,* **91**, (2), 101–8.

Cetl, I. (1973). Mendel's hybridization experiments with other plants than *Pisum. Folia Fac. sci. nat. Universitatis Purkynianae Brunensis, Biologia, Brno,* **41**, 2–42.

Churchill, F.B. (1979). Sex and the single organism: biological theories on sexuality in the mid-nineteenth century. In *Studies in the History of Biology,* pp. 139–77. Johns Hopkins University Press, Baltimore, MD.

Churchill, F.B. (1987). From heredity theory to *Vererbung*: the transmission problem, 1850–1915. *Isis,* **78**, 337–64.

Čižmář, J. (1979). Botanist Carl Theimer (1823–1870) in the cultural context of Brno during the period of Mendel's research. *Folia Mendeliana,* **14**, 265–70.

Cock, A.G. (1973). William Bateson, mendelism and biometry. *Journal of the History of Biology,* **6**, 1–36.

Cock, A.G. (1980). Faking and the intent to deceive. *British Medical Journal,* **281**, 1214–15.

Cock, A.G. (1982). Bateson's impression at the unveilling of the Mendel monument at Brno in 1910. *Folia Mendeliana,* **17**, 217–23.

Coleman, W. (1967). Ferdinand Schindler's letters to William Bateson, 1902–1909. *Folia Mendeliana,* **2**, 9–15.

Colladon, J.A. (1821). Résumé travaux de la Soc. Cant. de Genève. *Notice des séances Soc. Helv. Sciences Nat. Bibl. Universelle,* **17**, 288–9. (cit. Stubbe 1972).

Comenius, J.A. (1668). *Via lucis—vestigata et vestiganda.* Amsterdam. English translation, *The way of light,* Liverpool University Press, 1938.

Corcos, A. and Monaghan, F. V. (1984). Mendel had no 'true' monohybrids. *The Journal of Heredity*, **75**, 499–500.

Corcos, A.F. and Monaghan, F.V. (1987*a*). Correns, an independent discoverer of mendelism? I. An historical/critical note. *The Journal of Heredity*, **78**, 330.

Corcos, A.F. and Monaghan, F.V. (1987*b*). Role of de Vries in the rediscovery of Mendel's paper. II. Did de Vries really understand Mendel's paper? *The Journal of Heredity*, **78**, 275–6.

Corcos, A.F. and Monaghan, F.V. (1990). Mendel's work and its rediscovery: a new perspective. *Plant Science*, **9**, 197–212.

Corcos, A.F. and Monaghan, F.V. (1993). *Gregor Mendel's experiments on plant hybrids: a guided study*. Rutgers University Press, New Brunswick, NJ.

Correns, C. (1899). Untersuchungen über die Xenien bei Zea mays, (vorläufige Mittheilung). *Berichte der Deutschen Botanischen Gesellschaft*, **17**, 410–17.

Correns, C. (1900*a*). Gregor Mendel. *Botanische Zeitung*, **58**, 229–30.

Correns, C. (1900*b*). G. Mendel's Regel über das Verhalten der Nachkommenschaft der Bastarde. *Berichte der Deutschen Botanischen Gesellschaft*, **8**, 158–68. English translation, *G. Mendel's law concerning the behavior of progeny of varietal hybrids*, in Stern and Sherwood (1966), pp. 119–32).

Correns, C. (1900*c*). Über Levkojenbastarde. Zur Kenntniss der Grenzen der Mendelschen Regeln. *Botanisches Centralblatt*, **84**, 97–113.

Correns, C. (1905). Gregor Mendel's Briefe an Carl Nägeli, 1866–1873. *Abhandlungen der Mathematisch–Physikalischen Klasse der königlich Sächsischen Gesellschaft der Wissenschaften*, Leipzig, **29**, 189–265.

Correns, C. (1922). Etwas über Gregor Mendels Leben und Wirken. *Die Naturwissenschaften*, **10**, 623–31.

Crew, F.A.E. (1966*a*). *The foundation of genetics*. Pergamon, Oxford.

Crew, F.A.E. (1966*b*). Mendelism comes to England. In *G. Mendel memorial symposium 1865–1965* (ed. M. Sosna), pp. 15–30. Academia, Prague.

Czihak, G. (1984). *Johann Gregor Mendel (1822–1884). Dokumentierte Biographie und Katalog zur Gedächtnissausstellung anlässlich des hundertjährigen Todestages mit Fachsimile seines Hauptwerkes: 'Versuche über Pflanzenhybriden'*. University of Salzburg.

Czihak, G. (1986). Johann Gregor Mendel—seine Zeit und die "Wiederentdeckung" anlässlich der Gedächtnissaustellung zu seinem 100. Todestag. Wien, 1–3 Februar 1984. *Folia Mendeliana*, **21**, 5–7.

Czihak, G. and Sládek, P. (1991). Die Persönlichkeit des Abtes Cyrill Franz Napp (1792–1867) und die innere Situation des Klosters zu Beginn der Versuche Gregor Mendels. *Folia Mendeliana*, **26**, 29–34.

Darden, L. (1985). Hugo de Vries's lecture plates and the discovery of segregation. *Annals of Science*, **42**, 233–42.

Darlington, C.D. (1951). Mendel and the determinants. In *Genetics in the 20th century—essays on the progress of genetics during its first 50 years* (ed. L.C. Dunn), pp. 315–32. Macmillan, New York.

Darlington, C.D. (1966). *Genetics and man*. Pelican, Harmondsworth.

Darlington, C.D. and Mather, K. (1950). *The elements of genetics*. Macmillan, London.

Darwin, C. (1859). *The origin of species by means of natural selection*. Murray, London.

Darwin, C. (1863). Über die Entstehung der Arten im Tier- und Pflanzenreiche durch natürliche Züchtung, Stuttgart. Translation of *The origin of species by means of natural selection*, London, 1859.

Darwin, C. (1868*a*). *The variation of animals and plants under domestication* (2 vols). Murray, London.

Darwin, C. (1868*b*). *Das Variiren der Tiere und Pflanzen im Zustande der Domestication*, Stuttgart. Translation of *The variation of animals and plants under domestication*, 1868.

Darwin, C. (1877). *Die Wirkungen der Kreuz- und Selbstbefruchtung im Pflanzenreiche*, Stuttgart. Translation of *The effect of cross-and self-fertilisation in the vegetable kingdom*, London, 1877.

Darwin, E. (1805). *Zoonomie, oder Gesetze des organischen Lebens*. Pest. Translation from *Zoonomia, or the law of organic life*, London, 1794.

Dathe, G. (1866). Kurzer Bericht über die 14. Wanderversammlung deutscher Bienenwirthe in Brünn. *Centralblatt*, pp. 4–10. Hanover.

Daubenton, L.J.M. (1782). *Instructions pour les bergers et pour les propriétaires de troupeaux*. Paris.

Dick, A. (1986). Gründung und Ausstattung des k.k. physikalischen Institutes der Universität Wien. *Folia Mendeliana*, **21**, 41–8.

Diebl, F. (1829). Bemerkungen über die von Freiherrn v. Witten hinsichtlich der verschiedenen Weizensorten geäusserten Ansichten. *Mitteheilungen, Brünn*, **23**, 177–9.

Diebl, F. (1835). *Abhandlungen aus der Landwirtschaftskunde für Landwirthe, besonders aber für diejenigen, welche sich der Erlernung dieser Wissenschaft widmen. II. Von dem Pflanzenbau*. Rohrer, Brno.

Diebl, F. (1839). Ueber den nothwendigen Kampf des Landwirthes mit der Nature, um die Veredlung und Erhaltung seiner Kultur-Gebilde. *Mittheilungen, Brünn*, **34**, 270–1.

Di Trocchio, F. (1989). *Legge e caso nella genetica Mendeliana*. Angeli, Milan.

Di Trocchio, F. (1991). Mendel's experiments: a reinterpretation. *Journal of the History of Biology*, **24**, 485–519.

Dobzhansky, T. (1964). The Mendel centennial. *The Rockefeller Institute Review*, **2**, 1–6.

Dodson, E.O. (1979). Mendelian documents in Canada. *Folia Mendeliana*, **14**, 285–7.

Doppler, C. (1851). *Compendium der Artitmetik und Algebra. Mit besonderer Rücksicht auf die Bedürfnisse der technischen Wissenschaften und des praktischen Lebens*. W. Braunmüller, Vienna.

Doyle, G.G. (1968). Too many x^2 or hanky-panky in the monastery? in *An empirical introduction to statistics* (ed. J.L. Hodges *et al.*), pp. 228–9. McGraw-Hill, New York.

Dubec, K. and Orel, V. (1980). Gregor Mendel's activities in meteorology. *Folia Mendeliana*, **15**, 215–42.

Duhamel, M. (1758). *La physique des arbres, ou il est traité de l 'anatomie de plantes et de l'économie vegetale*. Paris.

Dunn, L.C. (ed.) (1951*a*). *Genetics in the 20th century—essays on the progress of genetics during its first 50 years*. Macmillan, New York.

Dunn, L.C. (1951b). Introduction. In *Genetics in the 20th century—essays on the progress of genetics during its first 50 years* (ed. L.C. Dunn), pp. ix–xi. Macmillan, New York.

Dunn, L.C. (1965a). Mendel, his work and his place in history. In *Commemoration of the publication of Gregor Mendel's pioneer experiments in genetics. Proceedings of the American Philosophical Society*, **109**, (4), 189–98.

Dunn, L.C. (1965b). *A short history of genetics.* McGraw-Hill, New York.

Dzierzon, J. (1854). Die Drohnen. *Der Bienenfreund aus Schlesien*, **8**, 63–4.

East, E.M. (1923). Mendel and his contemporaries. *Scientific Monthly*, **16**, 225–37.

Edwards, W.F. (1986). Are Mendel's results too close? *Biological Review*, **1**, 295–312.

Edwardson, J.R. (1962). Another reference to Mendel before 1900. *The Journal of Heredity*, **53**, 152.

Efron, N.J. and Fisch, M. (1991). Science naturalized, science denatured: an evaluation of Ronald Giere's cognitivist approach to explaining science. *History and Philosophy of the Life Sciences*, **13**, (2), 187–221.

Ehrenfels, J.M. (1831). Fortsetzung der Gedanken des Herrn Moritz Beyer über das Merinoschaf. *Mittheilungen, Brünn*, **18**, 137–42.

Ehrenfels, J. M. (1837). Schriftlicher Nachtrag zu den Verhandlungen der Schafzüchter-Versammlung in Brünn, am 10. Mai 1836. *Mittheilungen, Brünn*, **1**, 2–4.

Eichler, K. (1904). *P. Pavel Křížkovský, životopisný nástin.* Brno. In English, *Outline of the biography of Father P. Křížkovský*.

Eichling, G.W. (1942). I talked with Mendel. *The Journal of Heredity*, **33**, 243–5.

d' Elvert, C. (1870). *Geschichte der k.k. mähr. schles. Gesellschaft zur Beförderung des Ackerbaues, der Natur- und Landes- kunde, mit Rücksicht auf die bezüglichen Cultur-Verhältnisse Mährens und Oestrr. Schlesiens.* Rohrer, Brno.

Endlicher, S. and Unger, F. (1843). *Grundzüge der Botanik.* Vienna.

Esser, W. (1823). *System der Logik.* Büchlersche Verlag und Druckerei, Elberfeld.

Ettingshausen, A.V. (1826). *Die combinatorische Analysis als Vorbereitung zum Studium der theoretischen höheren Mathematik.* Wallishauser, Vienna.

Ettingshausen, A.V. (1827). *Vorlesungen über die höhere Mathematik.* Gerold, Vienna.

Ettingshausen, A.V. (1843). *Anfangsgründe der Physik.* Gerold, Vienna.

Falk, R. and Sarkar, S. (1991). The real objective of Mendel's paper. A response to Monaghan and Corcos. *Biology and Philosophy*, **6**, 447–51.

Ferdinand, O. (1966). Mendel's effort to find some mathematical laws in the derivation of names. *Folia Mendeliana*, **1**, 31–4.

Festetics, E. (1819). Weitere Erklärung des Herrn Grafen Emerich Festetics über Inzucht. *Oekonomische Neuigkeiten und Verhandlungen*, Prague, **22**, 85–6.

Festetics, E. (1820). Bericht des Herrn Grafen Emerich Festetics als Repräsentaten des Schafzüchter-Vereins in Eisenburger Comitate. *Oekonomische Neuigkeiten und Verhandlungen*, Prague, **19**, 25–8.

Fincham, J.R.S. (1990). Mendel—now down to the molecular level. *Nature*, **343**, 208–9.

Fisher, R.A. (1930). *The genetical theory of natural selection.* Clarendon Press, Oxford.

Fisher, R.A. (1936). Has Mendel's work been rediscovered? *Annals of Science*, **1**, 115–37. The paper was republished by Bennett (1965), pp. 59–85 and by Lerner and Sherwood (1966), pp. 132–72.

Fisher, R.A. (1955*a*). Introductory notes on Mendel's paper. In *Experiments in plant hybridisation—Gregor Mendel* (ed. J.H. Bennett), pp. 1–6. Oliver and Boyd, Edinburgh.

Fisher, R.A. (1955*b*). Marginal comments on Mendel's paper. In *Experiments in plant hybridisation—Gregor Mendel* (ed. J.H. Bennett), pp. 52–8. Oliver and Boyd, Edinburgh.

Focke, W.O. (1881). *Die Pflanzen-Mischlinge. Ein Beitrag zur Biologie der Gewächse.* Gebrüder Bornträger, Berlin.

Fraas, K.N. (1852). *Geschichte der Landwirtschaft, oder: Geschichtliche Übersicht der Fortschritte landwirtschftlicher Kenntnisse in den letzten 100 Jahren,* Vols. I and II. Calvé, Prague.

Franke, H. and Orel, V. (1983). Christian Carl André (1763–1831) as a mineralogist and an organizer of scientific sheep breeding in Moravia. In *Gregor Mendel and the foundation of genetics* (ed. V. Orel and A. Matalová), pp. 47–56. Moravian Museum, Brno.

Freudenberger, H. (1977). *The industrialization of a Central European city: Brno and the fine woollen industry in the 18th Century.* Edington, Wiltshire.

Frey, M. (1839). Abhandlungen Über die Kultur des Johannisbeer-strauches und dessen Nutzen. *Mittheilungen, Brünn,* **11**, 81–5; **12**, 89–95; **13**, 97–101.

Frey, T. (1869). Nekrolog auf Zawadski. *Verhandlungen des Naturforschenden Vereines, Brünn,* **7**, 22–5.

Fries, E. (1862). *Epicrisis generis* Hieracium. Epicrisis, Uppsala.

Fuss, J. (1795). *Anweisung zur Erlernung der Landwirtschaft in Königsreich Böhmen.* Prague.

Fux, J. (1839). *Vorlesungen über reine Mathematik.* Olomouc.

Gaissinovitch, A.F. (1936). *Izbrannye raboty o rastitelnych gibridach.* Izd. Med. literatury, Moscow.

Gaissinovitch, A.F. (1965*a*). *Gregor Mendel. Opyty nad rastitelnymi gibridami.* Nauka, Moscow.

Gaissinovitch, A.E. (1965*b*). Gregor Mendel—commentarii. In *Gregor Mendel. Opyty nad rastitenimi gibridami* (ed. A.E. Gaissinovitch), pp. 133–52. Nauka, Moscow.

Gaissinovitch, A.E. (1966). An early account of G. Mendel's work in Russia. (I.F. Shmalhausen, 1874). In *G. Mendel memorial symposium 1865–1965* (ed. M. Sosna), pp. 39–40. Academia, Prague.

Gaissinovitch, A.E. (1967). *Gregor Mendel, Charles Naudin, Augustin Sageret.* Izbrannye raboty medicina, Moscow.

Gaissinovitch, A.E. (1988). *Zarozdenie genetiky.* Nauka, Moscow.

Gaissinovitch, A.E. (1990). C.F. Wolf on variability and heredity. *History and Philosophy of the Life Sciences,* **12**, 179–201.

Gardner, M. (1977). Great fakes of science. *Esquire,* October, 88–92.

Gärtner, F. (1849). *Versuche und Beobachtungen über die Bastarderzeugung im Pflanzenreiche.* K.F. Herring, Stuttgart.

Gasking, E.B. (1959). Why was Mendel's work ignored? *Journal of history of ideas,* **20**, 60–84.

George, W. (1967). Mendel and the classification of mammals. *Folia Mendeliana,* **2**, 23–8.

George, W. (1975). *Mendel and heredity*. Priori Press, London.

George, W. (1983). The making of Mendel. In *Gregor Mendel and the foundation of genetics* (ed. V. Orel and A. Matalová), pp. 279–87. Moravian Museum, Brno.

Geschwind, R. (1864). Die Hybridation der Forstgehölze. *Oesterreichische Monatsschrift für Forstwesen*, **14**, 399–417.

Geschwind, R. (1885). *Die Hybridation und Sämmlingszucht der Rosen, ihre Botanik, Classification und Cultur nach den Anforderunge der Neuzeit*. H. Voigt, Leipzig.

Gickelhorn, R. (1969). Gregor Mendels Lehramtsprüfung und Studienzeit in Wien. *Biologische Rundschau*, **7**, 145–59.

Gickelborn, R. (1973). Wurde Gregor Mendel bei der Lehramtsprüfung an der Wiener Universität ungerecht behandelt? *Bilogogische Rundschau*, **11**, 73–84.

Giere, R. (1988). *Explaining science: a cognitive approach*, Chicago University Press.

Gistel, J. (1848). *Naturgeschichte des Tierreichs für höhere Schulen*. Stuttgart.

Godron, D.A. (1863). Des hybrides végétaux, considérés au point de vue de leur fécondité et de la perpétuité ou non-perpétuité de leur caractères. *Annales sciences naturales, 4ᵉᵐᵉ série, Botanique*, **19**, 135–79.

Goldschmidt, R. (1949). Research and politics. *Science*. Quoted by Koestler (1971).

Goldschmidt, R.B. (1951). The impact of genetics upon science. In *Genetics in the 20th century—essays on the progress of genetics in its first 50 years* (ed. L.C. Dunn), pp. 1–23. Macmillan, New York.

Goldschmidt, R. and Minami, S. (1923). Über Vererbung der sekundären Geschlechtscharaktere. In *Studia Mendeliana ad centesimum diem natalem Gregorii Mendelii a grata patria celebrandum* (ed. H. Iltis), pp. 65–77. Apud Typos, Brno.

Goss, J. (1824). On the variation in the colour of peas, occasioned by cross-impregnation. *Transactions of the Horticultural Society*, London, **5**, 234–6.

Grmek, M.D. (1981). A plea for freeing the history of scientific discovery from myth. In *On Scientific discovery* (ed. M.D. Grmek, R.S. Cohen, and G. Cimino), pp. 9–42. Reidel, Dordrecht.

Guédés, M. (1981). Qu'est-ce qu'un mendéelien? *Histoire et Nature*, **17–18**, 69–72.

Gustafsson, A. (1966). Gregor Mendel and 100 years of genetic research. *Acta Agriculturae Scandinavica*, Suppl. **16**, 27–32.

Habro, J. (1971). *Attempts at mating honeybees in confinement*, Bibliography 12. Bee Research Association, London.

Häckel, E. (1876). *Die Perigenesis der Platidüle oder die Wellenzeugung der Lebenstheilchen*, Reimer, Berlin.

Hackler, P.A. (1851). Bericht der Landwirtschaflichen Section an die Gesellschaft, betreffend die Preiswerbungsschrift: Geschichtliche Übersicht der Fortschritte landwirtschaftlicher Kenntnisse in den letzten Hundert Jahren. *Mittheilungen, Brünn*, **1**, 81–93.

Haecker, V. (1923). Einige Aufgaben der Phänogenetik. In *Studia Mendeliana cd centesimum diem natalem Gregorii Mendelii a grata patria celebrandum* (ed. H. Iltis), pp. 78–91. Apud Typos, Brno.

Hagedorn, A.C. and Hagedorn, A.L. (1923). Twenty years of genetics. In *Studia mendeliana ad centesimum diem natalem Gregorii Mendelii a grata patria celebrandum* (ed. H. Iltis), pp. 93–103. Apud Typos, Brno.

Haldane, J.B.S. (1928). *Possible worlds*. Chatto and Windus, London.

Harkenfeld, von Sedláček (1826). Zustand des mährischen Weinbaues und Vorschläge ihn durch Einführung der Rebenschulen von den edelsten Sorten zu Vervollkommen. *Mittheilungen, Brünn*, **21**, 161–7.

Hartl, D.L. and Orel, V. (1992). What did Gregor Mendel think he discovered? *Genetics*, **131**, 245–53.

Haškovec, L. (1925). Opening speech of the chairman. In *Memorial-volume in honor of the 100th birthday of J.G. Mendel (ed. A. Brožek, L. Haškovec, and V. Růžička), pp. 1–5. Borový, Prague.*

Haynes, R. (1989). Genetics and the unity of biology. *Genome*, **31**, (1), 1–7.

Hedwig, R. (1797). *Sammlungen seiner Abhandlungen und Beobachtungen über botanisch–ekonomische Gegenstände, Vol. II. Leipzig.*

Heimans, J. (1962). Hugo de Vries and the gene concept. *American Naturalist*, **96**, 93–104.

Heimans, J. (1969). Ein Notitzblatt aus dem Nachlass Gregor Mendels mit Analysen eines seiner Kreuzungsversuche. *Folia Mendeliana*, **4**, 5–36.

Heimans, J. (1970). A recently discovered note on hybridization in Mendel's handwriting. *Folia Mendeliana*, **5**, 13–25.

Heimans, J. (1971). Mendel's ideas on the nature of hereditary characters. *Folia Mendeliana*, **6**, 91–103.

Heinen, W. (1943). *Der junge genius J.G. Mendel.* W. Spael, Essen.

Hempel, G.C.L. (1818). Horticultural Societät in London. *Oekonomische Neuigkeiten und Verhandlungen*, Prague, **51**, 408.

Hempel, G.C.L. (1820). Ueber die Entstehung und Wichtigkeit der verschiedenen Sorten der Getreidearten. *Oekonomische Neuigkeiten und Verhandlungen*, Prague, **21**, 161–5.

Henninger, K.A. (1879). Ueber Bastarderzeugung im Pflanzenreich. *Flora*, **62**, 225–33.

Herčik, F. and Novák, L. (1952). Chybné základy mendelismu. *Čs. Biologie*, **1**, 254–62. In English, *Erroneous basis of Mendelism.*

His, W. (1874). *Unsere Körperform und das physiologische Problem ihrer Entstehung. Briefe an einen befreundeten Naturforscher*, Vogel, Leipzig.

Hoffmann, H. (1869). *Untersuchungen zur Bestimmung des Werthes von Species und Varietät: ein Beitrag zur Kritik der Darwinschen Hypothese.* J. Richter, Giessen.

Hoffmeister, D.F. and Henriksen, H. (1979). The collection of Mendeliana at the University of Illinois, Urbana-Champaign. *Folia Mendeliana*, **14**, 281–4.

Hofmeister, W. (1849). Ueber die die Fruchtbildung und Keimung der höheren Cryptogamen. *Botanische Zeitschrift*, **7**, 793–800.

Hofmeister, W. (1851). *Vergleichende Untersuchungen der Keimung, Entstehung und Fruchtbildung höherer Kryptogamen und der Samenbildung der Coniferen.* Leipzig.

Hončariv, R. (1971). The logical structure of Mendel's idea. *Folia Mendeliana*, **6**, 65–7.

Hoppe, B. (1971). Die Beziehung zwischen J.G. Mendel und C.W. Nägeli auf Grund neuer Dokumente. *Folia Mendeliana*, **6**, 123–38.

Hornschuh, Ch.F. (1848). Ueber Ausartung der Pflanzen. *Flora*, **2**, 17–28; **3**, 33–44; **4**, 50–64; **5**, 66–86.

Hrabětová-Uhrová, A. (1972). Herbarium relics after J.G. Mendel in the Old Brno monastery. *Folia Mendeliana*, **7**, 21–5.

Hrabětová-Uhrová, A. (1974). Joseph Veselý (Wessely), 1813–1892. *Folia Mendeliana*, **9**, 275–7.

Huber, K.A. (1978). Die Apostolische Visitation des St. Thomaskloster in Alt-Brünn. *Archiv für Kirchengeschichte von Böhmen–Mähren–Schlesien*, **5**, 190–236.

Hurst, C.C. (1911). Mendelian characters in plants, animals and man. *Verhandlungen des naturforschenden Vereines, Brünn*, **49**, 192–213.

Hutt, F.B. (1962). Restoration of the Mendel museum. *The Journal of Heredity*, **53**, 27–30.

Huxley, J. (1951). Genetics, evolution and human destiny. In *Genetics in the 20th century—essays on the progress of genetics during its first 50 years* (ed. L.C. Dunn), pp. 591–621. Macmillan, New York.

Iltis, A. (1954). Gregor Mendel 's autobiography. *The Journal of Heredity*. **45**, 231–4.

Iltis, H. (1911). Von Mendel Denkmal und seiner Enthüllung. *Verhandlungen des Naturforschenden Vereines, Brünn*, **49**, 335–63.

Iltis, H. (1912). Die Gescichte des Naturforschenden Vereines in Brünn in den Jahren 1862–1912. *Verhandlungen des Naturforschenden Vereines, Abhandlungen, Brünn*, **50**, 295–359.

Iltis, H. (ed.) (1923a). *Studia Mendeliana, ad centesimum diem natalem Gregorii Mendelii a grata patria celebrandum*. Apud Typos, Brno.

Iltis, H. (1923b). Die Mendel, Jahrhundertfeier in Brünn. In *Studia Mendeliana, ad centesimim diem natalem Gregorii Mendelii a grata patria celebrandum* (ed. H. Iltis), pp. 389–414. Apud Typos, Brno.

Iltis, H. (1924). *Gregor Johann Mendel. Leben, Werk und Wirkung*. Springer, Berlin. English translation, *Life of Mendel*, Norton, New York, 1932. 2nd edn: Hafner, New York, 1966.

Iltis, H. (1951). Gregor Mendel's life and heritage. In *Genetics in the 20th century—essays on the progress of genetics during its first 50 years* (ed. L.C. Dunn), pp. 25–34. Macmillan, New York.

Iltis, H. (1966). *See* Iltis (1924): used for citations of the English translation (2nd edn).

Jackson, B.D. (1881). *Guide to the literature of botany*. London. Reprinted by Hafner, New York, 1964.

Jahn, F., Lucas, E., and Oberdieck, G.G. 1850–60. *Illustriertes Handbuch der Obstkunde*, Vol. I. *Äpfel*. Ulmer, Ravensburg. Vol. II. *Birnen*. Ebner, Stuttgart, 1860.

Janko, J. and Orel, V. (1989–90). The cell in Purkyně's concept of procreation. *Folia Mendeliana*, **24–25**, 49–57.

Jindra, J. (1971a). A contribution to the logical nature of Mendel's Ideas. *Folia Mendeliana*, **6**, 69–70.

Jindra, J. (1971b). A possible derivation of the Mendelian series. *Folia Mendeliana*, **6**, 71–4.

Johannsen, W. (1913). *Elemente der exakten Erblichkeitslehre mit Grundsätzen der biologischen Variationsstatistik*. G. Fischer, Jena.

Joravsky, D. (1970). *The Lysenko affair*. Harvard University Press, Cambridge, MA.

Kabelík, J. (1910). *Korrespondence a zápisky Jana Helceleta*. Matice Moravská, Brno. In English, *Correspondence and notes of Jan Kabelík*.

Kalmus, H. (1978). A possible early attempt of Mendel's to find a mathematical theory for his hybridization experiments. *Folia Mendeliana*, **13**, 227–31.

Kalmus, H. (1983). The scholastic origin of Mendel's concept. *History of Science*, **21**, 61–83.

Kammerer, P. (1911*a*). Mendelsche Regeln und Vererbung erworbener Eigenschaften. *Verhandlungen des Naturforschen Vereines, Brünn*, **49**, 72–110.

Kammerer, P. (1911*b*). *Beweise für Vererbung erworbener Eigenschaften durch planmässige Züchtung*, Flugblatt der Deutschen Gesellschaft für Züchtungskunde, 12. Berlin.

Kammerer, P. (1914). *Die Bedeutung der Vererbung erworbener Eigenschaften für Erziehung und Unterricht*, Flugschrift der Sozialpedagogischen Gesellschaft, 4. Vienna.

Kammerer, P. (1924). *The inheritance of acquired chracteristics*. Boni and Liveright, New York.

Keller, A. (1828). Cantalupa Melonen. *Mittheilungen, Brünn*, **16**, 127.

Kerner von Marilaun, A. (1860). Niederösterreichische Weiden. *Verhandlungen des Zoologisch–botanischen Vereines in Wien. Abhandlungen*, **10**, 3–56 and 79–282.

Klácel, F.M. (1841). Rozwinutj wedectwa. *Časopis českéno musea, Praha*, **15**, 127–41. In English, *Development of scientific knowledge.*

Klácel, M.F. (1842). Náwrh. *Časopis českého musea, Praha*, **16**, 3–17. In Engliŝh, *Proposal.*

Klácel, F.M. (1843). O smrti. *Časopis českého musea, Praha*, **17**, 329–47. In English, *On death.*

Klácel, F.M. (1847). *Dobrovĕda*, Kronberg, Prague. In English, *Ethics.*

Kner, R. (1849). *Lehrbuch der Zoologie zum Gebrauch für höhere Lehramten*. Vienna.

Knievel, F. (1994). Mendelblüten–Blütenlese in Natur und Kunstblumen. *Biologie in unserer Zeit*, **24**, 43–7.

Knight, T.A. (1797). Letter to Sir Joseph Banks. Manuscript in British Museum, quoted by Mylechreest (1988).

Knight, T.A. (1799). An account of some experiments of the fecundation of vegetables. *Philosophical Transactions of the Royal Society of London*, **89**, 195–204.

Knight, T.A. (1800). Versuche über die Befruchtung der Pflanzen. *Oekonomische Hefte*, Leipzig, **5**, 322–40.

Knight, T.A. (1809). XXIII. On the comparative influence of male and female parents on their offspring. *Philosophical Transactions of the Royal Society of London*, **11**, 392–9.

Knight, T.A. (1823). Some remarks on the supposed influence of the pollen, in cross-breeding, upon the colour of the seed-coats of plants, and the qualities of their fruits. *Transactions of the Horticultural Society*, London, **4**, 278–80.

Knoll, F. (1967). Über einen noch unbekannten Sonderdruck der Hybriden-Arbeit Gregor Mendels. *Österreichische Akademie der Wissenschaften, Anzeiger der math. naturw. Klasse*, **9**, 226–9.

Koestler, A. (1971). *The case of the midwife toad*. Hutchinson, London.

Kölreuter, D.J.G. (1761–6). *Vorläufige Nachricht von einigen das Geschlecht der Pflanzen betreffenden Versuchen und Beobachtungen, nebst Fortsetzung 1, 2, 3*. Gledisch, Leipzig.

Kottler, M.J. (1979). Hugo de Vries and the rediscovery of Mendel's laws. *Annals of Science*, **36**, 517–38.

Krausse, E. (1983). Elucidation of the fertlization process in algae by N. Pringsheim (1823–1894). In *Gregor Mendel and the foundation of genetics* (ed. V. Orel and A. Matalová), pp. 259–67. Moravian Museum, Brno.

Křiwánek, L. and Suchánek, T. (eds) (1898). *Geschichte des mähr. Obst- und Wein- und Gartenbau Vereines in Brünn*. Verlag des Vereines, Brno.

Kříženecký, J. (1925). O dědičnosti získaných znaků a významu mendelismu pro teorii evoluční. In *Memorial-volume in honor of the 100th birthday of J.G. Mendel* (ed. A. Brožek, L. Haškovec, and V. Růžička), pp. 63–86. Borový, Prague.

Kříženecký, J. (1957). Přehled nauky o plemenitbě. *Polnopodárstvo*, Bratislava, **4**, 805–26. In English, *The survey of knowledge of animal breeding science*.

Kříženecký, J. (1963a). Genetische Abteilung Gregor Mendel im Mährischen Museum in Brünn. *Naturwissenschaftliche Rundschau*, **16**, 477–80.

Kříženecký, J. (1963b). Mendels zweite erfolgslose Lehramtsprüfung im Jahre 1856. *Sudhoffs Archiv für die Geschichte der Medizin und Naturwissenschaft*, **47**, 305–10.

Kříženecký, J. (1965a). *Fundamenta genetica. The revised edition of Mendel's classic paper with a collection of 27 original papers published during the rediscovery era*. Academia, Prague.

Kříženecký, J. (1965b). *Gregor Johann Mendel 1822–1884. Texte und Quellen zu seinem Wirken und Leben*. Leopoldina Akademie, Halle.

Kříženecký, J. (1974). The past of the G. Mendel memorial in Brno. *Folia Mendeliana*, **9**, 231–43.

Kříženecký, J. (1987). Wie J.E. Purkyně im mährischen Karst philosophierte— Purkyně und Gregor Mendel. *Folia Mendeliana*, **22**, 69–80.

Krumbiegel, I. (1957). *Gregor Mendel und das Schicksal seiner Vererbungsgesetze*. Wissenschaftliche Verlagsgesellschaft, Stuttgart. 2nd edn, 1967.

Kuhn, T.S. (1962). *The structure of scientific revolutions*. University of Chicago Press.

Kuhn, T.S. (1983). A function for thought experiments. In *Scientific revolutions* (ed. I. Hacking), pp. 6–27. Oxford University Press.

Kühne, F. (1881). Miscellen. *Ungarische Biene*, Temesvar, 1881, 2–8, 16–22, 25–6.

Kunzek, A. (1850). *Lehrbuch der Meteorologie*. W. Braunmüller, Vienna.

Lamarck, J.B. (1809). *Philosophie zoologique*. Paris.

Lamprecht, H. (1966). Die erste, bereits vom Mendel ausgeführte genanalytische Untersuchung der Testfarbe von *Phaseolus vulgaris*. *Phyton*, **11**.

Lamprecht, H. (1968a). Die neue Genenkarte von *Pisum* und warum Mendel in seinen Erbsenkreuzungen keine Genenkoppelung gefunden hat. *Arbeiten aus der Steiermarkischen Landesbibliothek am Joanneum in Graz*, **10**, 1–28.

Lamprecht, H. (1968b). *Die Grundlagen der Mendelschen Gesetze*. Parey, Berlin

Lauprecht, E. (1966). Zur Begegnung von Mendel mit dem bekannten Bienenzüchter Dathe. *Folia Mendeliana*, **1**, 19–22.

Lauprecht, E. (1982). Gregor-Mendel-Gedenktafel in Eystrup (Weser). *Folia Mendeliana*, **17**, 247–9.

Lebel, R.R. (1987). Fisher-Mendel data. *The Journal of Heredity*, **78**, 344.

Lerner, I.M. (1966). Genetics and animal breeding. In *G. Mendel memorial symposium 1865–1965* (ed. M. Sosna), pp. 189–97. Academia, Prague.

Leuckart, R. (1853). Zeugung. In *Handwörterbuch der Physiologie mit Rücksicht auf physiologische Pathologie*, Vol. 4 (ed. R. Wagner), pp. 707–1000. Brunswick.

Lints, F.A. and Delcour, J. (1968). Galton and the Mendelian ratios. *Heredity*, **23**, 153–5.

Littrow, J.J. (1825). *Populäre Astronomie*, J.G. Heubner, Vienna.

Littrow, J.J. (1835). *Die Wahrscheinlichkeitsrechnung in ihrer Anwendung auf das praktische Leben*. Becks Universitätbuchhandlung, Vienna.

Liznar, J. (1882). Die periodische Aenderung des Grundwasser-standes. *Meteorologische Zeitschrift*, Wien, **17**, 368–71.

Liznar, J. (1886). Über das Klima von Brünn. *Verhandlungen des Naturforschenden Vereins, Abhandlungen, Brünn*, **24**, 3–70.

Liznar, J. (1902). Über die Änderungen des Grundwasserstandes nach den vom Prälaten Gregor Mendel in den Jahren 1865–1880 in Brünn ausgeführten Messungen. In *Festschrift zur Erinnerung an die Feier des fünfzigsjährigen Bestandes der deutschen Staats-Oberreal-Schule in Brünn*, pp. 225–33. Winiker, Brno.

Loužil, J. (1972). Franz Thomas Bratranek Leben und Philosophie. *Jahrbuch des Collegium Carolinum München*, **13**, 182–210.

MacRoberts, M.H. (1984). L.B. Bailey's citations to Gregor Mendel. *The Journal of Heredity*, **75**, 500–1.

MacRoberts, M.H. (1985). Was Mendel's paper neglected or unknown? *Annals of Science*, **42**, 339–45.

Makowsky, A. (1863). Die Flora des Brünner Kreises. *Verhandlungen des Naturforschenden Vereines, Abhandlungen, Brünn*, **1**, 43–210. Also published separately under the enlarged title, with a reprint of the paper by Mendel (1863) added, *Die Flora des Brünner Kreises, Nach Pflanzengeographischen Principien. Mit einer meteorologischen Tabelle von Prof. G. Mendel*.

Makowsky, A. (1864*a*). Einige neue und interessante botanische Funde. *Verhandlungen des Naturforschenden Vereines, Brünn*, **2**, 37–8.

Makowsky, A. (1864*b*). Ueber Bienenarten etc. *Verhandlungen des Naturforschenden Vereines, Brünn*, **3**, 60.

Makowsky, A. (1866). Ueber Darwins Theorie der organischen Schöpfung. *Verhandlungen des Naturforschenden Vereins, Brünn*, **4**, 10–18.

Makowsky, A. (1868). *Geum urbano-rivale*. *Verhandlungen des Naturforschenden Vereines, Brünn*, **6**, 35.

Makowsky, A. (1869). Frisches Exemplar einer *Ficaria* welche vor 2 Jahren Herr Prälat Mendel bei Karthaus nächst Brünn gesammelt hat. *Verhandlungen des Naturforschenden Vereines, Brünn*, **7**, 28.

Marvanová, L. (1966). Mendels dischterische Versuche aus seinen Studienjahren. *Folia Mendeliana*, **1**, 15–18.

Marvanová, L. (1968). Le centenaire de l'élection abbatiale de Mendel. *Folia Mendeliana*, **3**, 13–20.

Marvanová, L. (1971). First impulse to Mendel's scientific education. *Folia Mendeliana*, **6**, 31–40.

Marvanová, L., Orel, V., and Sajner, J. (1965). *Iconographia Mendeliana*. Moravské Museum, Brno.

Matalová, A. (1973). A critical review of different editions of Mendel's *Pisum* paper. *Folia Mendeliana*, **8**, 243–54.

Matalová, A. (1974). Bibliographical note on Mendel's *Hieracium* paper. *Folia Mendeliana*, **9**, 225–30.

Matalová, A. (1979). A monument to F.M. Klácel (1809–1882) in the vicinity of the Mendel statue in Brno. *Folia Mendeliana*, **14**, 251–63.

Matalová, A. (1981). Published primary sources to Gregor Mendel's biography. *Folia Mendeliana*, **16**, 239–51.

Matalová, A. (1983). Mendel's confession in the ceiling paintings in the former Augustinian prelacy. *Folia Mendeliana*, **18**, 273–7.

Matalová, A. (1984). Response to Mendel's death in 1884. *Folia Mendeliana*, **19**, 217–21.

Matalová, A. (1985). Mendel's experimental plants decorate the Augustinian library ceiling. *Folia Mendeliana*, **20**, 5–8.

Matalová, A. (1990). Gregor Mendel als Wissenschaftler im Konflikte mit dem Staat. *Acta historiae rerum naturalium necnon technicarum*, Prague, **21**, 123–32.

Matalová, A. (1991). The laws on the origin of crystal forms and Mendel's theory on the origin and development of plant forms. *Acta historiae rerum naturalium necnon technicarum*, Prague, **23**, 79–92.

Matalová, A. and Kabelka, A. (1982). The beehouse of Gregor Mendel. *Folia Mendeliana*, **17**, 207–12.

Matoušková, B. and Matoušek, O. (1959). Darvinism v Tsechoslovakii. *Annaly Biologii, Moskva*, **1**, 31–52.

Mausbach, J. (1930). Wesen und Stufung des Lebens nach dem hl. Augustinus. In *Aurelius Augustinus. Die Festschrift des Gorres-Gesellschaft zum 1500. Todestages des heiligen Augustins* (ed. M. Grabmann and J. Mausbach), pp. 169–96. Cologne.

Mayr, E. (1982). *The growth of biological thought. Diversity, evolution and inheritance*. The Bellknap Press of Harvard University Press, Cambridge, MA.

Mayr, E. (1986). Joseph Gottlieb Kölreuter's contribution to Biology. *Osiris*, **2**, 135–76.

Mayr, E. and Provine, W.B. (1980). *The evolutionary synthesis: perspectives on the unification of biology*. Harvard University Press, Cambridge, MA.

Medvedev, Z.A. (1969). *The rise and fall of T.D. Lysenko*. Columbia University Press, New York.

Meijer, A.K. (1984). Gregor Johann Mendel (1822–1844). Ergänzende Dokumente zu seiner Abtswahl und seinem Tod. *Augustiniana*, Institutum Historicum Augustinianum Lovanii, **34**, 213–35.

Meijer, O. (1983). The essence of Mendel 's discovery. In *Gregor Mendel and the foundation of genetics* (ed. V. Orel and A. Matalová), pp. 123–78. Moravian Museum, Brno.

Meijer, O.G. (1985). Hugo de Vries no Mendelian? *Annals of Science*, **42**, 189–232.

Meijer, O.G. (1986). Hugo de Vries und Johann Gregor Mendel: die Geschichte einer Vereinigung. *Folia Mendeliana*, **21**, 69–93.

Meijknecht, J.C. (1950). *Gregor Mendel de omtdekker der erfelukheidswetten*. Paul Brand, Bussum.

Mendel, G. (1850). Autobiography. Manuscript in the Mendelianum (sig, no. 24). Kříženecký (1965*b*, pp. 74–7, explains its publication in German and later in English. This book uses the English translation by A. Iltis (1954), reprinted by Olby (1985), pp. 175–8. The facsimile was published by Sajner (1965).

Mendel, G. (1853). Ueber Verwüstung am Gartenrettich durch Raupen (*Botys margaritalis*). *Verhandlungen des Zool.–botanischen Vereines, Wien*, **3**, 116–18.

Mendel, G. (1854). Ueber *Bruchus pisi*, mitgeteilt von V. Kollar. *Verhandlungen des Zool.–botanischen Vereines, Wien*, **4**, 27–8.

Mendel, G. (1863). Bemerkungen zu der graphisch–tabelarischen Uebersicht der meteorologischen Verhältnisse von Brünn. *Verhandlungen des Naturforschenden Vereines, Abhandlungen, Brünn*, **1**, 246–9.

Mendel, G. (1864). Meteorologische Beobachtungen aus Mähren und Schlesien für das Jahr 1863. *Verhandlungen des Naturforschenden Vereines, Abhandlungen, Brünn*, **2**, 99–121.

Mendel, G. (1865). Meteorologische Beobachtungen aus Mähren für das Jahr 1864. *Verhandlungen des Naturforschenden Vereines, Abhandlungen, Brünn*, **3**, 209–20.

Mendel, G. (1866*a*). Versuche über Pflanzen-Hybriden. *Verhandlungen des Naturforschenden Vereines, Abhandlungen, Brünn*, **4**, 3–47. Editions in different languages were published by Matalová (1973). This book quotes the pagination of the original German paper. The pagination of the second English translation, published by Stern and Sherwood (1966), is very close to the pagination of the original German paper.

Mendel, G. (1866*b*). Meteorologische Beobachtungen aus Mähren und Schlesien für das Jahr 1865. *Verhandlungen des Naturforschenden Vereines, Abhandlungen, Brünn*, **4**, 318–30.

Mendel, G. (1867). Meteorologische Beobachtungen aus Mähren und Schlesien für das Jahr 1866. *Verhandlungen des Naturforschenden Vereines, Abhandlungen, Brünn*, **5**, 160–72.

Mendel, G. (1870*a*). Ueber einige aus künstlichen Befruchtung gewonnenen Hieracium-Bastarde. *Verhandlungen des Naturforschen Vereines, Abhandlungen, Brünn*, **8**, 26–31. Editions in different languages were published by Matalová (1974). This book quotes the pagination of the English translation, published by Stern and Sherwood (1966).

Mendel, G. (1870*b*). Meteorologische Beobachtungen aus Mähren und Schlesien für das Jahr 1869. *Verhandlungen des Naturforschenden Vereines, Abhandlungen, Brünn*, **8**, 131–43.

Mendel, G. (1871). Die Windhose vom 13. October 1870. *Verhandlungen des Naturforschenden Vereines, Abhandlungen, Brünn*, **9**, 54–71.

Mendel, G. (1879). Die Grundlage der Wetterprognosen. *Mittheilungen, Brünn*, **59**, 29–31.

Merton, R.K. (1973). *The sociology of science*. University of Chicago Press.

Miller, P. (1751). *Das englische Gartenbuch, oder Gärtner-Lexicon*. Nüremberg. Translation of the English version *The gardener's and the florist's dictionary or complete system of Horticulture*, published in eight editions, London, 1731–68.

Milovidov, P. (1968). Gregor Mendel's microscopic preparations. *Folia Mendeliana*, **3**, 35–53.

Molisch, H. (1934). *Erinnerungen und Welteindrücke eines Naturforschers*. Heim, Vienna.

Monaghan, F. and Corcos, A. (1986). Tschermak: a non-discoverer of Mendelism. An historical note. *The Journal of Heredity*, **77**, 468–9.

Monaghan, F.V. and Corcos, A.F. (1987). Reexamination of the fate of Mendel's paper. *The Journal of Heredity*, **78**, 116–18.

Monaghan, F.V. and Corcos, A.F. (1990). The real objective of Mendel's paper. *Biology and Philosophy*, **5**, 267–92.

Monaghan, F.V. and Corcos, A.F. (1993). The real objective of Mendel's paper: A response to Falk and Sarkar's criticism. *Biology and Philosophy*, **8**, 95–8.

de Morgan, A. (1847). *Formal logic or inference necessary and probable*. London.

Morgan, T.H. (1903). *Evolution and adaptation*. Macmillan, New York.

Morgan, T.H. (1932). The rise of genetics. *Science*, **76**, 261–7.

Mueller, H.J. (1951). The development of the gene theory. In *Genetics in the 20th century—essays on the progress of genetics during its first 50 years* (ed. L.C Dunn). pp. 77–99. Macmillan, New York.

Munzar, J. (1971). G. Mendel's erste, bis jetzt unveröffentlichte Abhandlung über Meteorologie. *Folia Mendeliana*, **6**, 185–91.

Mylechreest, M. (1988). Thomas Andrew Knight (1759–1838) and the Altenburg connection in the origin of mendelism. *Folia Mendeliana*, **23**, 27–32.

Nachstein, H. (1942). Gregor Johann Mendel. Zur 120. Wiederkehr seines Geburtstages am 22. Juli 1942. *Der Erbarzt*, **10**, 147–54.

Nachtweh, H. (1922). Zu Mendels 100jährigen Geburtsfeste, *Blätter für Obst-, Wein-, und Gartenbau und Kleintierzucht, Brünn*, **55**, 113–17.

Nägeli, C. (1841). Die Cirsien der Schweiz. *Neue Denkschrifte allg. schweiz. Gesellschaft der Naturwissenschaften, Bern*, **5**, 1–170.

Nägeli, C. (1845). Über einige Arten der Gattung *Hieracium. Zeitschrift der wissenscjaftlichen Botanik*, **1**, 103–20.

Nägeli, C. (1853). *Systematische Uebersicht der Erscheinungen im Pflanzenreich*, pp. 1–68. Freiburg im Breisgau. A copy of this lecture was sent to Mendel by Nägeli.

Nägeli, C. (1865a). Die Bastardbildung im Pflanzenreiche. *Sitzungsberichte der königl. bayer. Akademie der Wissenschaften. math. phys. Klasse. München*, **2**, 395–443.

Nägeli, C. (1865b). Entstehung und Begriff der naturhistorischen Art. *Sitzungsberichte der Königl, bayer. Akademie der Wissenschaften, math. phys. Klasse, München*, **1**, 1–53.

Nägeli, C. (1865c). Ueber die Bedingungen des Vorkommens von Arten und Varietäten innerhalb ihres Verbreitungsbezirkes. *Sitzungsberichte der königl. bayer. Akademie der Wissenschaften, math. phys. Klasse, München*, **1**, 367–95.

Nägeli, C. (1865d). Ueber den Einfluss äusserer Verhältnisse auf die Varietätenbildung im Pflanzenreiche. *Sitzungsberichte der Königl. bayer. Akademie der Wissenschaften. math. phys. Klasse, München*, **II**, 228–84.

Nägeli, C. (1866a). Die Theorie der Bastardbildung. *Sitzungsberichte der königl. bayer. Akademie der Wissenschaften, math. phys. Klasse, München*, **I**, 93–127.

Nageli, C. (1866b). Ueber die abgeleiteten Pflanzenbastarde. *Sitzungsberichte der königl. bayer. Akademie der Wissenschaften. math. phys. Klasse, München*, **I**, 71–93.

Nägeli, C. (1866c). Ueber die Zwischenformen zwischen den Pflanzenarten. *Sitzungsberichte der königl. bayer. Akademie der Wissenschaften, math. phys. Klasse. München*, **I**, 190–221.

Nägeli, C. (1866*d*). Aufklärung einiger Zwischenformen. *Sitzungsberichte der königl. bayer. Akademie der Wissenschaften, math. phys. Klasse, München,* **I**, 222–35.

Nägeli, C. (1866*e*). Ueber die systematische Behandlung der Hieracien rücksichtlich der Mittelformen. *Sitzungsberichte der königl. bayer. Akademie der Wissenschaften, math. phys. Klasse, München,* **I**, 324–52.

Nägeli, C. (1866*f*). Ueber die systematische Behandlung der Hieracien rücksichtlich des Umfanges der Species. *Sitzungsberichte der königl. bayer, Akademie der Wissenschaften, math. phys. Klasse, München,* **I**, 437–73.

Nägeli, C. (1884). *Mechanisch–physiologische Theorie der Abstammungslehre.* R. Oldenbourg, Munich. English translation, *A mechanico-physiological theory of organic evolution.* Open Court, Chicago, 1914.

Nägeli, C. (1891). Einleitung. In *Die europäischen Arten der Gattung* Primula (ed. E. Widmer), pp. 1–8.

Nägeli, C. and Peter, A. (1885). *Die Hieracien Mittel-Europas. Monographische Bearbeitung der Piloselloiden mit besonderer Berücksichtung der mitteleuropäischen Sippen.* Munich.

Nakazawa, S. (1976). Original reprint of Mendel's paper in Japan. *Folia Mendeliana,,* **11**, 47–52.

Nakazawa, S. (1978). *Mendel's discovery* [in Japanese]. Kyoritsu Co., Tokyo.

Napp, F.C. (1829). Kulturversuch mit dem weissen Melilotenklee, uneigentlich Hanfklee (*Melit. vulg.*). *Mittheilungen, Brünn,* **50**, 398–9.

Napp, F.C. (1831). Pomologische Versuche. *Mittheilungen, Brünn,* **31**, 321.

Napp, F.C. (1832*a*). Kulturresultate des Melilotenklees (*Melilotus vulgaris*). *Mittheilungen, Brünn,* **8**, 60 –1 .

Napp, F.C. (1832*b*). Mohar-Anbauversuch. *Mittheilungen, Brünn,* **23**, 184.

Napp. C. F. and Diebl, F. (1826). Saatverwüstung auf dem Stiftsherrschaft Scharditz in Mähren, durch die Larve der Wintersaateule (*Phalaema noctua segetum*) und den gemeinen Vielfuss (*Justus terrestis*). *Mittheilungen, Brünn,* **52**, 413–16.

Napp, C.F. and Diebl, F. (1828). Nachtrag. *Mittheilungen, Brünn,* **28**, 153–5.

Napp, C.F. and Diebl, F. (1838). Nachricht und Aufforderung zu einer Preisbewerbung vom mähr. schles. pomol. önol. Veriens-Ausschusse. *Mittheilungen, Brünn,* **20**, 157–8.

Nathusius, H. (1872). *Vorträge Viehzucht und Rassenkenntniss.* Wiegand–Hempel, Berlin.

Naudin, C. Nouvelles recherches sur 1 'hybridité dans les végétaux. *Annales sciences naturales. Botanique, 4-me série,* **19**, 180–203.

Nave, J. (1858). Ueber die Entwicklung und Fortpflanzung der Algen. *Jahresbericht der nat. wiss. Sektion der k.k. Mähr. Schles. Gesellschaft zur Beförderung des Ackerbaues, der Natur und Landeskunde, Brünn,* pp. 84–191.

Nave, J. (1864). *Anleitung zum Ansammeln, Präparieren und Untersuchen der Pflanzen mit besonderer Rücksicht auf die Kryptogamen.* H. Burdach, Dresden.

Nave, J. (1867). *A handy-book to the collection and preparation of freshwater and marine algae, diatoms, desmids, fungi, lichens, mosses and other of the lower cryptogamia with instructions for the formation of an herbarium.* R. Hardwicke, London. A second edition was published in the same year. A further three editions, without any changes, appeared in 1881, 1896, and 1904.

Neilreich, A. (1871). Kritische Zusammenstellung der in Oesterreich–Ungarn bisher beobachteten Arten, Formen und Bastarde der Gattung *Hieracium*. *Sitzungsberichte der Königliche und Kaiserliche Akademie der Wissenchaften, math. nat. Classe, Wien*, **63**, 424–500 (quoted by Weiling 1971).

Němec, B. (1925). Official speech. In *Memorial-volume in honor of the 100th birthday of J.G. Mendel* (ed. A. Brožek, L. Haškovec, and V. Růžička, pp. 15–29. Borový, Prague.

Nestler, J.K. (1829). Ueber den Einflusss der Zeugung auf die Eigenschaften der Nachkommen. *Mittheilungen., Brünn.*, **47**, 369–72; **48**, 377–80; **50**, 394–8; **51**, 401–4.

Nestler, J.K. (1837). Ueber Vererbung in der Schafzucht. *Mittheilungen. Brünn*, **34**, 265–9; **35**, 273–9; **36**, 281–6; **37**, 289–300; **38**, 300–3; **40**, 318–20.

Nestler, J.K. (1841). *Amts-Bericht des Vorstandes Über die vierte, zu Brünn von 20. bis 28. September 1840 abgehaltene Versammlung der deutschen Land- und Forstwirthe.* A. Skarnitze, Olomouc.

Nestler, J.K. and Diebl, F. (1836). *Die allgemeine Naturgeschichte.* Brno.

Neumann, A. (1930). *Acta et epistolae eruditor um monasterii ord. S. Augustini Vetero-Brunae*, Vol.1, (A) 1819–1850. Brno.

Niessl, G. (1863a). Über den golden Schnitt. *Verhandlungen des Naturforschernden Vereines, Brünn*, **1**, 14.

Niessl, G. (1863b). Über eine im Juli 1.J. unternomene Excursion nach Letowitz. *Verhandlungen des Naturforschenden Vereines, Brünn*, **1**, 76–7.

Niessl, G. (1865). Floristische Notizen, betreffend phanerogamische Pflanzen, welche aus den Floregebieten, oder doch zu den selteneren in denselben gehören. *Verhandlungen des Naturforschenden Vereines, Brünn*, **3**, 85–8.

Niessl, G. (1866). Berichte aus Excursionen, unternommenen nach Eisgrub und auf Polauerberge. *Verhandlungen des Naturforschenden Vereines, Brünn*, **4**, 80–4.

Niessl, G. (1867a). Exemplar von *Asplenium heufleri* Reicht. *Verhandlungen des Naturforschenden Vereines, Brünn*, **5**, 20–7.

Niessl, G. (1867b). Botanische Mittheilung. *Verhandlungen des Naturforschenden Vereines, Brünn*, **5**, 56–64.

Niessl, G. (1867c). Ueber Bastarde kryptogamischer Pflanzen. *Verhandlungen des Naturforschenden Vereines, Brünn*, **5**, 20–7.

Niessl, G. (1868a). Floristische Mittheilung aus Mähren und eine kurze Schilderung des Eislethen bei Frain. *Verhandlungen des Naturforschenden Vereines, Brünn*, **6**, 62–8.

Niessl, G. (1868b). Ueber *Asplenium adulterium* Milde und sein Vorkommen in Mähren und Böhmen. *Verhandlungen des Naturforschenden Vereines Brünn, Abhandlungen*, **6**, 165–76.

Niessl, G. (1869). Cirsiumbastarde. *Verhandlungen des Naturforschenden Vereines, Brünn*, **7**, 48–53.

Niessl, G. (1871). Obituary notice on Professor F. Unger. *Verhandlungen des Naturforschenden Vereines, Brünn*, **9**, 24–30.

Niesel, G. (1903). Jahresbericht. *Verhandlungen des Naturforschen Vereines, Brünn*, **41**, 18–21.

Niessl, G. (1906). Jahresbericht. *Verhandlungen des Naturforschen Vereines, Brünn*, **44** 5–9.

Niessl, G., Kalmus, H., and Nave J. (1864). Bericht über eine in den letzen Tagen des Juni gemeinschaftlich unternommene Excursion nach Zwittau. *Verhandlungen des Naturforschenden Vereines, Brünn*, **2**, 40–3.

Nilsson, N.H. (1930). *Linné, Darwin, Mendel. Three biografisca skisses*. Stockholm.

Norton, B. and Pearson, E.S. (1976). A note on the background to and refereeing of, R.A. Fisher's paper [On] The correlation between relatives on the supposition of Mendelian inheritance. *Notes and Records of the Royal Society*, **31**, 151–62.

Novitski, E. and Blixt, S. (1979). Mendel, linkage and synteny. *Bioscience*, **28**, 34–5.

Nowakowski, J. and Morse, R.A. (1971). Attempts at mating honeybees in confinement. *Gleanings in Bee Culture*. pp. 216–8, Medina, USA.

Obermajer, J. (1974). Medals and plaques bearing the portrait of Gregor Mendel. *Folia Mendeliana*, **9**, 235–62.

Obermajer, J. (1985). New medals with Mendel's portraits. *Folia Mendeliana*, **20**, 33–6.

Oborny, J. and Voigt, H. (1879). *Die Meteorologie und Wetter-Telegraphie im Dienste der Landwirtschaft*. Voigt, Berlin.

Olby, R.C. (1965). Francis Galton's derivation of Mendelian ratios in 1875. *Heredity*, **20**, 636–8.

Olby, R.C. (1966). *Origins of mendelism*. Constable, London. 2nd edn: University of Chicago Press, 1985.

Olby, R.C. (1967). Franz Unger and the *Wiener Kirchenzeitung*: an attack on one of Mendel's teachers by the editor of a Catholic newspaper. *Folia Mendeliana*, **2**, 29–37.

Olby, R.C. (1968). Galton's ratios. *Heredity*, **23**, 155–6.

Olby, R.C. (1971). The influence of physiology on hereditary theory in the nineteenth century. *Folia Mendeliana*, **6**, 99–103.

Olby, R.C. (1979). Mendel no Mendelian? *History of Science*, **17**, 55–72.

Olby, R. (1985). *See* Olby (1966).

Olby, R.C. (1986). Mendels Vorläufer: Kölreuter, Wichura und Gärtner. *Folia Mendeliana*, **21**, 49–67.

Olby, R. (1987). William Bateson's introduction of mendelism to England: a reassessment. *British Journal of the History of Science*, **20**, 399–420.

Olby, R. (1991). Invited editorial comment: Johann Gregor Mendel. *American Journal of Medical Genetics*, **40**, (26), 1.

Olby, R.C. (1993). Constitutional and hereditary disorders. In *Companion Encyclopedia to the History of Medicine* (eds. W.F. Bynum and R. Porter). The quotation in the text is from page 421. (Cited by Orel and Hartl 1994).

Oldroyd, D.R. (1980). The work of Mendel: the synthetic theory of evolution. In *Darwinian impacts* (ed. D.R. Oldroyd), pp. 158–73. Open University Press, Milton Keynes.

Orel, A. (1933). *Das Weltantlitz. Eine gemeinverständliche Natur-, Kultur-. Religions-. und Geschichtsphilosophie*. Grünewald, Mainz.

Orel, V. (1968). Will the story on 'too-good' results of Mendel's data continue? *BioScience*, **18**, 776–8.

Orel, V. (1969a). Abbot Mendel's expert opinion on the first weather forecasts for agriculture. *Folia Mendeliana*, **4**, 37–43.

Orel, V. (1969b). Die Stellungnahme G. Mendels zur ersten Herausgabe von Wetterprognosen für Landwirtschaftszwecke. *Wetter und Leben*, **21**, 208–216.

Orel, V. (1970a). Mendel and the central board of the Agricultural Society. *Folia Mendeliana*, **5**, 39–54.

Orel, V. (1970b). Die Auseinandersetzung um die Organisation der Brünner Naturforscher in der Zeit, da G. Mendel seine *Pisum*-Versuche durchführte *Folia Mendeliana*, **5**, 55–72.

Orel, V. (1971a). A reconstruction of Mendel's *Pisum* experiments and an attempt at an explanation of Mendel's way of presentation. *Folia Mendeliana*, **6**, 41–60.

Orel, V. (1971b). Mendel's publishing activities in the Agricultural Society *Folia Mendeliana*, **6**, 213–25.

Orel, V. (1971c). Unknown letters relating to Mendel's health. *Folia Mendeliana*, **6**, 265–70.

Orel, V. (1971d). Mendel and the evolution idea. *Folia Mendeliana*, **6**, 161–72.

Orel, V. (1972a). Professor Alexander Zawadski (1798–1868), Mendel's superior at the technical modern school in Brno. *Folia Mendeliana*, **7**, 13–20.

Orel, V. (1972b). Mendel and new scientific ideas at the Vienna University *Folia Mendeliana*, **7**, 27–36.

Orel, V. (1972c). Mendel's elder friar and teacher, Matthew Klácel (1808–1882). *The Quarterly Review of Biology*, **47**, 435–6.

Orel, V. (1973a). Response to Mendel's *Pisum* experiments in Brno since 1865. *Folia Mendeliana*, **8**, 199–211.

Orel, V. (1973b). The enigma of hybrid constancy in Mendel's paper perceived by Albert Blomberg in 1872. *Hereditas*, **73**, 41–4.

Orel, V. (1974a). The scientific milieu in Brno during the era of Mendel's research. *The Journal of Heredity*, **64**, 314–18.

Orel, V. (1974b). The prediction of the laws of hybridization in Brno already in 1820. *Folio Mendeliana*, **9**, 245–54.

Orel, V. (1975a). Das Interesse F. C. Napps (1792–1867) für den Unterricht der Landwirtschaftslehre und die Forschung der Hybridisation. *Folia Mendeliana*, **10**, 225–40.

Orel, V. (1975b). The building of greenhouses in the monastery garden of Old Brno at the time of Mendel's experiments. *Folia Mendeliana*, **10**, 201–8.

Orel, V. (1977). Selection practice and theory of heredity in Moravia before Mendel. *Folia Mendeliana*, **12**, 179–221.

Orel, V. (1978a). Newly found notes relating to Mendel's research. *Folia Mendeliana*, **13**, 225–7.

Orel, V. (1978b). The influence of T.A. Knight (1759–1838) on early plant breeding in Moravia. *Folia Mendeliana*, **13**, 241–60.

Orel, V. (1978c). Heredity in the teaching programme of Professor J.K. Nestler (1783–1841). *Acta universitatis Palackianae Olomucensis, fac. rer. nat.* **59**, 79–98.

Orel, V. (1979). The teaching of J.M. Schleiden and its possible influence on G. Mendel. *Janus*, **64**, 33–47.

Orel, V. (1981). Die Idee der Pflanzenentwicklung aus einer Pollenzelle im 19. Jahrhundert. *Acta historiae rerum naturalium necnon technicarum, Pragce*, **16**, 275–86.

Orel, V. (1983). Mendel's achievements in the context of the cultural peculiarities of Moravia. In *Gregor Mendel and the foundation of genetics* (ed. V. Orel and A. Matalová), pp. 23–46. Moravian Museum, Brno.

Orel, V. (1984). *Mendel*, Past Masters series. Oxford University Press. French translation: *Mendel, un inconnu célèbre.* Berlin, Paris. Indonesian translation: Bapakgenetika modern. Grafiti, Jakarta.

Orel, V. (1986). History of plant hybridization according to Mendel's contemporary Rudolf Geschwind. *History and Philosophy of the Life Sciences,* **8**, 251–63.

Orel, V. (1988). Commemoration of the N.I. Vavilov centennial in Brno. *Folia Mendeliana,* **23**, 37–50.

Orel, V. (1989). Genetic laws published in Brno in 1819. In *Proceedings of the Greenwood Genetic Center,* South Carolina, **8**, 81–2.

Orel, V. (1992). Jaroslav Kříženecký (1896–1964), tragic victim of Lysenkoism in Czechoslovakia. *The Quarterly Review of Biology,* **67**, 487–94.

Orel, V. (1994). Embryonální 'genetische Gesetze' zveřejněné v Brně před Mendelovým narozením. *Dějiny věd a techniky,* **28**, 15–23. [Embryonal genetic laws published in Brno before Mendel was born.]

Orel, V. (1996). Embryonal genetical laws published in Brno before Mendel was born. *The Journal of Heredity (in press).*

Orel, V. and Alpatov, V. (1979). An 1885 Moscow report on Mendel's research into acclimatization of tropical bees. *Folia Mendeliana,* **14**, 237–50.

Orel, V. and Cetl, I. (1973). *Secret of Mendel's discovery.* (In Japanese), Kyoikushuppan-kabushikikaisha, Tokyo.

Orel, V. and Čunderlík, I. (1985). What was Mendel's intention in preparing microscopic slides? *Folia Mendeliana,* **20**, 9–14.

Orel, V. and Fantini, B. (1983). The enthusiasm of the Brno Augustinians for science and their courage in defending it. In *Gregor Mendel and the foundation of genetics* (ed. V. Orel and A. Matalová), pp. 105–110. Moravian Museum, Brno.

Orel, V. and Gabriel, J. (1981). The impact of hermetic philosophy on early vine breeding. *Folia Mendeliana,* **16**, 263–7.

Orel, V. and Hartl, D.L. (1994). Controversies in the interpretation of Mendel's discovery. *History and Philosophy of the Life Sciences,.***16**, 423–64.

Orel, V. and Janko, J. (1988). Purkyně's concept of procreation and Mendel's research in heredity. In *Jan Evangelista Purkyně* in science and culture (ed. J. Purš), pp. 657–70. Československá akademie věd, Prague.

Orel, V. and Kuptsov, V.I. (1983). Preconditions for Mendel's discovery in the body of knowledge in the middle of the 19th century. In *Gregor Mendel and the foundation of genetics* (ed. V. Orel and A. Matalová), pp. 189–227. Moravian Museum. Brno.

Orel, V. and Marvanová, L. (1966). The Mendel centennial in Brno. *Folia Mendeliana,* **1**, 9–14.

Orel, V. and Matalová, A. (1990). Kříženeckého chápání Mendelova objevu pod vlivem teorie dědičnosti získaných vlastností. *Dějiny věd a techniky,* **23**, 79–91. In English, *The perception of Mendel's discovery by J. Kříženecký under the influence of the theory of heredity of acquired characteristics.*

Orel, V. and Vávra, M. (1968). Mendel's program for the hybridization of apple trees. *Journal of the History of Biology,* **2**, 219–24.

Orel, V. and Vávra, M. (1979). Pedagogue Johann Andreas Edmond Schreiber (1769–1850) evoked in Gregor Mendel first interest in natural science *Folia Mendeliana*, **14**, 243–50.

Orel, V. and Verbík, A. (1984). Mendel's involvement in the plea for freedom on teaching in the revolutionary year of 1848. *Folia Mendeliana*, **19**, 223–33.

Orel, V. and Wood, R. (1981). Early development in artificial selection as a background to Mendel's research. *History and Philosophy of the Life Sciences*, **3**, 145–70.

Orel, V., Rozman, J., and Veselý, V. (1965). *Mendel as a beekeeper*, Moravian Museum, Brno.

Orel, V., Czihak, G., and Wieseneder, H. (1983). Mendel's examination paper on the geological formation of the earth of 1850. *Folia Mendeliana*, **18**, 227–72.

Orel, V., Janko, J., and Geus, A. (1987). The enigma of generation in connection with heredity in the teaching of J.E. Purkyně (1787–1869). *Folia Mendeliana*, **22**, 7–33.

Ostenfield, C.H. (1904a). Zur Kenntnis der Apogamie in der Gattung *Hieracium*. *Berichte der Deutschen Botanischen Gesellschaft*, **22**, 376–81.

Ostenfeld, C.H. (1904b). Weitere Beiträge zur Kenntnis der Fruchtentwicklung bei der Gattung *Hieracium*, *Berichte der Deutschen Botanischen Gesellschaft*, **22**, 537–41.

van der Pas, P.W. (1972). The date of Gregor Mendel's birth. *Folia Mendeliana*, **7**, 7–12.

van der Pas, P.W. (1976). Hugo de Vries and Gregor Mendel. *Folia Mendeliana*, **11**, 3–16.

Pease, M. (1958). Research in poultry genetics at Cambridge *School of Agriculture, Cambridge University, Memoirs*, **30**, 5–9.

Peter, A. (1885). *Hieracia Naegeliana exsiccata*. Included are 28 specimens from Mendel, cited by Weiling (1971).

Peter, A. (ed.) (1885–6). *Hieracia Naegeliana*, 3 Centurien + 1 Supplement–Centurie. Munich.

Peters, J.A. (1959). *Classic papers in genetics*. Prentice-Hall, Englewood Cliffs, NJ.

Pettenkofer, M. (1862). Die Bewegung des Grundwassers in München von März 1856 bis März 1862. *Sitzungsberichte der königl. bayer. Akademie der Wissenschaften, München*, **I**, 272–90.

Piegorsch, W.W. (1986). The questions of fit in the Gregor Mendel controversy. *History of Science*, **24**, 173–82.

Pilgram, A. (1788). *Untersuchungen über das Wahrscheinliche der Wetterkunde durch vieljährige Beobachtungen*. Vienna.

Pilgrim, I. (1984). The too-good-to-be-true paradox and Gregor Mendel. *The Journal of Heredity*, **75**, 501–2.

Pilgrim, I. (1986). A solution to the too-good-to-be-true paradox and Gregor Mendel. *The Journal of Heredity*, **77**, 218–20.

Plesník, J. and Orel, V. (1949–50). Boj o uznání dědičnosti získaných vlastností v našej kulture. *Priroda* , 18–20, 33–4. In English, *Conflict relating to the acknowledgement of heredity of acquired characteristics in our culture.*

Pontecorvo, G. (1966). Template and stepwise processes in heredity. *Proceedings of the Royal Society of London, B*, **167**, 169.

Powers, L. and Rollins, R.C. (1945). Reproduction and pollination studies on guayule. *Parthenium argentatum* Gray and *P. incanum* H.B.K. *Journal of the American Society of Agronomy*, **37**, 184–93.

Priestley, J. (1797). Letter from Dr. Priestley to Sir John Sinclair, President of the Board of Agriculture. *Communication of the Board of Agriculture on subjects relative to the husbandry and internal improvement of the country*, **1**, 363–6. Reprinted in *Artificial selection and development of evolutionary theory* (ed. C.J. Bajema) (1982). Benchmark papers in systematic and evolutionary biology, 4, pp. 37–40. Hutchinson Ross, Stroudsburg.

Pringsheim, N. (1855). Über die Befruchtung und Keimung der Algen und das Wesen des Zeugungsaktes. *Monatsberichte der Königl. Preuse. Akademie der Wissenschaften, Berlin*, **III**, 1–33.

Procházka, L. (1985). Mikroskope von Johann Gregor Mendel. *Folia Mendeliana*, **20**, 15–27.

Proskowetz, E. (1902). Erinnerung an den österreichischen Forscher Gregor Mendel. *Neue Freie Presse, Wien*, no. 13619. Partly reprinted by Kříženecký (1965*b*, pp. 110–11).

Przibram, H. (1923). Artwandlung und Arterhaltung. In *Studia Mendeliana ad centesimum diem natalem Gregorii Mendelii a grata patria celebrandum* (ed. H. Iltis, L. Haškovec, and V. Růžička), pp. 175–86. Apud Typos, Brno.

Purkinje, J.E. (1825). *Symbolae ad ovi avium historiam ante incubationem*. Vratislaviae typis universitatis Bratislava University Press). English translation by G.W. Barthelmez in *Essays in biology in honour of H.M. Evans*, pp. 53–93, University of California Press, 1943.

Purkinje, J.E. (1834). Erzeugung (generation, genesis, procreation). *Encyclopädisches Wörterbuch der medicinischen Wissenschaften, Berlin*, **XI**, 515–48.

Purkyně, J.E. (1860). Buňka zvířecí. *Riegnův slovník naučný*, **1**, 988–9, Prague. In English: The animal cell.

Recht, H. (1976). *Die höhere Obst- und Gartenbauschule und das Mendeleum in Eisgrub*. Verlag des wissenschaftlichen Antiquariat, Vienna.

Redtenbacher, J. and Liebig, J. (1841). Ueber das Atomgewicht des Kohlenstoffs. *Annalen der Chemie*, **38**, 113–40.

Regel, E. (1853). Verwandlung von *Aegilops ovata* in *Triticum*. *Gartenflora*, **2**, 280–1.

Reye, T. (1872). *Die Wirbelstürme, Tornados und Wettersäulen in der Erdatmosphäre*. C. Rümpler, Hanover.

Richter, O. (1924). Ein kleiner Beitrag zur Biographie P. Gregor Mendels. In *Festschrift der Technischen Hochschule Brünn zu Feier ihres 75jährigen Bestandes*, pp. 123–41. Deutsche Technische Hochschule, Brno.

Richter, O. (1941). 75 Jahre seit Mendels Grosstat und Mendels Entdeckungen. *Verhandlungen des Naturforschenden Vereines, Brünn*, **72**, 109–73.

Richter, O. (1943). Johann Gregor Mendel wie er wirklich war. *Verhandlungen des Naturforschenden Vereines, Brünn*, **74**, II, 1–262.

Roberts, H.P. (1929). *Plant hybridization before Mendel*. Princeton University Press.

Romanes, G.J. (1881–95). Article on 'Hybridism'. In *Encyclopaedia Britannica* (9th edn). **Vol. 12**, pp. 422–6.

Römer, C. (1867). Verzeichnis von Pflanzen, welche in dem letzten Jahre in der Umgebung von Namiest gesammelt, und welche aus dieser Gegend bisher entweder noch gar nicht, oder nur von anderen Fundorten bekannt waren. *Verhandlungen des Naturforschenden Vereines, Brünn*, **5**, 55–6.

Root-Bernstein, R.S. (1983). Mendel and methodology. *History of Science*, **21** 275–95.

Roux, W. (1911). Ueber die bei der Vererbung blastogener und somatogener Eigenschaften anzunehmendien Vorgängen.*Verhandlungen des Naturforschenden Vereines, Brünn*, **49**, 270–323.

Růžička, V. (1914). *Nárys učení o dědičnosti pro studující, lékaře, učitelstvo škol středních a hospodářských s 60 vyobrazeními*. Prague. In English, *The outline of the science of heredity for students, physicians, and teachers of secondary and agriculture schools*, with 60 illustrations.

Růžička, V. (1925). Official speech. In *Memorial-volume in honor of the 100th birthday of J.G. Mendel* (ed. A. Brožek *et al.*), pp. 31–49. Borový. Prague.

Sageret, M. (1826). Considération sur la production des hybrides, des variantes et des variétés en général, et sur celles de la famille des Cucurbitacées en particulier. *Annales science naturelle, Prem. Série*, **8**, 294–314.

Sajner, J. (1963). Gregor Mendel's Krankheit und Tod. *Sudhoff Archiv für die Geschichte der Medizin und Naturwissenschaften*, **47**, 377–82.

Sajner, J. (1965). *Gregorii Mendel autobiographia iuvenilis*. Universitas Purkyniana Brunensis, Brno.

Sajner, J. (1967). Gregor Johann Mendel und Znaim. *Forschung, Praxis und Fortbildung, Organ für die gesammte praktische und theoretische Medizin*, **18**, 677–85.

Sajner, J. (1968). Gregor Johann Mendels Erkrankung im Jahre 1849. Eine pathographische Studie zu Mendels Persöhnlichkeit. *Clio Medica*, **3**, 59–63.

Sajner, J. (1974). *Johann Gregor Mendel Leben und Werk*. Augustinus Verlag, Würzburg (2nd edn, 1976).

Salm and Andre, (1814). An die Freunde der vaterländischen Industrie und der inländischen Schafzucht insbesondere. *Oekonomische Neuigkeiten und Verhandlungen*, Prague, **15**, 113–14.

Sandler, I. and Sandler, L. (1985). A conceptual ambiguity that contributed to the neglect of Mendel's paper. *History and Philosophy of the Life Sciences*, **7**, 3–70.

Sapp, J. (1990). The nine lives of Gregor Mendel. In *Experimental Inquiries*, (ed. H.E. Le Grand). pp. 137–66. Kluwer, Amsterdam.

Saunders, G. (1906). Gregor Johann Mendel. In *Report of the conference on genetics and applied sciences, London* (ed. G. Saunders), pp. 85–8. London.

Schams, F. (1836). 'Beschreibung der Weinberge Mährens. *Landwirthschafts-Kalender auf das Schalt-Jahr 1836*, pp. 5–18. Brno.

Schindler, A. (1902). *Gedenkrede auf Prälat Gregor Joh, Mendel—anlässlich der Gedenkenthüllung in Heinzendorf, Schlesien, am 20. Juli 1902*. Private print of the author. Reprinted by Kříženecký (1965*b*), pp. 77–100. The English translation by A. Iltis was reprinted by Olby (1966), pp. 175–8.

Schleiden, M.J. (1848). *Die Pflanze und ihr Leben*. Engelmann, Leipzig.

Schleiden, M.J. (1849–50). *Grundzüge der wissenschaftlichen Botanik nach einer methodologischen Einleitung als Einleitung zum Studium*. Engelmann, Leipzig.

Schmalhausen, I.F. (1874). *O rastitelnych pomesjach- nabljudenija iz petersburskoy flory*. Dissertation, St. Petersburg University.

Schneider, G. (1888–95). Die Hieracien der Westsudeten. *Das Riesengebirge in Wort und Bild*, **8**, 75–80. Trautenau.

Schreiber (1801). Nachricht von der Insdustrialschule zu Kunewald. *Patriotisches Tagesblatt für k.k. Erblande, Brünn*, 42–5.

Schubert, E. (1899). Stručné dějiny ovocnictví moravského. *Moravský hospodář, orgán Českého odboru zemědělské rady moravské*, 152–4, Brno. In English, *A short history of Moravian horticulture*.

Schultz, F. (1854–5). Plantes hybrides. *Archives de Flore Wissenbourg-Deidesheim*, **I**, 254–6.

Scourfield, D.J. (1911). Sketch of Mendel's life and work. An address delivered to the Mendel Society on June 6th, 1910. *The Mendel Journal, February*, 5–23.

Serre, J.L. (1981). Mendel's rejection of the concept of blending inheritance. *Fundamenta Scientiae*, **2**, 55–66.

Seton, A. (1824). On the variation in the colour of peas from cross-impregnation. *Transactions of the Horicultural Society, London*, **5**, 236–7.

Shull, G.H. (1911). Defective inheritance-ratios in *Bursa* hybrids (mit 6 Tafeln). *Verhandlungen des Naturforschenden Vereines, Brünn*, **49**, 157–68.

Sinoto, Y. (1971). Mendel's two papers on genetics, considered from the standpoint of evolution. *Folia Mendeliana*, **6**, 151–6.

Smith, A.M. (1988). Major differences in isoforms of starch-branching enzyme between developing embryos of round- and wrinkled-seeded peas (*Pisum sativum* L.). *Planta*, **175**, 270–9.

Sootin, H. (1959). *Gregor Mendel: father of the science of genetics*. Vanguard Press, New York.

Sosna, M. (ed.) (1966). *G. Mendel memorial symposium 1865–1965*. Academia, Prague.

Soyfer, V.N. (1989). *Vlast i nauka. Istoria razgroma genetiky v SSSR*. Hermitage, Tenalfy.

Spencer, H. (1864). *The principles of biology*. London.

Stern, C. (1965). Mendel and human genetics. In *Commemoration of the publication of Gregor Mendel's pioneer experiments. Proceedings of the American Philosophical Society*, **109**, (4), 216–26.

Stern, C. (1966). Foreword. In *The origin of genetics. A Mendel source book* (ed. C. Stern and E.R. Sherwood), pp. v-xiii. W.H. Freeman, San Francisco, CA.

Stern, C. (1972). The continuity of genetics. In *The twentieth century sciences* (ed. G. Holton), pp. 171–97. Norton, New York.

Stern, C. and Sherwood, E.R. (1966). *The origin of gentics. A Mendel source book*. W.H. Freeman, San Francisco, CA.

Stern, C. and Stern, E. (1978). A note on the 'Three rediscoverers' of Mendelism. *Folia Mendeliana*, **13**, 237–40.

Stoletov, B.H., Sirotin, A.M., and Obbedkov, S.K. (1948). *O polozenii v biologiceskoy nauke*. Ozig-Selchozgiz, Moscow.

Stomps, T.H. (1954). On the rediscovery of Mendel's work by Hugo de Vries. *The Journal of Heredity*, **45**, 293–4.

Strumpf, F.L. (1853). *Die Fortschritte der Chemie in ihrer Anwendung auf Gewerbe, Künste und Pharmazie*. Berlin.

Stubbe, H. (1963). *Kurze Geschichte der Genetik bis zur Wiederentdeckung der Vererbungsregeln Gregor Mendels*. G. Fischer Jena. 2nd enlarged ed published in 1865. English trans. in 1972 as *History of genetics*, MIT Press, Cambridge, MA.

Sturtevant, A.H. (1965). *A history of genetics*. Harper and Row, New York.

Sturtevant, A.H. (1967). Mendel and the gene theory. In *Heritage from Mendel* (ed. R.A. Brink), pp. 11–15. University of Wisconsin Press, Milwaukee, WI.

Tatum, E.L. (1977). A case history in biological research. In *Nobel lectures in molecular biology* (ed. D. Lewis and J.R.S. Fincham), pp. 67–80. Elsevier, New York.

Taufer, J. (1914). Praktický význam Mendlových zákonů dědičnosti v chovu hospodářských zvířat. *Chov hospodářských zvířat*, Brno, pp. 6–9. In English, Practical significance of Mendel's laws in animal breeding.

Teindl, (1822). Vortrag gehalten bei der am 1822 stattgefundenen Schafzüchter-Vereines-Versammlung. *Mittheilungen, Brünn*, **44**, 345–51; **45**, 353–60; **46**, 361–8; **47**, 369–74; **48**, 377–80.

Teindl, Hisrch and Lauer, J.K. (1836). Protokol über die Verhandlungen bei der Schafzüchter-Versammlung in Brünn am 9.und 10. Mai 1836. *Mittheilungen, Brünn*, **38**, 303–9; **39**, 311–17.

Theimer, C. (1863). Über Bastardbildung im Pflanzernreiche. *Verhandlungen des Naturforschenden Vereines, Brünn*, **1**, 19–21.

Theimer, C. and Wallaschek, E. (1865). Bericht über einen gemeinschaftlich gemachten botanischen Ausflug nach Napajedl, Hradisch und Göding. *Verhandlungen des Naturforschenden Vereines, Brünn*, **3**, 68–73.

Thoday, J.M. (1966). Mendel's work as an introduction to genetics. *Advancement of Science*, **23**, 120–4.

Timoféev-Ressowsky, N.V. (1981). On Mendel. *Folia Mendeliana*, **16**, 229–37.

Tjebbes, K. (1923). Die Existenzmöglichkeit des Kommunismus. In *Studia Mendeliana ad centesimum diem natalem Gregorii Mendelii a grata patria celebrandum* (ed. H. Iltis), pp. 192–200. Apud Typos, Brno.

Tomaschek, A. (1871). Ueber Culturen der Pollenschlauchzelle. *Programm des k.k. deutschen Ober-Gymnasiums in Brünn*. pp. 3–17.

Tomaschek, A. (1879). Ein Schwarm der amerikanischen Bienenart *Trigona lineata*, Lep. lebend in Europa. Part I. *Zoologischer Anzeiger*, **2**, 782–7. Part II. *Zoologischer Anzeiger*, **3**, 60–5 (1880).

Traugott-Thieme, M.K. (1795). *Erste Nahrung für den gesunden Menschenverstand*. Leipzig.

Trautmann, L. (1814). *Versuch einer wissenschaftlichen Einleitung zum Studium der Landwirtschaft*. Vienna.

Tschermak, E. (1900a). Ueber künstliche Kreuzung bei *Pisum sativum*. *Berichte der Deutschen Botanischen Gesellschaft*, **18**, 232–9.

Tschermak, E. (1900b). Ueber künstliche Kreuzung bei *Pisum sativum*. *Zeitschrift für das landwirtschaftliche Versuchswesen in Österreich*, **3**, 465–555.

Tschermak, E. (1901). Anmerkungen des Herausgebers. In *Versuche über Pflanzenhybriden. Zwei Abhandlungen (1865 und 1869) von Gregor Mendel* (ed. E. Tschermak), pp. 54–62. Ostwald's Klassiker, Leipzig.

Tschermak, E. (1940). *Versuche über Pflanzenhybriden von Gregor Mendel*, p. 66. Ostwald's Klassiker, Leipzig.

Tschermak, E. (1958). *Leben und Wirken eines österreichischen Pflanzenzüchters*. Parey, Berlin.

Twrdy, J. (1839). Ueber die Johannisbeerstrau-Kultur. *Mittheilungen, Brünn*, **17**, 129–30.

Unger, F. (1846). *Grundzüge der Anatomie und der Physiologie der Pflanzen*. Vienna.

Unger, F. (1851). *Anatomie und Physiologie der Pflanzen*. Vienna.

Unger, F. (1852). *Botanische Briefe*. C. Gerold, Vienna. English trans. by B. Paul, *Botanical letters to a friend*, London, 1863.

Unger, F. (1860). *Die physiologische Bedeutung der Pflanzencultur. Aus den populären Vorträgen der k. k. Gartenbau-Gesellschaft*. Gerold, Vienna.

Uschmann, G. (1974). Evolution und Genetik zur Zeit Mendels. In *Folia fac. sci. nat. Universitatis Purkynianae Brunensis, Biologia*, **XV**, 34–41.

Valentin, G.G. (1835). *Handbuch der Entwicklungsgeschichte des Menschen*. Berlin.

Vavilov, N.I. (1935). Mendelism i ego znacenie v biologii i agronomii. In *Gregor Mendel opyty nad rastitelnymi gibridami*. (ed. N.I. Vavilov), pp. 5–12. Ogizselchozgiz, Moscow. Reprinted in *Opyty nad rastitelnym gibridami* (ed. A.E. Gaissinovitch), pp. 98–106. Nauka, Moscow, 1965. English trans. *Mendelism and its significance in biology and agronomy*, published by Orel (1988), pp. 37–41.

Vávra, M. (1984). Mendel's cooperation with the fuchsia breeder J.N. Tvrdý. *Folia Mendeliana*, **19**, 251–6.

Vavra, M. (1985). F.M. Klácel (1808–1882) as inspirer of students in botany. *Folia Mendeliana*, **20**, 29–32.

Vávra, M. and Matalová, A. (1983). Cultivation of new varieties of ornamental plants in Moravia in the period of Mendel's research. In *Gregor Mendel and the foundation of genetics* (ed. V. Orel and A. Matalová), pp. 77–86. Moravian Museum, Brno.

Vávra, M. and Orel, V. (1971). Hybridization of pear varieties by Gregor Mendel. *Euphytica*, **29**, 60–7.

Vávra, M. and Orel, V. (1976). Did Mendel grow the Florentine or Steyer vine? *Folia Mendeliana*, **11**, 33–5.

Volodin, B. (1968). *Mendel*. Molodaya gvardya, Moscow.

de Vries, H. (1870). *De invloed der temperatuur op de Levensverschijnsenlen der planten*. Martinus Nijhoff.

de Vries, H. (1889). *Intracellulare pangenesis*. Fischer, Jena. Republished by A. Oosthoek (1920) in *Hugo de Vries opera e periodicis collata*, Vol. V, pp. 1–148. English trans.: *Intracellular pangenesis*. The Open Court, Chicago, 1910.

de Vries, H. (1899). Sur la fécondation hybride de l'albumen. *Comptes Rendus de l'Academie des Sciences*, **129**, 973–5.

de Vries, H. (1900a). Sur la loi de disjonction des hybrides. *Comptes Rendus de l'Academie des Sciences*, **130**, 845–7.

de Vries, H. (1900b). Das Spaltungsgesetz der Bastarde.*Berichte der Deutschen Botanischen Gesellschaft*, **18**, 83–90.

de Vries, H. (1900c). Sur les unités des caractères spécifiques et leur application á l'étude des hybrides.*Revue générale de botanique*, **12**, 257–71.

de Vries, H. (1901–3). *Die Mutationstheorie.* Veit, Leipzig.

Vybral, V. (1968). Die leitende Funktion des Abtes Gregor Mendel in der Mährischen Hypothekenbank und ihr politischer Hintergrund. *Folia Mendeliana,* **3**, 21–33.

Vybral, V. (1971). Der Streit Mendels mit der Staatsverwaltung über die Beitragspflicht des Klosters zum Religionsfonde. *Folia Mendeliana,* **6**, 231–7.

van der Waerden, B.L. (1968). Mendel's experiments. *Centaurus,* **12**, 275–88.

Wagner, R. (1853). Nachtrag zu dem voranstehenden Artikel Zeugung and Nachtrag zum Nachtrag des Artikels Zeugung. In *Handwörterbuch der Physiologie mit Rücksicht auf physilogische Pathologie,* 4th vol. (ed. R. Wagner), pp. 1001–18a–d. F. Vieweg, Brunswick.

Wahlberg, T. (1810). *Neueste Beobachtungen zur Veredlung des Feldbaues und der Forstwirtschaft.* Wien.

Wegener, A. (1917). *Wind- und Wasserhosen in Europa.* F. Vieweg, Brunswick.

Weiling, F. (1966a). Hat J.G. Mendel bei seinen Versuchen zu genau *gearbeitet?—Der* X^2 Test und seine Bedeutung für die Beurteilung genetischer Spaltungsverhältnisse. *Der Züchter,* **36**, 359–65.

Weiling, F. (1966b). J.G. Mendels 'Versuche über Pflanzen-Hybriden' und ihre Würdigung in der Zeit bis zu ihrer Wiederentdeckung. *Der Züchter,* **36**, 273–32.

Weiling, F. (1967). J.G. Mendels Wiener Studienaufenthalt 1851–1853. *Sudhoffs Archiv für die Geschichte der Medizin und Naturwissenschaften,* **51**, 260–6.

Weiling, F. (1968). F.C. Napp und J.G. Mendel. Ein Beitrag zur Vorgeschichte der Mendelschen Versuche. *Theoretical and Applied Genetics,* **38**, 144–8.

Weiling, F. (1969). Die *Hieracium*-Kreuzungen J.G. Mendels sowie ihr Niederschlag in der Literatur und Herbarien. (Zu 100. Jahrestag von Mendels *Hieracium*-Vortrag vom 9.Juni 1869.) *Zeitschrift für Pflanzenzüchtung,* **62**, 63–9.

Weiling, F. (1970a). Die Meteorologie als die wahrscheinliche Quelle der statistischen Kenntnisse J.G. Mendels. *Folia Mendeliana,* **5**, 73–85.

Weiling, F. (1970b). *Kommentar zu Gregor Mendel: Versuche über Pflanzenhybriden.* Ostwalds Klassiker der exakten Wissenschaften, NF 6. Vieweg, Brunswick.

Weilung, F. (1971). Der Niederschlag der Arbeiten J.G. Mendels in der Literatur bis 1900. *Folia Mendeliana,* **6**, 139–42.

Weiling, F. (1972). Über die von Max Pettenkofer angeregten Untersuchungen des Zusammenhanges von Cholera–und Typhus-Massenerkrankungen mit dem Grundwasserstand, im Hinblick auf die langjährigen Grundwasserstandsbeobachtungen J.G. Mendels. (The lecture on the meeting of the German Region of the International Biometrical society in 1972.) The copy of the lecture is preserved in the Mendelianum, sign. no. 9441.

Weiling, F. (1975a). Rudolf Geschwind (1829–1910), ein wenig bekannter Forstpflanzen- und Rosenzüchter des ehemaligen Böhmisch–Ungarischen Raumes. *Folia Mendeliana,* **10**, 209–24.

Weiling, F. (1975b). J.G. Mendel sowie die von Max Pettenkofer angeregte Untresuchungen mit dem Grundwasserstand. *Archiv Zeitschrift Wissenschaftsgeschichte,* **59**, 1–19.

Weiling, F. (1976). Die Auseinandersetzung um die Besetzung des Lehrstuhles für Landwirtschaftslehre in Brno (Brünn) in den Jahren 1825 bis 1828 und Ihr

Zusammenhang mit der damaligen Schafzucht in Mähren. *Folia Mendeliana*, **11**, 17–32.

Weiling F. (1982). Die Philosophische Lehranstalt in Brünn (1808–1849) und die österreichische Bildungspolitik jener Zeit. Ihre Bedeutung für Entdeckungstätigkeit Johann Gregor Mendels. *Mittheilungen des Österreichchischen Staatsarchives*, **35**, 110–33.

Weiling, F. (1983). J.G. Mendels 'Versuche über Pflanzenhybriden' und die damals herrschende Theorie der Befruchtung höherer Pflanzen. In *Gregor Mendel and the foundation of genetics* (ed. V. Orel and A. Matalová), pp. 237–58. Moravian Museum, Brno.

Weiling, F. (1984*a*). Fünf weitere Sonderdrücke der 'Versuche über Pflanzenhybriden' J.G. Mendels aufgetaucht. *Folia Mendeliana*, **19**, 257–63.

Weiling, F. (1984b). Die Übernahme der Brünner meteorologischen Station durch J.G. Mendel im Jahre 1878 im Rahmen seiner längjährigen meteorologischen Tätigkeit. *Folia Mendeliana*, **18**, 235–50.

Weiling, F. (1985). Zur Frage der 'überzufällig grossen Genauigkeit' der Versuche J.G. Mendels. *Mittheilungen der Österreichischen Gesellschaft für Geschichte der Naturwissenschaften*, **5**, 1–25.

Weiling, F. (1986). Das Wiener Universitätssetudium 1851–1853 des Entdeckers der Vererbungsregeln Johann Gregor Mendel. *Folia Mendeliana*, **21**, 9–40.

Weiling, F. (1989). Which points are incorrect in R.A. Fisher's statistical conclusion: Mendel's experimental data agree too closely with his expectation? *Angewandte Botanik*, **63**, 129–43.

Weiling, F. (1991). Historical study: Johann Gregor Mendel 1822–1884. *American Journal of Medical Genetics*, **40**, (26), 1–25.

Weiling, F. and Orel, V. (1967). Wo erhielt J. G. Mendel Anregung zu seiner "Graphische–tabelarischen Übersicht der meteorologischen Verhältnisse von Brünn? *Folia Mendeliana*, **2**, 17–22.

Weinstein, A. (1958). Did Nägeli fail to understand Mendel's work? *Proceedings 10th International Congress of Genetics*, **2**, 369.

Weismann, A. (1885). *Die Continuität des Keimplasma's als Grundlage einer Theorie der Vererbung*. G. Fischer, Jena.

Weismann, A. (1892). *Aufsätze über Vererbung*. Fischer, Jena.

Weldon, W.F.R. (1902). Mendel's laws for the alternative inheritance in peas. *Biometrica*, **1**, 228–54.

Wichura, M. (1854). Ueber künstlich erzeugte Weidenbastarde. *Flora*, **1**, 1–8.

Wichura, M. (1865). *Die Bastardbefruchtung im Pflanzenreich erläutert an den Bastarden der Weiden*. Morgenstein, Breslau (Wroclaw).

Widmer, E. (1891). *Die europäischen Arten der Gattung* Primula. Mit einer Einleitung von C.v. Nägeli. Oldenbourg, Munich.

Wiegmann, A.F. (1828). *Über die Bastarderzeugung im Pflanzenreiche*. Vieweg, Brunswick.

Wieseneder, H. (1983). Commentary on Mendel's paper on the geological formation of the earth of 1850. In Orel *et al.* (1983), *Folia Mendeliana*, **18**, 230–4.

Wiesner, J. (1901). Gustav Theodor Fechner und Gregor Mendel. *Wiener Abendpost*, No. 269.

Wilkie, J.S. (1962). Some reasons for rediscovery and appreciation of Mendel's work in the first years of the present century. *The British Journal for History of Science*, **1**, 5–17.

Witten, V. (1828). Ueber Weizenarten. *Mittheilungen, Brünn*, **40**, 317–18.

Wolf, C.F. (1759). *Theoria generationis*. Halae and Salam (Halle).

Wood, R.J. (1973). Robert Bakewell (1725–1795), pioneer breeder, and his influence on Charles Darwin. *Folia Mendeliana*, **8**, 231–42.

Wood, R.J. (1989–90). Martin Barry (1802–1855) and his theory of blood corpuscles as determinants of the organism derived from both parents. *Folia Mendeliana*, **24–25**, 45–8.

Wright, S. (1966). Mendel's ratios. In *The origin of genetics. A Mendel source book* (ed. C. Stern and E.R. Sherwood), pp. 173–5. W.H. Freeman, San Francisco, CA.

Wunderlich, R. (1982). Der wissenschaftliche Streit über die Entstehung des Embryos der Blütenpflanzen im zweiten Viertel des 19. Jahrhunderts (bis 1856) und Mendel's 'Versuche über Pflanzenhybriden'. *Folia Mendeliana*, **17**, 225–42.

Zawadski, A. (1854). Ueber die Anforderungen der Naturforschung in der jetzigen Zeit. In *Program der Oberrealschule*, pp. 1–16. Brno.

Zawadski, A. (1857). Die Section für Botanik und Pflanzenphysiologie. *Mittheilungen, Brünn*, **3**, 19–23.

Zeising, A.Z. (1854). *Neue Lehre von den Proportionen des menschlichen Körpers*. Leipzig.

Zirkle, C. (1949). *Death of a science in Russia. The fate of genetics as described in Pravda and elsewhere*. University of Pennsylvania Press.

Zirkle, C. (1951). The knowledge of heredity before 1900. In *Genetics in the 20th century—essays on the progress of genetics during its first 50 years* (ed. L.C. Dunn), pp. 35–57, Macmillan, New York.

Zirkle, C. (1964). Some oddities in the delayed discovery of mendelism. *The Journal of Heredity*, **55**, 65–72.

Zirkle, C. (1966). Some anomalies in the history of mendelism. In *Mendel memorial symposium 1865–1965* (ed. M. Sosna), pp. 31–5. Academia, Prague.

Živansky, F. (1874). *Krátký návod průmyslového včelařství pro údy moravského spolku včelařského*. Brno. In English, *Short instruction on industrial bee-keeping for members of the Moravian Apicultural Society*.

Živansky, F. (1873). *Kurze Anleitung zum Betriebe der Vernunfmässigen Bienenzucht*. Brünn (Brno).

Zlámal, B. (1937). Cyrill František Napp (1792–1867). Augustiniánský opat na Starem Brně. Thesis, Olomouc University. The copy of the manuscript is preserved in the Mendelianum, sign. no. 1740. In English, *Cyrill František Napp (1792–1867), abbot of the Augustinian monastery at Old Brno*.

Zlámal, B. (1991). Cyrill Franz Napp (1792–1867) Augustiner Abt in Altbrünn. Biographische Skitze. *Folia Mendeliana*, **26**, 35–95.

Zumkeller, A. (1971). Recently discovered sermon sketches of Gregor Mendel. *Folia Mendeliana*, **6**, 247–56. Reprinted in German by Sajner (1974), pp. 108–22.

Index